THE SIGNIFICANCE TEST CONTROVERSY

THE SIGNIFICANCE TEST CONTROVERSY

edited by
Denton E. Morrison and
Ramon E. Henkel

 Routledge
Taylor & Francis Group

LONDON AND NEW YORK

First published 1970 by Transaction Publishers

Published 2017 by Routledge
2 Park Square, Milton Park, Abingdon, Oxon OX14 4RN
711 Third Avenue, New York, NY 10017, USA

Routledge is an imprint of the Taylor & Francis Group, an informa business

Library of Congress Catalog Number: 2006042891

Library of Congress Cataloging-in-Publication Data

The significance test controversy : a reader / Denton E. Morrison and Ramon
  E. Henkel, editors.
        p. cm.
  Originally published: Chicago : Aldine Pub. Co., 1970.
  ISBN 0-202-30879-0 (alk. paper)
  1. Statistical hypothesis testing. 2. Social sciences—Statistical methods.
  I. Morrison, Denton E. II. Henkel, Ramon E., 1931-

HA33.M67   2006
519.5'6—dc22   2006042891

ISBN 13:978-0-202-30879-1 (pbk)

Contents

i

PART FOUR. CRITICISM FROM OTHER QUARTERS

PART FIVE. EPILOGUE

Contents

PART FOUR: CRITIQUES AND FURTHER QUARRELS

PART FIVE: AFTERGLOW

Preface

OUR INTEREST in compiling this volume stems from our concern with the considerable amount of indiscriminate use of significance tests in behavioral research. Even their strongest proponents agree that there is much misuse, misinterpretation, and meaningless use of the tests. More important than the question of how the tests are correctly used, however, is the question of whether the tests are useful, and why or why not. We are concerned that many users lack an understanding of the latter questions. We do not know how much of this lack is because an extensive literature critical of past and current practice in the use of significance tests is unknown to researchers and how much results from a failure to heed the criticism. But we hope that collecting a substantial portion of this literature in one volume will help make researchers more mindful of both the practical problems and philosophical pitfalls involved in using the tests.

While the essential tone of this volume is critical of the tests, our broader purpose is to document the controversy over use of the tests in behavioral research. We have used the term "controversy" to characterize the literature on significance tests, but the sense in which a "controversy" exists must be understood in a special way. "Controversy" implies dialogue over points of disagreement, and in view of the fact that such dialogue has not always occurred, the term may be an overstatement. Both in sociology and psychology critics of the tests have reacted to what they view as erroneous research practice based on misguided statistical training. Essays that respond specifically to this criticism by explicitly defending the tests have appeared only in the sociological literature, however, so that the controversy has only in part taken the form of an extended debate or dialogue. In the behavioral sciences in general the overwhelming practice by both researchers and those responsible for statistical training has been to ignore the issues raised by the critics and to continue doing things as before. Thus the preponderance of the negative side of the "debate" in this volume does not represent so much bias as redress, since the amount of behavioral science writing that implicitly supports the tests is far greater than that which is critical.

Although most of the literature critical of significance tests has been written by sociologists and psychologists, the book is addressed to all behavioral scientists and to both professionals and students. We have thus selected readings for a group that as a whole is not extensively trained in mathematics, logic, and statistics. We hope that readers who are learned in these subjects will also find something of value in the readings, though the selections are mainly nontechnical. The readings do not provide computational instruction, nor do they compare one test with another. Dozens of standard statistical texts have done, if not overdone, this. The readings necessarily require a grasp of the broad and basic technical matters dealing with the nature and meaning of the sampling assumptions, probability statements, and parameter estimates connected with use of the tests. We assume, therefore, that the reader has some knowledge of statistics, at least that obtained through an introductory course.

The general failure to incorporate a critical perspective on significance tests into research training and practice has brought a considerable reiteration of the critical points in the literature, and this reiteration is reflected in substantial repetition of the same criticism in the essays of this volume. While we realize this repetition has its negative aspects, our view is that it will be more beneficial than harmful in a book of this sort. And though we readily confess to a desire to proselytize a skeptical perspective on the tests, our rationale is not simply that the sheer force of repetition will help drive home some of the points made, for the critical points are made at varying levels of statistical and philosophy of science difficulty in the readings. Thus, we hope the repetition will both provide something of value for a considerable range of reader sophistication and encourage step-wise growth in many readers' level of understanding. Also, the criticism often appears to have been made without the knowledge that similar criticism was made elsewhere; the independent occurrence of similar critical points in the literature adds to their persuasiveness and validity. It is impossible to overestimate the extent to which use and misuse of the tests is ingrained in the behavioral sciences. Researchers will profit from knowing that writers with diverse substantive and methodological interests have offered carefully reasoned and often basically similar criticism of the tests. Consequently, we have deliberately avoided deleting repetitious material from the selections and have attempted to be relatively complete in documenting the controversy by including both the major and minor essays in the main behavioral sciences where they have occurred—sociology and psychology.

This volume is not, however, intended as a general history of either the concept of statistical significance or the broader controversy surrounding this notion. Such a history would require inclusion of an extensive literature by such writers as Fisher, Neyman, E. S. Pearson, Jeffreys, and Wald, at the very least. While such a collection would be of considerable value, we believe that, since most of this literature is quite technical, most behavioral scientists would neither read it nor gain as much from it as from the selections we have included. The essays in the first part of this reader provide an ample, if critical, general historical context on the tests for the present purpose.

We wish to be clear that the volume is not intended as a blanket condemnation of statistics. The focus of our critical concern is inferential and *not* descriptive statistics, both in terms of how statistical inference is done in behavioral research and, more important, whether, given *basic* (in contrast with *applied*) scientific goals, it is worth doing. In essence, our view is that the significance test as typically employed in behavioral science is bad statistical inference, and that even good statistical inference in basic research is typically only a convenient way of sidestepping rather than solving the problem of scientific inference. Although we do not pretend that the selections of this volume go far toward solving the problems of scientific inference, we are convinced that the diversion of energy away from the rituals of significance testing in basic scientific research will be a worthy first step toward this goal and will, in fact, be one difference in behavioral science that *is* significant. Thus, we hope that this volume will be a modest contribution not only to more sensible statistical practice but to a more adequate philosophy of social science.

For this perspective and, indeed, for making the volume possible we are indebted to the contributors, both those who share our views and those who disagree. We have had a considerable, though largely vicarious, dialogue with all the contributors, a dialogue from which we have profited and which we trust will be profitable for our readers.

S. F. Camilleri is acknowledged for his special influence on our thinking and on our motivation to compile this volume. We also gratefully acknowledge the useful suggestions and encouragement of Richard Hill and Alexander J. Morin, the editorial help of Nancy Hammond and Joan Levinson, and the secretarial assistance of Mary Wilson and Ann Ries. Our wives provided encouragement and understanding, which also made a significant difference.

Denton E. Morrison
Michigan State University

Ramon E. Henkel
University of Maryland

INTRODUCTION
A Preview of the Issues and an Overview of the Readings

A RESEARCHER is typically able to study only part of the cases in which he is interested. In statistical language, the part he studies is his "sample," and the larger body of cases his "population" or, synonymously, his "universe." Any datum from a sample is called a "statistic." A statistic, like a mean or correlation coefficient or percentage, describes some characteristic of the sample and is thus called a "descriptive statistic."

The extent to which credence should be placed in a given statistic as a description of the population is the problem of inferring from the part to the whole, sometimes called the problem of inductive inference, or the problem of generalization. In empirical research this problem is often treated as one of statistical inference, to be answered by what has emerged in behavioral science as the central tool of this mode of making inferences, the test of significance. In its simplest terms, then, a test of significance is a method used to infer knowledge about a population (a population "parameter," technically) on the basis of a statistic gained from a sample.[1]

The readings of this volume rarely question whether tests of significance,

---

1. Technical descriptions of significance tests are readily available in any statistics text that covers statistical inference. More detailed nontechnical descriptions of the tests are available in many of the selections in this reader, for instance in our paper on "Significance Tests Reconsidered" (1969 [Chapter 21]). (Note: Detailed references to items briefly cited in parentheses appear in the consolidated list of "References," pp. 313–324. References to this volume and Editors' notes are given in brackets.)

correctly used, make a meaningful contribution to solving a certain set of problems of inductive inference. The readings, however, raise and debate many questions as to what constitutes correct use of tests of significance, and whether such use is practiced or practicable in behavioral research. Issues of this type we call "statistical issues." These issues revolve around both the assumptions of mathematical theory on which tests of significance are based and conventional practices that, although not dictated by the mathematical theory, affect applications or interpretations of the tests.

While some writers represented in this volume (and most of the authors of conventional works on statistical inference) seem to assume that statistical inference is coterminous with or at least a fundamental aspect of scientific inference, many of our readings implicitly or explicitly question the contribution of statistical inference to scientific inference. The latter writers are concerned, then, with a set of philosophy of science issues connected with the use of significance tests. Typically these writers base their arguments on close consideration of some of the statistical issues; but in addition to whatever questions they raise about the use and interpretation of the tests, they argue that even correct use and interpretation of the tests do not match up with the ideal model of science that they favor.

The model of science favored by those who raise philosophy of science issues is that of basic or pure science in contrast with applied science. The central feature of basic science, in their view, is a concern with the development of a body of general theory; in contrast, applied science is directed toward providing information for the solution of practical problems. More specifically, their view of basic science has the following crucial elements: (1) the notion that the hypotheses of science are constructed to hold whenever and wherever instances of the substantive phenomena in question occur under specified conditions (in contrast with hypotheses that are thought to hold only in some time- and place-bound existent population), and (2) the notion that belief in a scientific hypothesis develops gradually on the basis of the deductive system that produced it, the abstractness and fertility of its concepts, and the strength and consistency of its empirical support over replication (in contrast with the discrete labeling of every hypothesis as "rejected" or "accepted" in each research situation, to provide guides to immediate action based on this judgment). Typically those who raise philosophy of science issues argue that many of the users of the tests implicitly favor this model of basic science also, but that the users are led astray by the tests.

To some extent, of course, the dichotomy of statistical and philosophy of science issues is bound to be arbitrary, since the two sets are intimately related. So, moreover, are the issues within each set. Nevertheless, the following breakdown provides a useful framework for orientation to the readings. We employ it both in introducing the main points in the sections and selections and in presenting our conclusions in the final chapter. Further, at the end of this chapter a detailed tabular guide to the readings according to the issue categories is provided.

An Outline of the Specific Issues

THE STATISTICAL ISSUES

1. Sampling. The issue of how the method by which the data are generated is related to the applicability of significance tests to the data. The question of the legitimacy and meaning of significance tests applied to data not generated by random sampling or random assignment (randomization).

2. Population. The issue of the population to which inferences based on significance tests apply. Whether and how the population to which significance tests allow inferences is determined, given the method of sampling.

3. Meaning. The issue of the meaning of the probability statements resulting from significance tests. How statistical and substantive significance differ.

4. Level. The issue of how and why a given significance level is chosen. How conventional practices in the choice of a significance level are related to the varying purposes of the research.

5. Power. The issue of the role of the power of significance tests in statistical inference. How sample size and other power considerations influence the ability of a test to reject false hypotheses.

6. Technique. The issue of the role of computation and data analysis techniques in significance testing. How the level of significance reported can be influenced by the selection of computational formulas and the manipulation of data.[2]

7. Causality. The issue of the role of significance tests in assessing causality. The design and analysis conditions under which significance tests provide information of value for inferring causal relationships.

THE PHILOSOPHY OF SCIENCE ISSUES

8. Scope. The issue of whether significance tests contribute to obtaining the desired population scope of scientific inference (obviously related to the issue of population described above).

9. Form. The issue of whether the logical characteristics of the hypotheses involved in significance testing measure up to the requirements of hypotheses that characterize basic science. Whether the null hypotheses common in behavioral science contribute to the development of scientific theory.

---

2. None of the many articles that deal mainly or only with issues of technique (for instance, Lewis and Burke, 1949) is included in this reader.

10. Process. The issue of whether the process by which hypotheses are evaluated in significance testing is appropriate to the creation of scientific knowledge. Whether the fixed criterion for making decisions about hypotheses in significance testing contributes to scientific inference.

11. Purpose. The issue of the purpose of making inferences with significance tests as compared with the purpose of scientific inference. Consideration of the place of firm decision and action on isolated hypotheses in the scientific enterprise.

Organization and Overview of the Readings

Although it might be advantageous, it is not possible to organize and order the readings neatly around the issues. Many of the readings, as implied above, attack a variety of statistical and philosophy of science issues because their authors quite correctly see these two types of issues as fundamentally interrelated. Also, there is merit to an organization that shows how the issues unfold and develop and—at least in Part Two—how issues are raised, debated, and then re-raised and re-debated. Consequently, the book is organized primarily around the two main behavioral disciplines in which the literature has appeared, psychology and sociology. To some extent this mode of organization also serves to group the readings according to the broad areas of experimental (psychology) and nonexperimental (sociology) research designs. The readings will make clear that the tests have not been immune from criticisms in either context.

The literature critical of significance tests in psychology and sociology has appeared mainly in the last 15 years. The controversy over the tests has a much longer and deeper history, however, and Part One, "Critical Historical Context," which contains three chapters from Lancelot Hogben's book *Statistical Theory* (1957), gives a basis for understanding the historical antecedents of the more recent criticisms in behavioral science. Hogben's analysis also deals critically with many of the issues that are discussed in the subsequent selections. Thus we include Hogben at length not only because of his penetrating historical and critical analysis, but because his work on this topic is generally unknown to American researchers despite his influence on a number of the writers whose work appears in subsequent sections (Camilleri, 1962 [Chapter 16]; Bakan, 1966, 1967 [Chapter 25]; Meehl, 1967 [Chapter 26]; Morrison and Henkel, 1969 [Chapter 21]). Hogben is delightful and rewarding reading, but his style is abstruse. Consequently, we advise that only the statistically hardy start with him; others will profit more from Hogben after obtaining a familiarity with the issues in the other readings.

Part Two, "The Controversy in Sociology," which follows Hogben's essays, documents a debate that began over a decade ago and continues today. While the contestants are often talking past each other on the issues or hardly acknowledge that they are responding to one another, to follow the controversy in complete detail is instructive. While some of the sociologists are involved in the discussion of philosophy of science issues, most of the essays in sociology constitute vigorous debate mainly of statistical issues.

The "Criticism by Psychologists" section, which constitutes Part Three of the book, does not represent a genuine debate as in the sociological literature. These essays touch firmly on several statistical issues, but the focus of their concern is on broader philosophy of science issues, with generally negative conclusions about the contribution of the tests to scientific inference.

Part Four, "Criticism from Other Quarters," presents three relative early and diverse critical essays in behavioral science that are of value, but they do not fit into either the sociology or psychology parallel mainstreams of critical literature on the tests.

In Part Five, "Epilogue," we state our position on each issue and indicate the basis for our belief in the benefits of less dependence on the tests in our essay, "Significance Tests in Behavioral Research: Pessimistic Conclusions and Beyond."

Section, Chapter, First Author, Date	Statistical Issues							Philosophy of Science Issues			
	Sampling	Population	Meaning	Level	Power	Technique	Causality	Scope	Form	Process	Purpose
Critical Historical Context											
1. Hogben (1957)			×						×	×	×
2. Hogben (1957)	×	×	×	×	×			×	×	×	×
3. Hogben (1957)	×	×	×	×			×	×	×	×	
The Controversy in Sociology											
4. Hagood (1941)	×	×	×				×	×		×	×
5. Zeisel (1955)			×							×	×
6. Lipset (1956)	×	×	×	×		×	×	×	×	×	×
7. Kendall (1957)	×		×	×		×			×	×	×
8. Davis (1958)										×	
9. Selvin (1957)	×	×	×	×		×	×	×	×	×	×
10. Gold (1958)			×		×		×				
11. Selvin (1958)					×		×	×			
12. Beshers (1958)	×						×				
13. Selvin (1958)							×				
14. McGinnis (1958)	×	×					×	×			×
15. Kish (1959)	×	×	×	×	×	×	×	×	×	×	
16. Camilleri (1962)	×	×	×	×			×	×	×	×	×
17. Skipper (1967)			×	×	×					×	
18. Duggan (1968)			×	×		×					
19. Labovitz (1968)			×	×	×	×	×			×	
20. Gold (1969)	×	×	×			×		×		×	
21. Morrison (1969)	×	×	×	×	×	×	×	×	×	×	×
22. Winch (1969)	×	×	×			×	×	×		×	×
Criticism by Psychologists											
23. Chandler (1957)			×		×						
24. Rozeboom (1960)			×	×	×	×			×	×	×
25. Bakan (1966, 67)	×	×	×	×	×	×		×	×	×	×
26. Meehl (1967)			×		×				×	×	
27. Lykken (1968)			×		×			×	×	×	×
Criticism from Other Quarters											
28. Berkson (1942)			×	×	×				×	×	×
29. Sterling (1959)				×	×			×		×	
30. Tullock (1959)								×		×	

PART ONE
Critical Historical Context

INTRODUCTION

Hogben's book, *Statistical Theory* (1957), is a systematic and damaging attack on various probability practices in research. In the three chapters reprinted here Hogben presents a critical and historical discussion of some major issues concerning tests of significance. Hogben is thus a logical starting point for this book, but because of both the scope of erudition he assumes the reader possesses and his somewhat elusive style of writing, his chapters will be a practical starting point only for those readers with considerable familiarity with the issues and their history. Others will probably want to defer reading Hogben until they become more familiar with the issues by reading certain of the subsequent papers, particularly Bakan's [Chapter 25], but also those by Rozeboom [Chapter 24] and Camilleri [Chapter 16].

In "The Contemporary Crisis or the Uncertainties of Uncertain Inference," the first of the chapters that follow, Hogben delineates four separate uses of probability in theoretical statistics. His "Calculus of Judgments," the term he uses for statistical inference, is most relevant here.[1] He elaborates his views

---

1. In a less direct sense Hogben's "Calculus of Aggregates" is relevant to Gold's [Chapter 20] and Winch and Campbell's [Chapter 22] use of tests of significance. The latters' random process models are essentially what Hogben would refer to as a Calculus of Aggregates—the attempt to see whether random aggregate behavior could account for the phenomenon observed.

on this topic in the two subsequent chapters we have reprinted. Most of the initial chapter is given to demonstrating that there are long-standing and deep-seated differences among statisticians on probability in general and on the mathematical bases for and interpretation of significance tests in particular. He points out that these differences have not prevented statisticians from making somewhat pompous claims that have been uncritically accepted by researchers. The contemporary crisis is, then, that despite widespread use in research, many uncertainties surround the mathematical assumptions, meaning, interpretation, and relevance of statistical theory.

In "Statistical Prudence and Statistical Inference" [Chapter 2], Hogben deals with the issue of the meaning of significance tests from a historical perspective and relates his discussion to the issues of the process and purpose of scientific inference. Hogben attempts to demonstrate that the term "statistical inference" is a misnomer for what is in fact simply a method of making prudent judgments in research situations where such judgments are necessary and possible. To understand this idea requires the recognition that Hogben wishes to reserve the term "inference" (also "interpretation") exclusively for what is involved in arriving at a valid assessment of whether a hypothesis should or should not take its place in a body of scientific knowledge, that is, scientific inference. He argues that the school of R. A. Fisher, the earlier and more influential among scientists, mistakenly employs significance tests as a mode of scientific inference. In contrast, he maintains that J. Neyman, E. S. Pearson, and A. Wald correctly use the tests as *decision* tests.

Throughout *Statistical Theory* Hogben tends to identify Fisher's approach as one example of what he calls the "Backward Look" (the other key example is the Bayesian approach), while the Neyman-Pearson-Wald approach is labeled the "Forward Look." The terms "Forward" and "Backward" imply as much (or more) about Hogben's attitudes toward the two practices as they do about the differences actually involved. Briefly, the Backward Look, as applied to Fisher's notion, involves retrospective interpretation of the information produced by a significance test to infer the general validity of a hypothesis in probabilistic terms. On the other hand, the Forward Look involves the application of probability theory to a given empirical outcome to determine the ratio of correct decisions about future outcomes of tests of a hypothesis.

The difference can be better understood by considering the meaning of a rare event—for example, a difference between two groups that a significance test indicates would occur less than five times in 100 if the null hypothesis of no difference between the groups were true. According to the Fisherian approach we would attempt to infer from a single occurrence like the one above whether the difference is "significant," i.e. whether the difference is such that we may infer the null hypothesis to be invalid ("rejected"). Although Fisher preferred the five percent level as a "convenient" criterion for making this inference, he did not dictate that it be a firm criterion, nor did he think it necessary to state the criterion level in advance of the test. The problem of judging whether a given finding is rare enough to warrant rejection of the null hypothesis is a matter of inference and interpretation for the researcher after he has performed the test.

The Neyman-Pearson-Wald approach, in contrast, does not, on the basis of any particular occurrence, attempt to judge the validity of the hypothesis under consideration. Rather, it is a procedure for judgment involving a firm rule (level) stated in advance that yields a specified proportion of valid judgments in the long run. In the short run, that is, in any particular instance, the procedure yields a decision. Such a procedure is appropriate because the research context in which such a *decision* test (in contrast with a *significance* test) is used requires a decision as a guide to actions and provides the required fixed framework for repetition of the test. Thus, making decisions in this way is prudent, but such decisions do not commit the researcher to an inference about the validity of the hypothesis in each instance; the hypothesis is simply rejected or accepted on each test, not "interpreted" with regard to its credentials as scientific knowledge.

Hogben favors the Neyman-Pearson-Wald approach for the following reason: Test procedure tells the researcher that a particular event would be one of a class of events that is collectively rare (for instance $p < .05$) if the null hypothesis were true, but the occurrence of such rare events is, by definition, perfectly compatible with the truth of the null hypothesis. Therefore there is no more basis for interpreting any particular event as outside the rare class (and the null hypothesis as false) than there is for interpreting the event as part of the rare class (and the null hypothesis as true). The probabilities involved in the tests refer only to classes of events, not to individual events, and the best one can hope for in using the tests is a certain proportion of correct decisions about hypotheses; meeting a given probability level does not allow an inference about the "significance" of the particular finding for the validity of the hypothesis.

To treat the probabilities of the Calculus of Judgments as indicative of significance in the Fisher tradition is, according to Hogben, to locate probability "in the mind"—the notion that a high degree of belief is warranted about the invalidity of null hypotheses rejected by "significant" findings. Hogben strenuously objects to this practice, preferring instead what he calls a "behaviorist" approach, which treats probability completely in terms of relative frequencies exhibited by observable events. Further, Hogben objects to the negative aspects of Fisher's stress on the rejection of elementary (no-difference) null hypotheses (rather than emphasizing positive knowledge through the development of specific and informative alternative hypotheses). Hogben is not, however, against scientists individually or collectively having beliefs about the validity of hypotheses, and he fully recognizes that making positive inferences must be a part of scientific activity; his criticism is directed generally at attempts to calculate and express such beliefs as probabilities, specifically with significance tests involving elementary null hypotheses.[2]

Hogben thinks the main application of the Calculus of Judgments is in practical applications, as in quality control in manufacturing. In this sense, then, he clearly is not endorsing the Neyman-Pearson-Wald approach as a prescription for scientific inference. He does allow the possibility that the tests may play an occasional minor role in the scientific enterprise as a screen against rash decisions or as advice to the researcher concerning which hunches are worth following. However, he quickly qualifies even this possible contribution of the tests:

> An experienced investigator, with no illusions about the practicality of formulating risks relevant to further effort in numerically intelligible terms consistent with the professional ethic of scientific research, may accordingly prefer to rely on common sense, if statistical theory has nothing better to confer [Chapter 2: 25].

Moreover, Hogben later points out that such a screening convention is worthless unless one considers the power of a test, and that the question of power cannot be resolved outside the context of practical situations requiring prudent research decisions.

In addition to the above issues, "Statistical Prudence and Statistical Inference" touches on an interrelated package of issues dealing with sampling, population, form, and scope, mainly in the context of Hogben's attempt to demonstrate the general inadequacy of the Fisherian approach to significance tests. In "Significance as Interpreted by the School of R. A. Fisher" [Chapter 3], Hogben deals with this group of issues in more detail and with increasing invective. He also documents important inconsistencies in Fisher's position, although some scholars will doubtless claim that both his selection and interpretation of Fisher's statements are biased. Both Fisher's position and Hogben's analysis of these issues are complex but can be briefly summarized as follows.

According to Fisher, the purpose of a significance test is to determine whether two groups, treated and untreated, can be considered samples from the same infinite hypothetical population. Thus, the null hypothesis takes the familiar form that the groups do not differ on some characteristic of interest. If the null hypothesis is rejected on the basis of a significance test, the appropriate inference is that the treatment made a significant difference and the

2. The reader is urged to compare Hogben's views with the notions of Rozeboom [Chapter 24] and Camilleri [Chapter 16]. While the latter writers both favor attempts to assess the probable validity of hypotheses, they are in complete agreement with Hogben in his negative view of using tests of significance for doing so.

groups did not come from the same infinite hypothetical population. Thus, the population involved in the Fisherian view of significance is the conceptual *resultant* of the test procedure.

Hogben points out that the "no-difference" form of the Fisher null hypothesis (sometimes also called the "point null" hypothesis) is insufficiently informative for scientific work, that a random process must be assumed in the formation of the groups, that such a process must be possible in a conceptual framework of infinite repetition on a *particular* population, and that in no sense can an infinite or any other population be the resultant of the test. His critical points all have direct relevance to the use of significance tests in behavioral research, since no-difference nulls and nonrandom samples from what researchers casually claim are infinite hypothetical populations are deeply imbedded practices in these disciplines. Hogben does not point out, however, that in the behavioral science disciplines the inference features to the Fisher approach have to some extent been amalgamated in practice with the features of the decision test approach that involve stating a level in advance and sticking to it unequivocally. It is doubtful that Hogben would look with favor on either this marriage or its offspring.

1. The Contemporary Crisis or the Uncertainties of Uncertain Inference

LANCELOT HOGBEN, F.R.S.

IT IS NOT WITHOUT REASON that the professional philosopher and the plain man can now make common cause in a suspicious attitude towards statistics, a term which has at least five radically different meanings in common usage, and at least four in the context of *statistical theory* alone. We witness on every side a feverish concern of biologists, sociologists and civil servants to exploit the newest and most sophisticated statistical devices with little concern for their mathematical credentials or for the formal assumptions inherent therein. We are some of us all too tired of hearing from the pundits of popular science that natural knowledge has repudiated any aspirations to absolute truth and now recognises no universal logic other than the principles of statistics. The assertion is manifestly false unless we deprive all purely taxonomic enquiry of the title to rank as science. It is also misleading because statistics, as men of science use the term, may mean disciplines with little connexion other than reliance, for very different ostensible reasons, on the same algebraic tricks.

This state of affairs would be more alarming as indicative of the capitulation of the scientific spirit to the authoritarian temper of our time, if it were easy to assemble in one room three theoretical statisticians who agree about the fundamentals of their speciality at the most elementary level. After a generation of prodigious proliferation of statistical techniques whose derivation is a closed book to an ever-expanding company of avid consumers without access to any sufficiently simple exposition of their implications to the producer-mathematician, the challenge of J. Neyman, E. S. Pearson, and Abraham Wald is provoking, in Nietzsche's phrase, a transvaluation of all values. Indeed, it is not too much to say that it threatens to undermine the entire superstructure of statistical estimation and test procedure erected by R. A. Fisher and his disciples on the foundations laid by Karl Pearson, Edgeworth, and Udny Yule. An immediate and hopeful consequence of the fact that the disputants disagree about the factual credentials of even the mathematical theory of probability itself is that there is now a market for textbooks on probability as such, an overdue awareness of its belated intrusion in the domain of scientific research and a willingness to re-examine the preoccupations of the Founding Fathers when the topic had as yet no practical interest other than the gains and losses of a dissolute nobility at the gaming table.

Since unduly pretentious claims put forward for statistics as a discipline derive a spurious cogency from the protean implications of the word itself, let us here take a look at the several meanings it enjoys in current usage. First, we may speak of statistics in a sense which tallies most closely with its original connotation, i.e. figures pertaining to affairs of state. Such are *factual* statistics, i.e. any body of data collected with a view to reaching conclusions referable to recorded numbers or measurements. We sometimes use the term *vital* statistics in this sense, but for a more restricted class of data, e.g. births, deaths, marriages, sickness, accidents and other happenings common to individual human beings and more or less relevant to medicine, in contradistinction to information about trade, employment, education and other topics allocated to the social sciences. In a more restricted sense, we also use expressions such as vital statistics or economic statistics for the exposition of *summarising* procedures (life expectation, age standardisation, gross or net reproduction rates, cohort analysis, cost of living or price indices) especially relevant to the analysis of data so described. By analysis in this context, we then mean sifting by recourse to common sense and simple arithmetical procedures what facts are more or less relevant to conclusions we seek to draw, and what circumstances might distort the true picture of the situation. Anscombe (1951) refers to analysis of this sort as statistics in the sense in which "some continental demographers" use the term.

If we emphatically repudiate the unprovoked scorn in the remark last cited, we must agree with Anscombe in one particular. When we speak of analysis in the context of demography, we do not mean what we now commonly call *theoretical statistics*. What we do subsume under the latter presupposes that our analysis invokes the calculus of probabilities. When we speak of a calculus of probabilities we also presuppose a single formal system of algebra; but a little reflection upon the history of the subject suffices to remind us that: (a) there has been much disagreement about the relevance of such a calculus to everyday life; (b) scientific workers invoke it in domains of discourse which have no very obvious connexion. When I say this I wish to make it clear that I do not exclude the possibility that we may be able to clarify a connexion if such exists, but only if we can reach some agreement about the relevance of the common calculus to the world of experience. On that understanding, we may provisionally distinguish between four domains to which we may refer when we speak of the *Theory of Statistics:*

(i) *A Calculus of Errors*, as first propounded by Legendre, Laplace, and Gauss, undertakes to prescribe a way of combining observations to derive a preferred and so-called *best* approximation to an assumed *true* value of a dimension or constant embodied in an independently established law of nature. The algebraic theory of probability intrudes at two levels: (a) the attempt to interpret empirical laws of error distribution referable to a long sequence of observations in terms consistent with the properties of models suggested by games of chance; (b) the less debatable proposition that unavoidable observed net error associated with an isolated observation is itself a sample of elementary components selected randomwise in accordance with the assumed properties of such models.

Few current treatises on theoretical statistics have much to say about the Gaussian Theory of Error; and the reader in search of an authoritative exposition must needs consult standard texts on *The Combination of Observations* addressed in the main to students of astronomy and geodesics. In view of assertions mentioned in the opening paragraph of this chapter, it is pertinent to remark that a calculus for combining observations as propounded by Laplace and by Gauss, and as interpreted by all their successors, presupposes a putative true value of any measurement or constant under discussion as a secure foothold for the concept of error. When expositors of the contemporary reorientation of physical theory equate the assertion that the canonical form of the scientific law is statistical to the assertion that the new physicist repudiates absolute truth, cause, and effect as irrelevant assumptions, it is therefore evident that they do not use the term statistical to cover the earliest extensive application of the theory of probability in the experimental sciences. I cannot therefore share with my friend Dr. Bronowski the conviction that the statistical formulation of particular scientific hypotheses subsumed under the calculus of aggregates as defined below has emancipated science from an aspiration so old-fashioned as the pursuit of absolute truth. Still less do I derive from so widely current a delusion any satisfaction from the promise of a new Elizabethan era with invigorating prospects of unforeseen mental adventure.

(ii) *A Calculus of Aggregates* proceeds deductively from certain axioms about the random behaviour of subsensory particles to the derivation of general principles which stand or fall by their adequacy to describe the behaviour of matter in bulk. In this context our criterion of adequacy is the standard of precision we commonly adopt in conformity with operational requirements in the chosen field of enquiry. Clerk Maxwell's Kinetic Theory of Gases is the *fons et origo* of this prescription for the construction of a scientific hypothesis and the parent of what we now call statistical mechanics. Beside it, we may also place the Mendelian Theory of Populations.

So far as I know, the reader in search of an adequate account of statistical hypotheses of this type will not be able to find one in any standard current treatise ostensibly devoted to Statistical Theory as a whole. This omission is defensible in so far as physicists and biologists admittedly accept the credentials of such hypotheses on the same terms as they accept hypotheses which make no contact with the concept of probability, e.g. the thermodynamic theory of the Donnan membrane equilibrium. We assent to them because they *work*. Seemingly, it is unnecessary to say more than that, since all scientific workers agree about what they mean in the laboratory, when they say that a hypothesis works; but such unanimity has no bearing on the plea that statistical theory works, when statistical theory signifies the contents of contemporary manuals setting forth a regimen of interpretation now deemed to be indispensable to the conduct of research in the biological and social sciences.

(iii) *A Calculus of Exploration*, which I here use to label such themes as regression and factor analysis, is difficult to define without endorsing or disclaiming its credentials. The expression is appropriate in so far as the ostensible aim of procedures subsumed as such is definable in the idiom of Karl Pearson as *concise statement* of unsuspected *regularities of nature*. This again is vague, but less so if we interpret it in terms of Pearson's intention, and of Quetelet's belief that social phenomena are subject to quantitative laws as inexorable as those of Kepler. Procedures we may designate in this way, more especially the analysis of covariance, may invoke significance tests, and therefore intrude into the domain of the calculus of judgments; but the level at which the theory of probability is ostensibly relevant to the original terms of reference of such exploratory techniques is an issue *sui generis*.

The pivotal concept in the algebraic theory of regression, as in the Gaussian theory of error, is that of *point-estimation*; and the two theories are indeed formally identical. In what circumstances this extension of the original terms of reference of the calculus of errors took place is worthy of comment at an early stage. We shall later see how a highly debatable transference of the theory of wagers in games of chance to the uses of the life table for assessment of insurance risks whetted the appetite for novel applications of the algebraic theory of probability in the half-century before the more mature publications of Gauss appeared in print. The announcement of the Gaussian theory itself coincided with the creation of new public instruments for collection of demographic data relevant to actuarial practice both in Britain[1] and on the Continent. Therewith we witness the emergence of the professional statistician in search of a theory. Such was the setting in which Quetelet obtained a considerable following, despite the derision of Bertrand (1889: 172) and other mathematicians *au fait* with the factual assumptions on which the Gaussian theory relies.

Since an onslaught by Keynes, the name of Quetelet, so explicitly and repeatedly acknowledged by Galton, by Pearson and by Edgeworth as the parent of regression and cognate statistical devices, has unobtrusively retreated from the pages of statistical textbooks; but mathematicians of an earlier vintage than Pearson or Edgworth had no illusions about the source of his theoretical claims nor about the relevance of the principles he invoked to the end in view. So discreet a disinclination to probe into the beginnings of much we now encounter in an up-to-date treatise on statistical theory will not necessarily puzzle the reader, if sufficiently acquainted with the unresolved difficulties of reaching unanimity with respect to the credentials of the remaining topics there dealt with; but we shall fail to do justice to the legitimate claims of its contents, unless we first get a backstage view of otherwise concealed assumptions. In what seems to be one of the first manuals setting forth the significance test drill, Caradog Jones correctly expounds as follows the teaching of Quetelet and its genesis as the source of the tradition successively transmitted through Galton, Pearson and R. A. Fisher to our own contemporaries:

It is almost true to say, however, that, until the time of the great Belgian, Quetelet (1796–1874), no substantial theory of statistics existed. The justice of this claim will be recognized when we remark that it was he who really grasped the significance of one of the fundamental principles—sometimes spoken of as the *constancy of great numbers*—upon which the theory is based. A simple illustration will explain the nature of this important idea: imagine 100,000 Englishmen, all of the same age and living under the same normal conditions—ruling out, that is, such abnormalities as are occasioned by wars, famines, pestilence, etc. Let us divide these men at random into ten groups, containing 10,000 each, and note the age of every man when he dies. Quetelet's principle lays down that, although we cannot foretell how long any particular individual will live, the ages at death of the 10,000 added together, whichever group we consider, will be practically the same. Depending upon this fact, insurance companies calculate the premiums they must charge, by a process of averaging mortality results recorded in the past, and so they are able to carry on business without serious risk of bankruptcy. . . . In his writings he visualizes a man with qualities of average measurement, physical and mental (*l'homme moyen*), and shows how all

1. The Office of the Registrar-General of England and Wales came into being in 1837.

other men, in respect of any particular organ or character, can be ranged about the mean of all the observations. Hence he concluded that the methods of Probability, which are so effective in discussing errors of observation, could be used also in Statistics, and that deviations from the mean in both cases would be subject to the binomial law (1921).

If we are to do justice to the claims of a Calculus of Exploration, we must therefore ask in what sense probability is indeed so effective in discussing errors of observation and in what sense, if any, are Quetelet's authentic deviations from a nonexistent population mean comparable to the Guassian deviation of a measurement from its putatively authentic value. We shall then envisage the present crisis in statistical theory as an invitation to a more exacting re-examination of its foundations than contemporary controversy has hitherto encompassed. After the appearance of his treatise on probability by Keynes, who dismisses Quetelet as a charlatan with less than charity towards so many highly esteemed contemporaries and successors seduced by his teaching, an open conspiracy of silence has seemingly exempted a younger generation from familiarity with the thought of the most influential of all writers on the claims of statistical theory in the world of affairs. Since his views will occupy our attention again and again in what follows, a few remarks upon his career, culled from Joseph Lottin's biography and from other sources, will be appropriate at the outset.

From 1820 onwards Quetelet was director of the Royal Belgian Observatory which he founded. In the 'twenties he professed astronomy and geodesy at the *Ecole Militaire*. The year following the publication of his portentous *Essai de Physique Sociale* (1835), their uncle King Leopold committed to his care Albert of Saxe-Coburg and his brother Ernest for a brief course of instruction on the principles of probability. Correspondence continued between Quetelet and the Prince, who remained his enthusiastic disciple, affirming his devotion to the doctrine of the *Essai* both as president of the 1859 meeting of the British Association in Aberdeen when Maxwell first announced his stochastic interpretation of the gas laws and in the next year as president of the International Statistical Congress held (1860) in London. As official Belgian delegate, Quetelet himself had attended (1832) the third annual meeting of the British Association at Cambridge. There he conferred with Malthus and Babbage, then Lucasian professor of the Newtonian succession and the inventor of the first automatic computer, also famous as author of the *Economy of Manufacture* (1832) and of the *Decline of Science in England* (1830). The outcome of their deliberations was the decision of the General Committee to set up a statistical section.

An incident in the course of Quetelet's relations with Albert is revealing and not without entertainment value. Gossart tells of it thus in the *Bulletin de la Classe des Lettres*, etc., of the Royal Belgian Academy.

Quetelet peu après la publication de ses lettres; presenta à l'Academie un mémoire *Sur la Statistique morale et les principes qui doivent en former la base*. . . . Par une fâcheuse coincidence, le volume dans lequel les théories de Quetelet touchaient à l'art de gouvernement paraissait à Paris au moment on éclatait la révolution de février 1848 et allait "se perdre au milieu des barricades" si bien que quelques exemplaires seulement furent alors distribués. En voyant ce que se passait en France et bientôt dans une partie de l'Europe, le prince Albert ne put s'empêcher de remarquer avec une certaine pointe de malice que le système social était "bien dérangé," que les "causes accidentelles" jouaient un grand rôle. "Le malheur," ajoutait il, "est que la loi qui les gouverne n'a pas été decouverte jusqu'à ce moment" (1919).

Quetelet's belief in eternal laws of human society was *en rapport* with a social philosophy unruffled by such mishaps as the Commune; and its Calvinistic temper is hard to reconcile with the libertarian *credo* for which Eddington finds sanction by appeal to the principle of uncertainty lately propounded by the exponents of statistical mechanics. "Tout en déplorant les maux que font à la société 'les changements brusque et les théories des rêveurs'," says Gossart, he attained *la resignation*. To be sure, "des fleaux frappent l'humanité au morale comme au physique," as he admits; but "quelque destructifs que soient leurs effets, il est au moins consolant de penser qu'ils ne peuvent altérer en rien les lois éternelles qui nous régissent. Leur action est passagere et le temps a bientôt cicatrisé les plaies du corps social."

(iv) *A Calculus of Judgments*, as here defined, ostensibly embraces a regimen of correct inference with respect to the credentials of hypotheses. This form of words is advisedly vague, because it is impossible to prescribe its terms of reference more explicitly without prejudging the outcome of the contemporary controversy we are about to examine. If one states that it includes both the theory of significance and of decision tests and the theory of interval estimation in terms of confidence or fiducial limits, the reader will infer all that we need say definitively and with propriety at this stage.

As such, the calculus of judgments subsumes a programme which is almost exclusively a product of our own century; but the emergence of the aspiration the programme endorses is traceable to the doctrine of inverse probability adumbrated in the posthumous publication of Bayes (1763) and propounded more explicitly by Laplace (1812). The end Laplace himself had in view was to vindicate the credentials of inductive reasoning conceived in retrospective terms consistent with his own cosmogony and hence likewise in terms of dubious relevance to verification in the domain of experimental design. Most statisticians now reject the doctrine in its original form; but its essentially retrospective orientation is profoundly relevant to the contemporary crisis in statistical theory.

At what level the theory of probability can appropriately intrude in a prescription for reasoning rightly is an issue which we can discuss with profit if, and only if, we can arrive at agreement about the relevance of the theory of probability to induction in the traditional sense of the term. This is not the exclusive prerogative of the mathematician. It is the birthright and duty of every self-respecting scientific worker who subjects his data to the type of analysis prescribed by one or other school of statistical inference. That fundamental differences with respect to the relevance of the mathematical theory to the world of real decisions have come into focus so lately is not surprising, when we reflect on the circumstances that its application to the technique of interpretation basks in the reflected glory of the pragmatic triumphs of Maxwell, Mendel, and their successors. In the deductive unfolding of a theory which must stand or fall with the operational requirements of laboratory practice, we are entitled to start from any axioms however arbitrary. We should therefore scrutinise with some suspicion the following remarks about statistical methods by Wilks:

> The test of the applicability of the mathematics in this field as in any other branch of mathematics consists in comparing the predictions as calculated from the mathematical model with what actually happens experimentally (1943: 1).

This assertion would be unexceptionable, if statisticians invoked the algebra of Professor Wilks only in connexion with the genetical theory of populations, Brownian movement of visible particles, the collisions of gas molecules, the emission of photons, and with cognate themes which constitute the scope of a calculus of aggregates; but such topics have no direct relevance to what we imply by statistical theory in the context of a calculus of judgments. In the calculus of aggregates we invoke the theory of probability to prescribe a hypothesis; but a Calculus of Judgments does not undertake to prescribe hypotheses. It claims only to prescribe a rule which will entitle us to arbitrate on their merits. One may hence ask with propriety what controlled experiment prosecuted on either side of what iron curtain over what number of centuries would settle the dispute between Jeffreys and Fisher concerning the legitimacy of Bayes's postulate or the contest between Fisher and Neyman over test procedure. When the terrain of combat is the realm of means, experience and experience alone should dictate the outcome. When it is in the realm of ends we cannot invoke pragmatic sanctions with the assurance of an acceptable decision.

My subtitle [to Hogben's *Statistical Theory: An Examination of Contemporary Crisis in Statistical Theory from a Behaviourist Viewpoint*] does not overstate what is a real intellectual dilemma of our time. *Crisis* is a word which has become tarnished by misuse; and some of my readers may well wish me to justify the statement that there is indeed a contemporary crisis in statistical theory. Poincaré cites a remark that everyone believes in the normal law of error, the physicists because they think that the mathematicians have proved it to be a logical necessity, the mathematicians because they believe that physicists have established it by laboratory demonstration. The gap between theory and practice has vastly deepened since his time, as is evident from the concluding remarks of E. S. Pearson in the following excerpt:

> That the frequency concept is not generally accepted in the interpretation of statistical tests is of course well known. With this characteristic forcefulness R. A. Fisher (1945) has recently written: "In recent times one often repeated exposition of the tests of significance, by J. Neyman, a writer not closely associated with the development of these tests, seems liable to lead mathematical readers astray, through laying down axiomatically, what is not agreed or generally true, that the level of significance must be equal to the frequency with which the hypothesis is rejected in repeated sampling of any fixed population allowed by hypothesis. This intrusive axiom, which is foreign to the reasoning on which the tests of significance were in fact based, seems to be a real bar to progress. . . ."
>
> But the subject of criticism seems to me less an intrusive mathematical axiom than a mathematical formulation of a practical requirement which statisticians of many schools of thought have deliberately advanced. Prof. Fisher's contributions to the development of tests of significance have been outstanding, but such tests, if under another name, were discovered before his day and are being derived far and wide to meet new needs. To claim what seems to amount to patent rights over their interpretation can hardly be his serious intention. Many of us, as statisticians, fall into the all too easy habit of making authoritative statements as to how probability theory should be used as a guide to judgment, but ultimately it is likely that the method of application which finds greatest favour will be that which through its simplicity and directness appeals most to the common scientific user's understanding. *Hitherto the user has been accustomed to accept the function of probability theory laid down by the mathematicians; but it would be good if he could take a larger share in formulating himself what are the practical requirements that the theory should satisfy in application"* (1947: 142 [Hogben's italics]).

Meanwhile the user, as Pearson calls him, continues to perform an elaborate ritual of calculations quite regardless of the fact that there are now at least three schools of theoretical doctrine with no common ground concerning what justification we have for applying a calculus of probability to real situations and with little agreement about how we should proceed to do so. Lest some readers should regard this as an overstatement, it will not be amiss to quote from a recent symposium following a paper read before the Royal Statistical Society by Anscombe (1948) who undertook the courageous assignment of an impartial appraisal of the views respectively advanced during the last decade by R. A. Fisher, by H. Jeffreys and by J. Neyman and E. S. Pearson. Dr. J. O. Irwin who opened the discussion said:

> I think all students of statistics should learn something about probability from a frequency point of view. When teaching students with mature minds who are

yet new or almost new to the subject, I usually give an outline of the different theories of the subject, tell them that they will find the frequency theory the most useful in practice, and to suspend judgment on which theory they will ultimately prefer as a basis until they have had more opportunity of study.

Practically minded people with no great taste for logical and philosophical speculation need not probe too deeply. They will probably be just as good statisticians if they don't. More theoretically profound minds will gain much insight if they do and will be able to help the others on critical occasions. But we must admit that what level we agree to call axiomatic is largely a matter of taste (1948: 201).

Professor G. A. Barnard, who followed Dr. Irwin, said in a more explicitly accommodating vein:

All three main theories of statistical inference seem to have their proper sphere of application. What we should ask is, not so much which is right, but to what sort of field each theory should be applied; which framework is better in certain circumstances. For example, in considering industrial inspection Dodge and Romig, two engineers, evolved the notions of producer's risk and consumer's risk, and these have been practically useful in statistical inspection. They are identical in content with the Pearson notion of errors of first and second kind. Again, during the war we had occasion to deal with other sorts of inspection problems, and in this connection we introduced the idea of the process curve, which is nothing but the *a priori* distribution of Professor Jeffreys. I must admit that in my own experience so far, cases where Professor Fisher's theory would have been most suitable have not been very frequent. I think that is because most of my problems have been those where it is necessary to make an administrative decision, rather than those in which one is concerned to establish or disprove a scientific theory. Our work is more concerned with immediate practical decisions, but I do not doubt that with wider experience I should have been able to quote practical cases for that also.

We are, then, left with three theories—the Jeffreys theory, the Pearson theory, and the Fisher theory. I think when we are discussing the foundations of statistics we should draw attention to the fact that this discussion is really, from a practical point of view, a discussion of the fine points of detail. All statisticians agree about what should be done in practical problems. The situation in statistics is really quite like the problem in mathematics. The foundations of mathematics have been discussed and queried for a long time. These discussions are now so broad and widespread that there is a journal devoted to them entirely. Yet no mathematician doubts that any of the mathematics are sound. . . . We have also to remember that a significance test, interpreted somewhat narrowly, as it must be, only allows us to say what is not true, but that does not involve proving a general proposition. In this connection a remark of Professor Jeffreys is worth quoting —that the methods of significance tests used in this way seem to enable us to disprove a great deal but never to prove anything. . . . I should like finally to make it clear that I disagree with all four parties to the controversy. The snag in

Professor Jeffrey's theory is that to work it one has to specify a probability distribution for a class of alternative hypotheses and the whole of the probability has to be distributed. One must when interpreting one's experiments be able to think of all possible explanations of the data, and that, I think, none of us believe that we can do. It is always possible for someone to produce later an entirely new explanation we had never thought of, and which would not be represented in the hypothesis nor in the alternatives we had tested.

Taking that criterion it does suggest that in talking about inference from the probability point of view, leaving aside the rigorous ground of the randomization test, we ought to formulate our ideas, not in terms of probability, but in terms of odds. Bayes's theorem, in terms of odds, says that *a posteriori* odds = likelihood ratio (A) × *a priori* odds. We can separate out the two factors on the right-hand side, and it seems to me that along these lines it is possible to reconcile to some extent the various theories. Professor Jeffreys takes the second factor as equal to one by a special axiom or assumption. Professor Fisher seems to say we ought to neglect the second factor; but that is equivalent to saying that A times 1 is equal to A. Finally, Neyman and Pearson say that A itself is a frequency probability of errors, and this is so provided that the reference set used is that of the sequential probability ratio test (1948).

Reported in *oratio obliqua,* Professor E. S. Pearson said with more insight into the consumer's viewpoint:

It has yet to be shown that a mathematical theory could make possible their assimilation into the process of inference on a numerical basis. In balancing these elements to reach conclusions leading to action the power of judgment was called into play; it was something which might be intuitive, a quality which scientific training aimed at developing, but whose possession was no monopoly of the statistician. The judgment might be an individual one or it might be a collective one according to the magnitude of the problem. The question which he raised was whether it was possible to lay down usefully any formal rules of induction, specifying how the various aspects of the problem needing review could be brought together to reach a balanced decision (1948b).

None of the contributors to this symposium advanced the most usual excuse for renouncing the traditional obligation of the man of science to understand what he is doing, i.e. the assertion that the procedures described by statistical theories work [*sic*] in spite of the fact that there is so exiguous a basis of agreement about their credentials. It is gratifying to record doubts about its cogency expressed by one of them in another context, if only because the consumer with no appetite for methodological disputation all too readily succumbs to reassurance on such terms. Acceptability of a statistically *significant* result of an experiment on animal behaviour in contradistinction to a result which the investigator can repeat before a critical audience naturally promotes a high output of publication. Hence the argument that the techniques *work* has a tempting appeal to young biologists, if harassed by their seniors to produce results, or if admonished by editors to conform to a prescribed ritual of analysis before publication. A reminder that the plea for justification by works derives its sanction from a different domain of statistical theory is therefore likely to fall on deaf ears, unless we reinstate reflective thinking in the university curriculum. Meanwhile, the views of E. S. Pearson on the teaching of statistics will commend themselves to the reflective few who entertain a pardonable scepticism about the allegedly useful contribution of current theories in so-called operational research:

> Probably there are several of us who can recall a considerable number of re-
> ports, or appendices to reports, written on both sides of the Atlantic by mathe-
> matically trained statisticians, which were hardly more than a waste of the paper
> on which they were written. There were cases where the results of statistical
> analyses were simply put on one side because the practical man, whether scientist
> or service technician, shrewdly sensed that the theoretical treatment was either
> not needed, or was actually leading to conclusions which the data could not
> possibly warrant. The trouble usually arose because the mathematical enthusiast
> had allowed his theory to run away with his common sense, or, perhaps, because
> he had never received an adequate training in the application of theory. It was
> true that biologists did extremely well in operational research; but their success
> often seemed due to the way in which an experimental training had taught them
> to handle data rather than to the fact that they mastered statistical technique
> quickly (1948a: 218).

I hope that these citations will dispel any doubt about whether there are very fundamental differences within the hierarchy of theoretical statistics concerning what the theory of probability can contribute to a regimen of scientific inference. They also disclose a widespread disposition on the part of the makers of the theory to disclaim at all costs any relevance of their differences to the requirements of those who use it. Kendall, who is deeply disturbed by the clamour the contemporary controversy has assumed, attempts in a recent paper "On the Reconciliation of Theories of Probability" to resolve disagreement by making explicit on what axioms widely current techniques in use depend; but his approach is essentially that of the pure mathematician seeking to remove *internal* inconsistencies of an otherwise satisfactory calculus. From the standpoint of the user, this accomplishes nothing unless he can also show that otherwise arbitrary postulates have any verifiable foundation in *external* experience. The engaging humour of his final remarks, which I shall now quote, suggests that Kendall himself is not wholly satisfied with the outcome of his pacific negotiations:

> A friend of mine once remarked to me that if some people asserted that the
> earth rotated from east to west and others that it rotated from west to east, there
> would always be a few well-meaning citizens to suggest that perhaps there was
> something to be said for both sides, and that maybe it did a little of one and a

little of the other; or that the truth probably lay between the extremes and perhaps it did not rotate at all (1949: 115).

It would be less necessary to insist that the issues at stake are of the utmost importance to the user, especially to the vast class of users who take the techniques on trust, if theoretical statisticians were content to arbitrate on the value of conclusions advanced by research workers on the basis of enquiries designed *ad hoc* and with due regard to background knowledge of the enquiry. Of late, more especially during the last fifteen years, they have in fact advanced claims with much wider terms of reference, as illustrated by the following citation from R. A. Fisher:

> Inductive inference is the only process known to us by which essentially new knowledge comes into the world. To make clear the authentic conditions of its validity is the kind of contribution to the intellectual development of mankind which we should expect experimental science would ultimately supply. . . .
>
> It is as well to remember in this connection that the principles and method of even *deductive* reasoning were probably unknown for several thousand years after the establishment of prosperous and cultured civilisations.
>
> . . . The liberation of the human intellect must, however, remain incomplete so long as it is free only to work out the consequences of a prescribed body of dogmatic data, and is denied the access to unsuspected truths, which only direct observation can give. . . .
>
> . . . The chapters which follow are designed to illustrate the principles which are *common to all experimentation*, by means of examples chosen for the simplicity with which these principles are brought out. Next, to exhibit the principal designs which have been found successful in that field of experimentation, namely agriculture, in which questions of design have been most thoroughly studied, and to illustrate their applicability to other fields of work (1935a, 5th ed.: 7–9 [Hogben's italics]).

This passage is instructive more because of what it implies than because of what it explicitly asserts. We get the impression that recourse to statistical methods is prerequisite to the design of experiments of any sort whatever. In that event, the whole creation of experimental scientists from Gilbert and Hooke to J. J. Thomson and Morgan has been groaning and travailing in fruitless pain together; and the biologist of today has nothing to learn from well-tried methods which have led to the spectacular advances of the several branches of experimental science during the last three centuries. Nor is this all. We learn that we shall find the pattern of new and more powerful methods in a procedure for carrying out agricultural field trials prescribed, though Fisher does not say so, by one school of statisticians, and one alone.

What we then naturally ask is whether consequent advances of our knowledge of the soil, if any, have been commensurate with such a claim. In the parallel domain of marine biology, our knowledge of how to reproduce all relevant conditions for the culture of sea creatures has made great strides by recourse to entirely traditional principles of experimentation, while our theoretical knowledge of the growth needs of plants has not conspicuously broadened as the outcome of experiments designed in conformity with the demands of Greco-Latin Squares or randomised blocks. In the latter domain the claim that the theoretical statistician knows better than the man on the job how to do it is one which derives its sanction from a particular theory of statistical inference; and Fisher himself would be first to admit this. If the theory turns out to be false, the result of increasingly widespread use of methods prescribed to design experiments must result both in curbing the ingenuity of the investigator at stupendous cost of time and in deterioration of standards of good workmanship in the laboratory. I believe we can already detect signs of such deterioration in the growing volume of published papers —especially in the domain of animal behaviour—recording so-called significant conclusions which an earlier vintage would have regarded merely as private clues for further exploration. Be that as it may, the fact that such methods are now in general use signifies that it is not merely an academic exercise to clarify the credentials of current views on statistical inference. Least of all is it merely a matter of moment to the trained mathematician at a time when trained mathematicians cannot reach agreement about them among themselves.

Should our adjudication lead us to embrace, with all its as yet half-formulated implications, the viewpoint of the new American school, the consequences will be far more drastic than many of our island contemporaries as yet recognise. In the closing words of his essay *Of the Academical or Sceptical Philosophy* Hume asks: "when we run over libraries persuaded of these principles, what havoc must we make?" Such havoc I suggest that little if anything in the cookery books will remain. We may have to reinstate statistics as continental demographers use the term. Laboratory experiments will have to stand on their own feet without protection from a façade of irrelevant computations. Sociologists will have to use their brains. In my view, science will not suffer.

2. Statistical Prudence and Statistical Inference

LANCELOT HOGBEN, F.R.S.

AT THE CORE of contemporary controversy in statistical theory is the following question: What bearing, if any, has the rarity of an observable occurrence as prescribed by an appropriate stochastic hypothesis on our legitimate grounds for accepting or rejecting the latter when we have already witnessed the former? The form of the answer we deem to be appropriate will define what we here conceive to be the proper terms of reference of a Calculus of Judgments, i.e. *statistical inference* as some contemporary writers use the term. Such is the theme of this chapter.

At the outset, it will forestall misunderstanding if we concede that some contemporary writers use the term statistical inference in a wider sense than as defined above, embracing any sort of reasoning which takes within its scope considerations referable to a calculus of probability. Such usage is regrettable, because the form of words suggests a more radical difference. By inference in the traditional sense of the term as used by the logicians of science from Bacon to Mill and Jevons, we signify rules which *unreservedly* lead to correct conclusions. By statistical inference, or, as we shall later say, *stochastic induction*, we here imply rules of reasoning which lead to correct conclusions subject to a reservation expressible in stochastic terms. We relinquish the

Reprinted from *Statistical Theory* by Lancelot Hogben, F.R.S. By permission of W. W. Norton & Company, Inc. Copyright All Rights Reserved 1957, 1968 by W. W. Norton & Company, Inc.

claim that any such rule *always* leads to a correct conclusion. All we claim for such a rule is that it guarantees an assignable probability of correct or false assertion, if we apply it consistently to the relevant class of situations.

By an assignable probability we then mean the proportion of true or false assertions to which the rule will lead us in an endless sequence of such situations, and the theory of probability can endorse such a rule if, and only if, the assemblage of data for our observational record is the outcome of random-wise sampling. We thus specify the proportion of false assertions as the *uncertainty safeguard* (P_f) of the rule with an upper acceptable limit $P_f \leqslant \alpha$. Alternatively, we may assign a lower acceptable limit $P_t = (1 - P_f) \geqslant (1 - \alpha)$ to the proportion of true statements which consistent application of the rule will endorse in the long run. We may then speak of such a lower limit as the *stochastic credibility* of the rule. What limit we deem to be acceptable in this context is not a logical issue. If we commonly set $\alpha = 0.05$ for illustrative purposes in the course of the chapters which follow [in Hogben's book], we do so because it is a widely current convention which will help some readers to feel at home in familiar territory.

In pursuing our theme on this understanding and historically as heretofore, we may conveniently distinguish between two sorts of statistical inference as *test procedure* and *interval estimation*. The dichotomy is provisional. For we shall later see that interval estimation embraces test procedure as a special case. Meanwhile, two difficulties beset our task. One arises from the circumstance that a strictly behaviourist formulation of the terms of reference of a stochastic Calculus of Judgments, as in the preceding paragraph, is wholly consistent with the attitude to test procedure adopted by only one of two different schools of doctrine; and its implications have come clearly into focus only since the contemporary controversy has forced the contestants to make explicit latent assumptions lazily embraced by their predecessors. Contemporaneously, claims put forward by exponents of the alternative school have changed in response to criticism not previously anticipated. Thus our foothold on fact is insecure when we seek to evaluate the original intentions of the authors with due regard to the way in which the consumer has meanwhile interpreted them.

Since emphasis on differences of test prescription and on what each type of test can accomplish has shifted in the turmoil of discussion, it is not easy to circumscribe acceptably the connotation of the term *decision test* as defined by Wald on the basis of views first advanced by J. Neyman and E. S. Pearson in contradistinction to the term *significance test* as prescribed by the school of R. A. Fisher in the older tradition of Karl Pearson and Udny Yule. In this chapter, we shall therefore disclaim any attempt to verbalise the distinction with finality. Instead, we shall restrict our attention to the historical background of the controversy. Against this background, we may formulate a preliminary statement of what a significance test is and of what are its claims to usefulness.

We may trace the beginnings of a view of test procedure widely current during the twenties and the thirties of this century to the practice of citing with point-estimates in surveying and in the physical sciences an *empirical* assessment *either* of the precision index (h) prescribed by the Gaussian, as we now say *normal*, law of error *or* of some parameter of the distribution related thereto, more especially the so-called *probable error*. For a normal error distribution of known precision, the probable error ($p.e.$) defines a continuous class of corresponding negative and positive deviations from the mean bounding half its total area. In sampling randomwise, equal probability thus attaches to the occurrence of deviations numerically less than and deviation numerically greater than the $p.e.$ The probability that a deviation from the mean will numerically exceed no more than three times the $p.e.$ is about 0.95.

Throughout the nineteenth century, those who followed this procedure did so with little disposition to claim that it embodies a major innovation of scientific reasoning. Nor did they need to do so. In the domain of precision instruments and of mechanical defects, it would be easy to cite a variety of situations in which the specification of a deviation from the putative true value of a dimension or constant has a utility none the less admissible because no professional logician would endorse its title to disclose a new formula for scientific induction. A single fictitious and over-simplified example from the contemporary setting of quality control (statistical inspection) at its most elementary (p-chart) level will suffice to clarify what we can legitimately and usefully say about a probable error in the Gaussian domain.

We shall suppose that: (*a*) a machine turns out lengths of metal wire of thickness guaranteed to lie within 0.98 and 1.02 mm.; (*b*) the variation of the thickness under normal working conditions is approximately normal with a probable error 0.004 mm. about mean 1.0 mm. Thus values which numerically exceed the mean by more than the guaranteed 0.02 mm. will turn up in only about one four hundredth of a very long sequence of samples during the normal production process. So long as the machine delivers the guaranteed product there is no need to interfere with it; but it may well run off a length of wire 1.025 mm. thick. The production engineer has then to face a dilemma. In which of two ways is he to interpret the event? He will know that the mishap may well be an extreme example of the uncontrollable vagaries of the mechanical set-up. He may also legitimately suspect that the machine has developed a fault which calls for repair. Without incurring the risk of having a large consignment of defective products on his hands, he can immediately settle the issue by an overhaul; but this spells needless loss of time if his suspicion is groundless. A plausible rationalisation of his choice therefore presupposes the possibility of balancing the cost of taking a wise precaution which may be unnecessary against the possible penalty of failing to do so.

His predicament is indeed on all fours with that of the reflective house-holder whose dog does not habitually bark in the night. There may be no burglar, when the dog does indeed bark. On the other hand, the risk may be worth the effort of getting out of bed, if there is enough portable property of value to conserve. Neither the householder who justifies his decision to make a search on these terms, i.e. as a wise rule of conduct, nor the engineer, who justifies the procedure of overhauling the machine as the less calamitous of two acts of choice, necessarily commits himself to an *inference* in the sense in which logicians traditionally use the term. To be sure, we might say of the householder and of the production engineer that each is testing the truth of alternative hypotheses; but the recognition of the rarity of the relevant event as a danger signal rather than the interpretation of its outcome dictates the choice of the test procedure; and the finality of the selected test has nothing to do with the rarity of the event which prompts its choice. If the dog's bark evokes the decision of the householder to go downstairs, he will admittedly be able to infer the truth or falsity of the hypothesis that a burglar is on the premises; but if he decides to remain in bed, he will also know before breakfast whether there has been an entry.

It suffices to speak of the engineer's rule as a rule of *statistical prudence;* but we may reasonably conclude that the theoretical statistician means something more than this when he uses a form of words so portentous as *statistical inference*. Indeed, some contemporary writers pinpoint such a distinction by using the term *conditional*, in contradistinction to unconditional, inference for what we here refer to as statistical prudence. Contrariwise, others make no such distinction and seemingly claim for what a test procedure can accomplish little more than the prescription of a wholesome discipline. If so, the research worker who embraces a test procedure presumably undertakes what may be a time-consuming programme of laborious computations in the mistaken belief that the statistician has much more to offer.

When we speak of testing a hypothesis, the form of words suggests a procedure for endorsing a valid verdict for or against its title to subsume new information worthy to take its place in the enduring corpus of scientific knowledge. The consumer has therefore the right to know that he is doing nothing of the sort, if the test is merely a *screening* convention to check rash decisions or to arbitrate on the advisability of following up a plausible hunch. An experienced investigator, with no illusions about the practicability of formulating risks relevant to further effort in numerically intelligible terms consistent with the professional ethic of scientific research, may accordingly prefer to rely on common sense, if statistical theory has nothing better to confer. The truth is that the distinction between statistical inference as an interpretative device defined at the beginning of this chapter and statistical prudence conceived as a discipline to forestall rash judgments did not become clear till Fisher repudiated the scholium of Bayes (1763) in his "Mathematical Foundations of Theoretical Statistics" (1921). This evoked a vigorous rejoinder from Jeffreys (see 1939: §7.4), whose confessedly idealistic approach to the theory of probability would not otherwise fall within the scope of an assessment of the present crisis from a behaviourist viewpoint.

Jeffreys propounded what Anscombe (1948) calls a *proper theory of induction*, i.e. a system of which the terms of reference are at least intelligible to those who find the initial assumptions acceptable. In controversy with him, Fisher consistently refused to commit himself explicitly to a definition of probability located in the mind, and, by so doing, enlisted the sympathy of contemporaries to whom such a formulation would be repugnant; and nothing in his earlier publications unequivocally advances the claims of test procedure as an innovation of logical technique. As controversy over the scholium sharpened, we can trace the emergence of a new *motif* in his writings. In his "Mathematical Foundations" referred to above, he elaborates the formal algebra of a test procedure essentially *en rapport* with that of his predecessors in the tradition of Karl Pearson, and in an idiom which does not provoke the professional logician to assent or denial. Only when controversy with Jeffreys had clarified some of the cruder implications of abandoning the doctrine of Laplace did his assertion of the claims of *uncertain inference* explicitly annex a procedure which the logician might reasonably have regarded as the more fitting preserve of a disciplinary precaution.

The treatise of Jules Gavarret (1840) signalises what seems to be the earliest intrusion of the probable error into the domain of natural variation. Therein Gavarret propounds a statistical approach to the evaluation of the efficacy of a treatment. It is in essence the significance test as commonly conceived in the twenties and thirties of our own century; but it seems to have exercised little influence on medical research in the author's own lifetime, perhaps because overshadowed by spectacular contemporaneous advances in experimental physiology. Meanwhile, there was a soil more favourable than medicine for the seed of the word. In the domain of natural variation, comparative anatomy was in the ascendant. Under the impact of the Darwinian doctrine in the setting of Galton's racialist creed and of the controversy over slavery in America, anthropometry became a fashionable academic playground. Thus we find one of the earliest examples of the use of the terms *test* and *significant* in their now current meaning in a memoir by J. Venn in the *Journal of the Anthropological Institute:*

> But something more than this must be attempted. When we are dealing with statistics, we ought to be able not merely to say vaguely that the difference does or does not seem significant to us, but we ought to have some test as to what

difference would be significant. For this purpose appeal must be made to the theory of probability. Suppose that we have a large number of measures of any kind, grouping themselves about their mean in the way familiar to every statistician, their degree of dispersion about this mean being assigned by the determination of their "probable error." . . . For instance, the difference in the mean length of clear vision between the A's and the C's is about an inch and a quarter; that between the same classes, of the age of 24, is slightly more, viz. about an inch and one-third. But the former is the difference between the means of 258 and 361, the latter that between means of 25 and 13. By the formula above given we find that the respective probable errors of the differences between these means are one-twelfth and one-third of 3.7 inches, i.e. about 0.3 inches and 1.2 inches. The latter is almost exactly the observed difference, which is therefore seen to be quite insignificant. The former is about one-quarter of the observed difference, which is therefore highly significant; for the odds are about 25 to 1 that a measure of any kind shall not deviate by three times its probable error.

The above remarks are somewhat technical, but their gist is readily comprehensible. They inform us which of the differences in the above tables are permanent and significant, in the sense that we may be tolerably confident that if we took another similar batch we should find a similar difference; and which of them are merely transient and insignificant, in the sense that another similar batch is about as likely as not to reverse the conclusion we have obtained (1888:147–148).

All this adds up to little in terms of the world's work, because anthropometry had then (as now) scanty, if any, practical value other than for manufacturers of ready-made wearing apparel or of school furniture. That reliance on test procedures so conceived suddenly becomes so universally *de rigueur* among experimental biologists in the twenties and thirties of our own century is an enigma which admits of no wholly satisfactory solution; but one clue to the mystery may be the fact that the influence of men such as Farr had already popularised interest in descriptive medical statistics in anticipation of public legislation to enforce immunisation procedures. By the end of the nineteenth century, the vaccination issue had indeed become the focus of a vehement public debate which provided a propitious setting for the reception of the views Gavarret had earlier expressed. Had it been true that men of science were themselves of one accord in the polemics of the last decade of the nineteenth and of the first decade of the twentieth century, the outcome might have been otherwise; but the opponents of vaccination could themselves claim enthusiastic supporters among biologists, for instance Alfred Russel Wallace. In such circumstances, a counsel of moderation could prevail against reckless washing of dirty linen in public only by taking the issue to a higher court of appeal with a better prospect of assembling a unanimous bench of respected judges.

Such was the situation when Yule and Greenwood published a memoir which was the curtain-raiser to the appearance on the stage of the statistician as arbitrator on matters about which trained observers disagree. The publication of the last named authors bore the title "The Statistics of Anti-typhoid and Anti-cholera Inoculations and the Interpretation of Such Statistics in General" (1915). It appeared in the middle of a world catastrophe, at the end of which the notion of significance becomes for the first time prominent in a new *genre* of statistical textbooks, such as those of Bowley and of Caradog Jones (1921). Meanwhile a generation suckled on Karl Pearson's *Grammar of Science* (1892) and attuned to the controversies of *Biometrika* were rising to positions of influence in the biological world. Of such were Raymond Pearl in America and in England Major Greenwood himself.

By then, food shortage at the end of the First World War had catalysed interest both in deficiency diseases and in agricultural output. The former merits comment because research on the vitamins enlisted methods of bio-assay which rely on group averages in contradistinction to one-to-one corre-spondence of stimulus and response of one and the same individual. Food production is especially relevant because there was now a ready audience for the honest broker of the vaccination controversy, when R. A. Fisher (1925a), then engrossed in fertiliser records, published a well-known textbook osten-sibly addressed to research workers in general, though in fact almost exclu-sively concerned with the agricultural field trial. Almost overnight it became a best seller. In America, agricultural statisticians, there led by Snedecor (1937) became the most enthusiastic converts to an evangel which is indeed a still unanswered challenge to Bernard's teaching and to the Baconian recipe.

Pari passu throughout the thirties, the use of statistical tests in the experi-mental sciences became more fashionable; but there was little inclination to probe their several claims in the domains of error and of natural variation. Meanwhile medicine remained aloof from the statistical approach to the therapeutic or to the prophylactic trial until unprecedented production of new synthetic drugs and of antibiotics on the eve of, and during, a second world war. An informed public now eagerly awaited a verdict on their merits. By that time, controversy concerning the current statistical recipes had become vocal and vigorous in mathematical circles; but its echoes did not penetrate the pathological laboratory. It would be unnecessary to write this book, if many men of science did as yet realise how little the College of Cardinals can agree about the rubric.

I have deliberately used the term *field trial* in the foregoing remarks. Such at first, and rightly, was the designation of what the statistician of a later vintage refers to as *experiment*; but the acceptability of the new title is instruc-tive, because it focuses attention on a confusion of aims. The twenties and thirties witnessed rapid advances in techniques of bio-assay first as tools of pure research in dietetics and in endocrinology, later as the means of accredit-ing the reliability of products in commercial production. Juxtaposition of statistical tests employed in comparable circumstances in the milieu of the new pharmacy seemingly encouraged the belief that there is a common denominator for the objectives which a commercial corporation and a disinterested scientific worker pursue. As Abraham Wald explicitly defines it, and as several expositors of Fisher's methods for the use of pharmacologists define it by implication, statistical inference is this common denominator.

Perusal of the earliest prescriptions for the testing of hypotheses may well leave the reader in considerable doubt about what the pioneers did claim. What is clear is that they carried over from the proper domain of the Gaussian theory of point-estimation a body of concepts with disputable relevance to a new class of situations, when the immunisation controversy recruited the statistician as the arbitrator of truth and falsehood. The ill-fated marriage of the Gaussian theory of error with the empirical study of populations had already generated a sturdy progeny of misconceptions dealt with elsewhere. Hence we may record the event without bewilderment, even if the *ipsissima verba* of Karl Pearson, who played a leading role in the background of the debate, do not greatly help us to formulate a clear distinction between the scope of scientific induction as interpreted in the English tradition from Bacon to J. S. Mill and induction restated in terms of stochastic theory. We encounter an early use of the now familiar expression *significant difference* in Pearson's discussion on the "Probability that Two Independent Distributions of Frequency Are Really Samples from the Same Population"; and it would be difficult to select a more forceful illustration of the *Backward Look* than in the following citation therefrom:

> In a memoir (1900: 157), I have dealt with the problem of the probability that a given distribution of frequency was a sample from a *known* population. That investigation was the basis of my treatment of the "goodness of fit" of theory and frequency samples. The present problem is of a somewhat different kind, but is essentially as important in character. We have two samples, and *a priori* they may be of the same population or of different populations; we desire to find out what is the probability that they are random samples of the same population. This population is one, however, of which we have no *a priori* experience. It is quite easy to state innumerable problems in which such knowledge is desirable. We have two records of the number of rooms in houses where (i) a case of cancer has occurred, (ii) a case of tuberculosis has occurred; the number of cases of each disease may be quite different, and we may not be acquainted with the frequency distribution of the number of rooms in the given district. *What is the chance that there is a significant difference in the tuberculosis and the cancer houses?* Or again, we have a frequency distribution of the interval in days between bite and onset of rabies in two populations of bitten persons (i) who have been and (ii) who have not been inoculated in the interval. What is the probability that the inoculation has modified the interval? Many other illustrations will occur to those who are dealing with statistics, but the above will suffice to indicate the nature of the problems I have in view (1911: 250 [Hogben's italics]).

For Pearson, as for Venn (1866), a difference is *significant*, in the sense that it casts doubt on a hypothesis, if the chance of its occurrence is very small on the assumption that the hypothesis is true. Thus it would not be easy to confuse so many issues with so few words as those of the query: "what is the chance that there is a significant difference in the tuberculosis and the cancer houses?" The publication of Yule and Greenwood (1915)[1] cited above, defines significance in the same way. At the outset, the authors give their own interpretation of the question: "Is there a significant difference between the attack or fatality rates of the two classes?" To them—as to a subsequent generation of consumers—this is strictly equivalent to asking: "Is the observed difference greater than we could fairly attribute to the action of chance?" The answer they seek, but without attempting a definition of *fairly* in this context, is indeed an answer to a different question: how often "errors of sampling would lead to as great a discrepancy as or a greater discrepancy than that actually observed between *theory* and observation" [Hogben's italics].

One implication of the foregoing is that we can first look at our sample and then decide what rule to apply. This is wholly inconsistent with a behaviourist approach. Moreover there is another objection, which we can get into focus more readily if we examine test prescription through the spectacles of the Forward Look. In effect, we then say that we shall reject a particular hypothesis if the deviation (X) of the sample score from the mean (M_x) of the distribution *numerically* exceeds a certain value X_r, so chosen that the probability assigned by the same hypothesis to sample scores in the range $M_x \pm X_r$ is $P_t = (1 - P_f) = (1 - \alpha)$. The choice of a *score rejection criterion*, so defined, raises a debateable issue at the outset. If the rarity of an observed occurrence as prescribed by a particular hypothesis can indeed endorse any intelligible grounds for rejecting it, there is still no obvious reason, other than the culture-lag of the Quetelet mystique, to compel us to define a score rejection criterion in this *two-sided* way. If our real concern is with the possibility that inoculation may lower the attack rate, our criterion of rarity will be relevant to the end in view only if we define it in a *one*-sided way, i.e. in terms of score values less than one which we may denote alternatively as x_r if the origin of the distribution is the least value (here zero) x may have, or $- X_r$ if we transfer the origin to the mean for algebraic convenience.

The distinction is worthy of emphasis, because writers in the Yule–Fisher tradition, though commonly disposed to adopt the *modular* (two-sided) rejection criterion when the sampling distribution is symmetrical, otherwise

1. The first edition of Yule's *Introduction to the Theory of Statistics* (1911) devotes no more than three pages to illustrations of a method of testing the significance of a difference, i.e. the truth of the hypothesis that the universes from which different samples come are in all relevant particulars alike. The rest of the book deals with summarising indices devised in accordance with stochastic considerations in the manner of Quetelet with no explicit recognition of the need to distinguish between a rule of decision and a law of nature.

employ a *vector* (one-sided) rejection criterion without explicitly disclosing the relevance of the algebraic properties of a particular sampling distribution to the factual content of the decision involved. The use of the word *theory* in the foregoing citation from Yule and Greenwood helps us to trace this confusion of aim to its source. What they here *signify* by theory is the long-run mean value of the sample difference, i.e. *zero* if the universes from which two samples come are alike with regard to all relevant particulars; but in this context the word is an unwarranted intruder from the proper domain of precision instruments. We cannot here assume that any universe parameter has a *true* value which would suffice to define the sample structure if we did not make mistakes. The situation to which Yule and Greenwood refer is in no relevant respect comparable to the class of problems we discuss in the Gaussian domain. No parameter of the distribution has any special claim to dictate how to delimit from the class of all samples a particular sub-class to which we may be able to assign a very low frequency. Indeed, we shall see that the repudiation of such an arbitrary choice is an essential feature of the theory of test procedure expounded by E. S. Pearson, Neyman, and Wald.

It goes without saying that the adverb *fairly*, as used by Yule and Greenwood, begs the whole question at issue. The algebra which the test invokes tells us how often we should encounter a certain and arbitrarily defined range of observed values on the conditional assumption that the samples come from like universes. Thus the mere fact that a particular observation is one of an arbitrarily delimited class of values, themselves collectively rare, is *ipso facto* consistent with the possibility that the hypothesis is true. In short, the so-called laws of chance provide for the very contingency which Yule and Greenwood invite us to regard as inconsistent with the truth of the test hypothesis. This being so, one may find it difficult to sympathise with Greenwood's dignified expression of grief in response to the ensuing and pertinent comment of his doughty clinical opponent Sir Almroth Wright:

> It cannot be too clearly understood that the mathematical statistician has no such secret wells of wisdom to draw from, and that his science does not justify his going one step beyond the purely numerical statement that—as computed by him from the data he has selected as suitable for his purposes—the probabilities in favour of a particular difference being or not being due to the operation of chance are such and such. There need, therefore, be no hesitation in saying that when the mathematical statistician makes free with the terms *significant* and *non-significant*, he is simply taking upon himself a function to which he can lay no claim in his capacity as a mathematician.

It is thus equally impossible to extract any intelligible definition of test procedure from Karl Pearson's earlier pronouncements or to pin down the expressed views of Yule and Greenwood in the publication last cited to a decisive statement which would confer on the term statistical inference any intention not implicit in the suggested alternative statistical prudence. The same is true of the book which introduced statistical test procedure to a wider audience of investigators. In *Statistical Methods for Research Workers* (1925a), destined to be the parent of a large fraternity of manuals setting forth the same techniques with exemplary material for the benefit of readers willing—and as it transpired, only too anxious—to take them on trust, Fisher's formulation of the rationale of the significance test neither discloses a new out-look explicitly nor clarifies views expressed by his predecessors [see Chapter 3]. All that is novel is a refinement of the algebraic theory of the sampling distributions—with one notable exception embraced by Karl Pearson's (1895) system of moment-fitting curves.

Without reading into the words of R. A. Fisher or of the authors cited above more than they would have conceded on second thoughts, we may none the less fairly define in broad outline under three headings the essential features of the *significance*, in contradistinction to the *decision*, test:

(i) we set up a single, the so-called null, hypothesis that one or more samples come from a hypothetical infinite population whose random sampling distribution is specifiable;

(ii) we decide to *reject* the hypothesis whenever the deviation of the sample score (X) from the *mean* of the distribution so prescribed is such that $P_f = \alpha$, if P_f is the probability of meeting a score deviation numerically[2] equal to or greater than X.

(iii) our reliance on the foregoing procedure places on us *no* obligation to specify in advance the size (r) of the samples to which we propose to apply the test, and hence no pre-assigned rejection score criterion X_r consistent with the agreed value of α.

For relevant source material with reference to (i) and (ii) the reader may consult "Significance as Interpreted by the School of R. A. Fisher" [Chapter 3]. Here our concern will be the content of (i) and (ii) in so far as they involve a still largely inarticulate conflict of interest intensified by amendments explicitly adopted to forestall criticism of the theoretical foundations of the test procedure without reconsideration of the saleability of the final product to a consumer fully acquainted with what the producer can now guarantee.

From this viewpoint, an issue raised in (ii) claims prior attention. We have assumed that the test merely prescribes when to *reject* the null hypothesis. Actually, the earliest exponents of the significance concept are by no means definite about this; and what seems to be the first wholly unequivocal pro-nouncement of R. A. Fisher is in the *Design of Experiments* (1935a: 15–16).

---

2. Exponents of the test procedure are not wholly consistent with regard to the use of a two-sided criterion here implicit in the word *numerically* in accordance with the choice of the mean as origin; but they give no factual reasons for adopting a one-sided criterion, when they do so.

This first appeared after publications which prescribe another view of test procedure had already raised the question: When and in what sense can we legitimately accept the alternative to the null hypothesis? In the context referred to, Fisher's statement that the test outcome can sometimes disprove but never prove the truth of the null hypothesis is, to say the least, obscure; and it evades the sixty-four dollar question: What do we mean by proof in the domain of statistical inference?

If we do explicitly limit the terms of reference of our test procedure to rejection, it seems to me that we may interpret it in terms consistent with a behaviourist viewpoint in either of two ways:

(*a*) we shall assign an uncertainty safeguard to a rule which is not *comprehensive*. In effect we say: In some situations we shall reserve judgment and in others we shall make a decisive statement in accordance with a prescribed rule. We can then assign an uncertainty safeguard ($P_f \leqslant \alpha$) which defines the probability of erroneous decision in a restricted class of situations; but we do so with no means of knowing how often the test will fail in the sense that we arrive at no decision at all.

(*b*) we shall impose on ourselves the self-denying ordinance of relinquishing the prescribed hypothesis only in exceptional circumstances as a disciplinary precaution against too ready acceptance.

Only the first of the two views here stated merits to rank as a technique of statistical inference in contradistinction to what we might more appropriately designate statistical prudence. Either way, the utility claimed for the performance of the test raises the question: Have we any more reason for deciding when it is important to reject than for deciding when it is important to accept a particular hypothesis? This at least is an issue on which the consumer may rightly claim to voice an opinion; and it is scarcely deniable that the laboratory worker who invokes a test procedure conceived in terms of a unique null hypothesis commonly assumes that the test procedure justifies acceptance or rejection on equal terms. Indeed, the present writer can see no reason why the investigator should shoulder the responsibility of performing an elaborate drill of statistical computations, unless fortified by this belief. A few citations from the good books which have taught the laboratory worker to do so will show that he has ample encouragement for his faith [Hogben's italics]:

(1) Many scientific investigations involve the employment of the method of framing working hypotheses and testing them experimentally. As long as the experiments fail to disprove them, so long are the hypotheses accepted. This is the general method by which statistical inferences are made . . . the probability level of the observed difference is calculated accordingly. . . . The hypothesis is *accepted* if the level is fairly high and . . . if the level is low (say below 0.05) the hypothesis is rejected (Tippett, 1931: 69–70).

(2) In presenting the results of any test of significance the probability itself should be given. The reader is then in a position to form his own opinion as to the justification of the *acceptance or rejection* of the hypothesis in question (Mather, 1942: 21).

The last source is of special interest, since the book carries the *nihil obstat et imprimatur* of a "Foreword" by R. A. Fisher; and it is therefore pertinent to quote the author in a context[3] which exhibits the investigator in the act of interpretation:

(3) . . . which for 1 degree of freedom has a probability of 0.30–0.20, *showing that there is no interaction* between the classifications, i.e. that the type of water does *not* affect germination (Mather, 1942, rev. ed.: 194).

Snedecor, the most widely read exponent of the Fisher test battery, is less explicit, but does not dispel the belief that acceptance of something is the presumptive alternative to rejection of the null hypothesis:

(4) Statistical evidence is not proof. Even after extensive sampling the investigator may not reject the null hypothesis when in fact the hypothesis is false. For example, m may not be zero in the population, yet natural variation may be great enough to confine t to a moderate value in any practicable size of sample. Therefore when one fails to reject the hypothesis he does not thereby conclude that m is zero. He decides only that m is *so small as to be unimportant to his investigation* (Snedecor, 1937, 4th ed.: 47).

In a later chapter we shall see why we can legitimately draw no such conclusion as the one last stated without due regard to what Neyman calls the *power* of the test [see Hogben, 1957: Chapter 15]. In that event, what we deem to be small depends more on the size of the sample than on its importance to the investigator. The writer disowns any intention of carping criticism if seemingly implicit in citing twice from Mather's book. Mather writes less as a logician interested in the test credentials than as an investigator concerned with its utility. His attitude is of interest in this context mainly because his words make explicit the only terms in which a hard-headed investigator will presumably welcome a significance test procedure as an instrument for validifying conclusions reached in the laboratory or in the field. Should the laboratory worker turn to treatises written from the viewpoint of the mathematician, he will not necessarily find an interpretation of the use of the test inconsistent with the understanding that failure to reject is equivalent to acceptance:

(5) We begin by asserting that the hypothesis to be tested is true. . . . We may calculate the probability $P(D > D_0)$ that the deviation D will exceed any given quantity D_0. . . . Let us choose $P(D > D_0) = \epsilon$ where ϵ is so small that we are

---

3. Here the null hypothesis is that there is no interaction.

prepared to regard it as *practically certain that an event of probability ε will not occur in one single trial.* . . . If we find a value $D > D_0$ this means that an event of probability ε has presented itself. However, on our hypothesis such an event ought to be practically impossible in one single trial, and thus we must come to the conclusion that in this case our hypothesis has been *disproved by experience.* On the other hand, if we find a value $D \leqslant D_0$ we shall be willing to *accept the hypothesis* as a reasonable interpretation of our data . . . (Cramér, 1946).

(6) Improbable arrangements give clues to assignable causes; and excess of runs points to intentional mixing, a paucity of runs to intentional clustering. It is true that these conclusions are never foolproof. Even with perfect randomness improbable situations occur and may mislead us into a search for assignable causes. However this will be a rarity, and with an appropriate criterion we shall in actual practice be misled once in 100 times and *find* assignable causes 99 out of 100 times (Feller, 1950).

These citations suffice to show that expositors of the significance test give the laboratory or field worker enough encouragement for overlooking the reservation specified in (ii) above. That the investigator would indeed less readily embrace the type of test procedure they expound if fully aware of its implications will be clear enough if we now examine (i) against the background of the situation discussed by Yule and Greenwood in the publication already cited. The null hypothesis (H_0) which circumscribes the prescription of the test procedure in accordance with (i) above is that our two groups (treated and untreated) come from the *same infinite hypothetical population.* This is the negation of the assertion that prophylactic inoculation is efficacious; but the main preoccupation of the investigator, who will commonly approach the task of carrying out the trial with a good hunch about the outcome, a hunch derived from preliminary experiments on related animals, from laboratory culture of bacteria or of viruses or from clinical observation, is in practice to establish the affirmative, i.e. to vindicate the credentials of a new instrument of preventive medicine. For what reason then should he or she be eager to take advantage of a test which can merely assign a low probability to erroneously asserting that the treatment is useless, but with no guarantee that the most likely result of applying it will be an open verdict, i.e. no verdict at all? We can justify the choice of our null hypothesis on such terms only from the disciplinary viewpoint defined by (*b*) above but we are then using the idiom of statistical prudence rather than that of statistical inference. As Keynes remarks, the assumption

that it is a positive advantage to approach statistical evidence *without* preconceptions based on general grounds, because the temptation to "cook" the evidence will prove otherwise to be irresistible, has no *logical* basis and need only be considered when the partiality of an investigator is in doubt (1921: 300).

All procedures of the type under discussion, including the entire test battery of *analysis of variance* and *analysis of covariance* elaborated by R. A. Fisher and by his pupils, are referable in the last resort to the hypothesis that samples come from one and the same specified population in accordance with (i) above. If we ask why, the only reason offered is the reason given by Fisher in the context referred to above. So specified, the null hypothesis is seemingly unambiguous and on that account its algebraic formulation is tractable; but this butters no parsnips from the viewpoint of the laboratory worker who understands what he is doing. In an oblique reply to criticism of the significance test in these terms (1935a, 5th ed.: 188 [see: 50]), Fisher admittedly invites the research worker to be the arbiter of the choice of the null hypothesis; but the motives which have promoted his own exploration of sampling distributions and the models set forth by all his expositors are inconsistent with such freedom and with his own unequivocal expression of faith in the infinite hypothetical population as the keystone of the test theory edifice.

In the set-up of the clinical trial, choice of a null hypothesis appropriate to the operational intent is never consistent with (i), since it implies that the two samples come from different populations; and the only hypothesis consistent with (i) is that the mean (M_d) of the sample score difference (d_m) in randomwise extraction is zero $(M_d = 0)$. The intelligible alternative to our prescribed null hypothesis will be unambiguous, only if our main concern is to make a terminal statement of the form $M_d \geqslant k$, i.e. to the effect that prophylactic treatment lowers the proportion who succumb to attack by a target figure (k percent) deemed sufficient to justify its adoption. As we shall later see, we must then define our uncertainty safeguard in the limiting form $P_f \leqslant \alpha$ in contradistinction to the form specified in (ii) above; and this is true of any test prescription referable to a *discrete* sampling distribution, if we claim the right to prescribe a preassigned rejection criterion at any acceptable level.

The intelligible alternative $(M_d \geqslant k)$ to the null hypothesis that both treatment groups are samples from a single infinite hypothetical situation, signifies that each is a sample from a unique population; but Fisher's explicit statements proscribe the choice of a null hypothesis conceived in such terms. What thus emerges from the pivotal role of the infinite hypothetical population in the theoretical superstructure of the significance test is a curious restriction. The terminal statement which the test procedure ostensibly endorses provides an answer (if any) devoid of operational value in the context of an experiment rightly undertaken to confirm a positive assertion suggested by prior information. Since the test procedure merely endorses the negation of a null hypothesis conceived within the straitjacket of the single infinite hypothetical population, the outcome will thus be an irrelevant decision or no decision at all.

The clinical trial sheds light on the legitimate claims of a significance test for another reason. It encourages us to probe more deeply into the structure of the alleged single infinite hypothetical population from which we extract two groups subjected to equally efficacious treatments. The truth is that the attack rate of a group of persons does not merely depend on what we associate with the treatment criterion. It will depend on age, medical history, nutrition and innumerable other changing circumstances. Thus our assumed infinite hypothetical population is not a fixture; though we may here concede that we can extract a very large number of samples from a factually finite population during a period in which relevant change is negligible. If we merely propose to apply the conclusion which the test endorses to situations in which elevant change is indeed negligible, we shall indeed do no violence to the canon of the fixed historical framework of repetition. Unless new circumstances conspire to make operative an otherwise latent difference with respect to the efficacy of two treatments, we may therefore reasonably continue to believe that their efficacy is constant if previous experience of an infinite hypothetical population conceived in the foregoing terms has justifiably convinced us that this is true; but then we must ask ourselves what we mean by *justifiably* in this context. For the judgment it implies, we assume some rational basis outside the framework of the test procedure dealt with; and the test procedure itself disclaims the title to justify such a conclusion, if the prescribed alternatives are reservation of judgment and rejection of the null hypothesis that our two samples, in Fisher's own words, come from the same infinite hypothetical population.

A new issue arises if we take courage from Fisher's second thoughts (1935a, 5th ed.: 188 [see: 50]) and choose a null hypothesis appropriate to the end in view. We then formulate it in terms of a target value k on the assumption that: (*a*) we do not wish to relinquish lightly the benefit of substituting treatment B for treatment A; (*b*) a difference as great as k is our criterion of minimal efficacy when we use the word *lightly* in this context. With appropriate choice of a one-sided criterion of rejection we may adopt as our null hypothesis either that $M_d \geqslant k$ or that $M_d < k$. If we choose and reject the latter our equivalent assertion is the former. In the set-up of the Yule-Greenwood trial, we are then saying that the adoption of treatment B (*inoculation*) will ensure a reduction of the attack rate by at least $100k$ percent, the alternative (treatment A) being *no* inoculation. This at least would seem to be a useful statement, if true; but closer examination raises doubt both about its usefulness and about its credibility.

For the sake of argument we have conceded the claim that the biologist, if convinced that treatment B has no effect in one milieu, may have good grounds for dismissing the possibility that it will be efficacious in a new one; but his legitimate assurance that $M_d \geqslant k$ is not on the same footing. Experience has taught him that epidemic diseases may disappear dramatically in response to changes of the social environment without intervention of the sort here signified as treatment. Hence he may well find that the advantage of treatment B has become negligible after lapse of a comparatively short time interval. Though we may define a framework of repetition in terms which plausibly accommodate the formal credentials of the test procedure with the theory of a stochastic model, we are therefore on shifting sands, when we seek to define our framework of repetition in terms of future conduct.

Nor does the mere fact that treatment A in a prophylactic trial of the type dealt with by Yule and Greenwood (1915) commonly means *no treatment at all*, restrict the relevance of the difficulty here disclosed. In therapeutic trials now zealously conducted on the same prescription, the dilemma is at least equally real. We have much evidence that: (*a*) otherwise indistinguishable bacteria may be more or less resistant to the sulphonamides; (*b*) widespread use of sulphonamides—especially in low dosage—results in selection of resistant strains. Thus experience of sulphonamide therapy which was much more efficacious than $KMnO_4$ or $HgOCN$ for treatment of gonorrhoea at the time of its introduction in the mid-thirties proved to be highly disappointing when used for British troops (1944) in Italy, where the German authorities had already distributed sulfa-drugs freely as a preventive measure to the prostitute population.

If we dismiss all the foregoing considerations, we have still to dispose of a formidable objection to the use of a statistical test procedure in the conduct of the clinical trial, and it is an objection which confronts the decision test of Neyman, E. S. Pearson, and Wald no less than the Yule–Fisher significance test [see Hogben, 1957: Chapter 15]. It is now customary to assume that the problem of the *prophylactic* trial (e.g. whether vaccination is efficacious against attack), is formally identical with that of the *therapeutic* trial (e.g. is penicillin more efficacious than sulphonamides for the treatment of gonorrhoea?). The identification is admissible, only if we lose sight of the end in view. In the practice of preventive medicine our concern is with numbers, and the framework of our problem is essentially one of social accountancy. In the practice of curative medicine our concern is with the sick individual, and unreflective reliance on averages as a criterion of preference may lead us to recommendations inconsistent with the end in view.

That this is so will be immediately apparent if we look at the problem through the spectacles of Claude Bernard (1957). Let us therefore consider a fictitious situation. We may suppose: (i) that a disease D is incurable if untreated; (ii) that a clinical trial of the usual type leads us to assess the recovery-rate under treatment A as about 25 percent and the recovery-rate under treatment B as about 50 percent. In such circumstances we too easily then content ourselves with a recommendation to step up the recovery rate 25 percent by substituting treatment B for treatment A. If so, our preoccupation with averages has blinded us to biological realities. If we are alert to the manifold interaction of nature and nurture, the outcome invites us to ask the question: What peculiarities are common to individuals who respectively respond or fail to respond to one or other treatment?

For heuristic reasons, let us now assume that: (*a*) persons with grey or brown eyes invariably respond to treatment B and fail to respond to treatment A; (*b*) persons with blue eyes invariably respond to treatment A but do not respond to treatment B; (*c*) blue-eyed individuals and individuals with grey or brown eyes occur in the population in the ratio 25 : 50; (*d*) our clinical trial groups are representative in the sense that the ratio of blue-eyed persons to persons with grey or dark eyes is also close to 25 : 50. On these assumptions the recovery rate would be 100 percent if all D patients with blue eyes continued to receive treatment A and all D patients with grey or brown eyes henceforth received treatment B.

Doubtless, such reflections will not greatly trouble the mind of the true believer. If they do, the true believer may gain what reassurance the circumstances solicit from one of Fisher's most recent pronouncements. He states his matured views on the composition, location and duration of the infinite hypothetical population in terms which are worthy of citation because the choice of phraseology suggests that the concept is less the reason for the faith that is in them than the *result* of the application of his method by the militant church of his following:

> Briefly, the hypothetical population is a *conceptual resultant* of the conditions we study (1925b: 700).

If such considerations are not wholly negligible in the set-up of the therapeutic trial, they are of compelling relevance to the task of the field worker in the social sciences; and it is agreeable to be able to cite one exponent of social statistics alert to the pitfalls which beset reliance on the concept of an infinite hypothetical population. Margaret Jarman Hagood's discussion of its status in *Statistics for Sociologists* is worthy of citation at length:

> In any test of significance there is a testing of some hypothesis about a universe from which the set of observations (a limited universe itself in this case) may be considered a random sample. That is, the logical structure of a superuniverse and the variation expected in random samples from it is the same for the observer sociologist as for the experimentalist. Imagining any experimental counterpart of the logical model is a more difficult matter, however. It is easy enough for the experimentalist to imagine repeated experiments under identical conditions, whether or not he can actually perfect his technique to the degree that he can reproduce conditions identically. His universe of possibilities can therefore be put into meaningful terms; it can at least be imagined, even if it cannot actually be produced. It is not so easy for the sociologist to imagine a set of observations repeated under conditions identical with those of one date. The fact of change in

social and cultural phenomena renders unrealistic any conception of identical repetition of the complex of factors conditioning characteristics such as fertility and level of living. . . . To what, then, does the variation expected from random sampling from such a universe of possibilities correspond? Only a feat of imagination involving an infinite prolongation of a present moment, where conditioning factors remain the same but "chance" factors continue to produce random variation can supply the answer. With this done, the observer sociologist along with the experimentalist still faces the problem of interpretation of the chance variation—with the alternatives of ascribing it to the present limitations in knowledge or to the statistical nature of the occurrence of events. . . . It has been suggested that the limited universe of measures on all of a series of demographic units as of a certain date be considered a sample in time; that the random variations of sampling from a superuniverse have their counterpart in the fluctuations which would be observed if we made observations on successive days, or for successive years, while the general influencing conditions would not have altered appreciably. It is evident, however, that such successive fluctuations would not be independent, nor could they be thought of as being produced by forces independent of each other, and therefore they would not be expected to have the same distribution as those produced by chance factors in random sampling (as in fact can be shown to be the case). . . . Another suggestion is that the measures on demographic units may be conceived of as one of an infinite set of such measures secured by dividing the total area surveyed into different series of areal units by shifting of boundaries under certain conditions of contiguity and uniformity of size. The matter of the arbitrary nature of the "lumps" in which our demographic information is secured, and the possible variations to be expected by recombining the information into different lumps has not been explored adequately. While the matter needs attention, it is probably not the answer to the search for a realistic counterpart of the universe of possibilities and to the random variation expected in samples from such a universe. . . . At present, the sociologist must face the fact that the postulated, hypothetical, infinite universe of possibilities, concerning which he tests hypotheses to establish the "significance" of his results, is merely a logical structure, for which he can offer no real counterpart in his research situation. Then what is the utility of such a construct and of the tests of significance based upon it? The answer to this question is not perfectly clear at the present stage of the application of statistical methods to sociological research (1941: 429–432).

I shall not comment on Margaret Hagood's concluding remarks concerning the possibility that "a case for the use of such a construct may be made," because it is not the obligation of the research worker to bow to the dictates of statistical theory until he or she has conclusively established its relevance to the technique of enquiry. On the contrary, the onus lies on the exponent of statistical theory to furnish irresistible reasons for adopting procedures which have still to prove their worth against a background of three centuries of progress in scientific discovery accomplished without their aid.

3. Significance as Interpreted by the School of R. A. Fisher

LANCELOT HOGBEN, F.R.S.

OF THREE WIDELY DIVERGENT VIEWS* about the nature of statistical inference, two have hitherto attracted little attention except among professional mathematicians, and have had few protagonists among practical statisticians except —as is true of Neyman's school—in connexion with new recipes for statistical inspection in commerce and manufacture. Contrariwise, the overwhelming majority of research workers in the biological field (including medicine and agriculture), as also a growing body of workers in the social sciences, rely largely on rule of thumb procedures set forth in a succession of manuals modelled on *Statistical Methods for Research Workers* by R. A. Fisher (1925a). This publication, which disclaims any attempt to set forth comprehensive derivations to justify the rationale of the author's methods, some of them first announced therein without formal proof, has thus become the bible of a world sect. It has a special importance because the author refers to it in a highly controversial context as a source book of his own *Weltanschauung*.

Some of Fisher's ideas took shape in the context of the early developments of the theory of relativity; and he learnt to move with ease and grace in the empyrean of the hypersphere when abstract multidimensional geometry was

* [Hogben refers to the views of Bayes, Neyman-Pearson-Wald, and Fisher.—The Editors.]

still an uncharted domain to theoretical statisticians of an earlier vintage. None of his many self-confessed expositors has attempted the task of interpreting the mathematical techniques involved on a plane intelligible to the investigator with no better equipment than a good practical grasp of the elements of the infinitesimal calculus or to analyse the logical content of the concepts invoked *vis-à-vis* the classical theory of probability. The consumer who is not a trained mathematician has therefore to learn the jargon of *degrees of freedom*, an expression which is meaningful only to those so fortunate as to be familiar with such branches of applied mathematics as the theory of the gyrostat or the development of thermodynamics in the tradition of Willard Gibbs.

Thus the controversy provoked by the more recent writings of Neyman, Wald, von Mises and others is essentially a challenge to the adequacy of statistical procedures already invoked by most laboratory workers who make much use of statistics; and those who use them most have so far had little opportunity to contribute to a discussion the outcome of which is highly relevant to their day's work. A temperate appraisal of the content of the controversy therefore calls for a more detailed examination of the views of R. A. Fisher and of his disciples than of the more concisely and explicitly stated opinions of his opponents.

To do justice to an evolving system of thought with lexicographical punctiliousness is commonly difficult, if only because it is by no means to the discredit of any philosopher to say that he has changed his views in the course of a prolific career. Happily, we may sidestep this obstacle. Fate has not deprived us of an up-to-date assessment of his mature views, since no sentiment of false modesty has deterred R. A. Fisher from taking the unusual precaution of collecting what he regards as his hitherto major contributions in an impressive quasi-memorial volume with a biographical introduction, with a portrait in photogravure of the author and with his own respectfully retrospective comments on the outstanding importance of each of the literary landmarks re-erected therein.

In these collected *Contributions* Fisher refers explicitly to the core of the current controversy in the following words:

> Pearson and Neyman have laid it down axiomatically that the level of significance of a test must be equated to the frequency of a wrong decision "in repeated samples from the same population." This idea was foreign to the development of tests of significance given by the author in 1925, for the experimenter's experience does not consist in repeated samples from the same population, although in simple cases the numerical values are often the same; and it was, I believe, this coincidence of values in simple cases which misled Pearson and Neyman, who were not very familiar with the ideas of "Student" and the author (1950: 35.173a).

In the references which follow, Fisher quotes under his own name only a short paper (1937) "On a Point Raised by M. S. Bartlett," his book *Statistical*

Methods for Research Workers (1925a) and a third (1939) as above. Before we consult *Statistical Methods*, we may pause to anticipate a curious implication of the foregoing remarks. Our next citation asserts that the static population of the experimenter's experience is also infinite. If so, we exclude the treatment of sampling without replacement from a finite universe from the domain of discourse at the outset, a limitation which is certainly inconsistent with the practice of Fisher's disciples and with the much-quoted tea cup test (see below) in his later book *Design of Experiments* (1935a). Fisher's own analysis of the lady and the tea-cup problem involves a nonreplacement situation with reference to which the only conceivable meaning one can give to his infinite population is precisely the postulate he repudiates in the passage cited above, viz. an infinite succession of 4-fold trials without replacement from 2-class universes of 8 objects.

The index of the 1948 edition of the *Statistical Methods* (1925a, 10th ed.) does not cite any page reference against *significance, null* or *hypothesis*, but gives p. 41 under *Tests, Significance of*. With this clue and from elsewhere in the same source we may try to get a clear idea of a theory of decision test procedure and/or estimation alternative to that of Neyman and Pearson. I shall first cite the only relevant indication the index supplies:

> (i) The idea of an infinite population distributed in a *frequency distribution* in respect of one or more characters is fundamental to all statistical work. From a limited experience, for example, of individuals of a species, or of the weather of a locality, we may obtain some idea of the infinite hypothetical population from which our sample is drawn, and so of the probable nature of future samples to which our conclusions are to be applied. If a second sample belies this expectation we infer that it is, in the language of statistics, drawn from a different population; that the treatment to which the second sample of organisms had been exposed did in fact make a material difference, or that the climate (or the methods of measuring it) had materially altered. Critical tests of this kind may be called tests of significance, and when such tests are available we may discover whether a second sample is or is not significantly different from the first (41).

Since we have here no definitive criterion of when a sample score belies our expectation, we must seek further afield for an elucidation of what a decision test can accomplish. Three pages later we come to the next unequivocally relevant statement which follows a definition of a normal score deviation in any infinitesimal range dx.

> (ii) In practical applications we do not so often want to know the frequency at any distance from the centre as the total frequency beyond that distance; this is represented by the area of the tail of the curve cut off at any point. Tables of this total frequency, or probability integral, have been constructed from which, for any value of $(x - \mu)/\sigma$, we can find what fraction of the total population has a larger deviation; or, in other words, what is the probability that a value so

distributed, chosen at random, shall exceed a given deviation. Tables I and II have been constructed to show the deviations corresponding to different values of this probability. The rapidity with which the probability falls off as the deviation increases is well shown in these tables. A deviation exceeding the standard deviation occurs about once in three trials. Twice the standard deviation is exceeded only about once in 22 trials, thrice the standard deviation only once in 370 trials, while Table II shows that to exceed the standard deviation sixfold would need nearly a thousand million trials. The value for which $P=0.05$, or 1 in 20 is 1.96 or nearly 2; it is convenient to take this point as a limit in judging whether a deviation is to be considered significant or not. Deviations exceeding twice the standard deviation are thus formally regarded as significant. Using this criterion we should be led to follow up a false indication only once in 22 trials, even if the statistics were the only guide available. Small effects will still escape notice if the data are insufficiently numerous to bring them out, but no lowering of the standard of significance would meet this difficulty (44).

Evidently, it is difficult to confer any meaning on the expressions "about once in 22 trials" or "only once in 370 trials" unless we postulate the extraction as a repetitive process in conformity with the Neyman–Pearson axiom. We might thus conclude that Fisher's objection to the axiom cited arises from the circumstance that it interprets significance levels in terms of frequency of wrong decisions. There is in fact nothing in either of the preceding paragraphs to suggest an alternative view. The criterion which would lead us "to follow up a false indication only once in 22 trials" presupposes that the sample does in fact come from a particular infinite population; but has no bearing whatsoever on how often we should follow a false indication if the sample came from some other infinite population. The sort of decision to which the test leads us is therefore useful if, and only if, our sole concern is to set a fixed limit to the conditional risk of rejecting the unique proposition implicit in the null hypothesis, viz. that the infinite population from which the sample comes has a certain structure.

The choice of an appropriate null hypothesis then presupposes: (*a*) that the main preoccupation of the investigator is to safeguard himself or herself against the risk of rejecting as the source of the sample a population with a particular structure; (*b*) there exists a convenient distribution function definitive of repeated sampling from such a population. If so, the numerous manuals devoted to the exposition of the Fisher battery of tests display a truly astonishing prescience concerning the preoccupations of the research worker or indifference to what they are. Thus a prodigious literature on therapeutic trials records the outcome of a Chi square (with one degree of freedom) test for the 2×2 table in conformity with the null hypothesis that one treatment is as good as another. Actually, the risk the research worker may be most anxious to avoid is that of rejecting a new treatment which is better than the old one, in which event the procedure prescribed by Fisher's pupils and expositors has then no bearing on the presumptive reason for carrying out the test. Clearly, the considerations which dictate the null hypothesis implicit in the test procedures advocated in the many manuals which expound Fisher's methods are mainly referable to what is an essential desideratum of any test procedure, viz. that the sample distribution of the population specified by the hypothesis chosen is definable and amenable to tabulation for ready reckoning. No less evidently, the fact that it may have this desirable property has no necessary connexion with any risk the laboratory or field worker is unwilling to incur.

There is another puzzling feature of the passage first cited, when we interpret it side by side with the second. In the former Fisher refers to an infinite population as a fundamental postulate of his theory of statistical work. In the latter he elucidates his views about significance levels in terms of a particular population which is *ex hypothesi* infinite; but we could pair off each statement with appropriate verbal changes with a corresponding statement about the risk of following up a false indication about the structure of an urn containing 25 balls some red and some black on the basis of our knowledge that 3 out of 5 balls extracted from it without replacement are red. Surely the reason for introducing the concept of infinity has therefore no special relevance to the population of objects to which the score values of the frequency distribution refer. Its convenience can arise only in a conceptual framework of infinite repetition indispensable to the formulation of a correspondence between assertions and the occurrences falsely or truly described by them.

The rest of the chapter beginning on p. 41 sheds no new light on what Fisher claims for a decision test or on his views about statistical inference in general. Later we come on the following statement which encourages us to hope for more:

> The treatment of frequencies by means of Chi square is an approximation, which is useful for the comparative simplicity of the calculations. The exact treatment is somewhat more laborious, though necessary in cases of doubt, and valuable as displaying the true nature of the inferences which the method of Chi square is designed to draw (96).

There follows an exposition of a treatment of 2×2 tables to test the hypothesis of proportionality, i.e. that two small samples are samples from populations with the same definitive parameter, p.[1] The conclusion is:

> Without any assumption or approximation, therefore, the table observed may be judged significantly to contradict the hypothesis of proportionality if
> $$\frac{18! \, 13!}{30!} (2992 + 102 + 1)$$

1. Fisher himself does not explicitly in this context make the important semantic distinction between the statement that the two samples come from the same infinite population and the statement that they come from populations which are alike in terms of the classificatory criterion relevant to the methods of sampling.

is a small quantity. This amounts to 619/1330665, or about 1 in 2150 showing that if the hypothesis of proportionality were true, observations of the kind recorded would be highly exceptional (97).

In this example, the reader will note that Fisher disclaims the necessity to formulate a criterion of rejection before undertaking a decision test; and calculates the frequency with which a particular range of score values, including the sample score itself will turn up, if the null hypothesis is true. No one can take exception to the deduction that "observations of the kind recorded are highly exceptional"; but the logician will wish to know if this conclusion has any necessary bearing on the type of inference relevant to the aim and nature of the test. Kendall's comment on a comparable distribution summarises the most we can say in such a situation:

If this probability is small we have the choice of three possibilities:
(*a*) An improbable event has occurred.
(*b*) The hypothesis is not true, i.e. the proportion of A's in the population is not $\bar{\omega}$.
(*c*) The sampling process is not random.

The calculation cited from *Statistical Methods* (1925a, 10th ed.: 97) refers to "the probabilities of the set of frequencies observed and the two possible more extreme sets of frequencies which might have been observed." If we have so far failed to clarify what role Fisher confers on a significance test as a tool of statistical inference, this *accouplement* at least gets into focus a fundamental difference between his own attitude to the theory of test procedure and that of Neyman, Pearson, Wald, and the Columbia Research Group. Their view is that a decision test embodies a rule with statistically specifiable consequences if followed consistently. The kingpin of the rule is a rejection criterion which is independent of the particular sample under examination. The criterion prescribes a range of sample scores deemed to be inadmissible if the population prescribed by the test hypothesis correctly describes the source of a sample. If the sample score (x_0) lies in a *critical region* defined by the vector rejection criterion $x > x_r$ any possible value of x greater than the observed one will also lie in it. If x_0 lies in a critical region defined by the vector rejection criterion $x < x_r$ any possible value of x less than x_0 will also lie in it.

The prestatement of the rejection criterion thus gives an intelligible meaning to the quadrature which defines the probability (α) of rejecting the hypothesis if true; but Fisher offers no alternative justification for the association with x_0 of all values of x greater than x_0 in the summation cited above, though the exposition of the test procedure is clearly inconsistent with any such prestatement of a uniform rejection criterion. This circumstance seemingly throws more light than does anything elsewhere stated on Fisher's insistent refusal to identify significance levels with frequency of wrong decisions. All he

offers in the source he himself cites as a basis for decision is a range of inadmissible score values specified *after* observing the sample on the basis of an
intuitive evaluation of its likeability.

The Design of Experiments (1935a), first published some ten years later than
Statistical Methods, gives us an opportunity for exploring the author's second
thoughts on significance and on estimation. In the index of the fifth edition
(1949) we find the following relevant entries:

> *Significance:* 13, 33, 55, 73, 105, 113, 182
> *Null Hypothesis:* 15–17, 182–208
> *Fiducial Probability:* 182, 195

Of the entries under significance only one explicitly throws further light on the
author's viewpoint. In the citation which follows, three different concepts
seemingly compete for mastery and somewhat inconclusively.

> It is open to the experimenter to be more or less *exacting* in respect of the
> smallness of the probability he would require before he would be willing to
> admit that his observations have demonstrated a positive result. It is obvious
> that an experiment would be useless of which no possible result would satisfy
> him. Thus, if he wishes to ignore results having probabilities as high as 1 in 20—
> the probabilities being of course reckoned from the hypothesis that *the phenom-
> enon to be demonstrated is in fact absent*—then it would be useless for him to
> experiment with only 3 cups of tea of each kind. For 3 objects can be chosen out
> of 6 in only 20 ways, and therefore complete success in the test would be achieved
> without sensory discrimination, i.e. by "pure chance," in an average of 5 trials
> out of 100. It is usual and convenient for experiments to take 5 percent as a
> standard level of significance, in the sense that they are prepared to ignore all
> results which fail to reach this standard, and, by this means, to eliminate from
> further discussion the greater part of the fluctuations which chance causes have
> introduced into their experimental results. No such selection can eliminate the
> whole of the possible effects of chance coincidence, and if we accept this con-
> venient contention, and agree that an event which would occur by chance only
> once in 70 trials is decidedly "significant," in the statistical sense, we thereby
> admit that no isolated experiment, however significant in itself, can suffice for the
> experimental demonstration of any natural phenomenon; for the "one chance
> in a million" will undoubtedly occur, with no less and no more than its appro-
> priate frequency, however surprised we may be that it should occur to us. *In order
> to assert that a natural phenomenon is experimentally demonstrable we need, not
> an isolated record, but a reliable method of procedure. In relation to the test of
> significance, we may say that a phenomenon is experimentally demonstrable when
> we know how to conduct an experiment which will rarely fail to give us a statistically
> significant result* (13–14 [Hogben's italics]).

Initial remarks about how exacting the experimenter needs to be re-echo the
impression that a significance test as conceived by the author is primarily a

disciplinary regimen to discourage rash judgments. The statement that it is *usual* for research workers to adopt a 5 percent significance level in the same context is true only of those who rely on the many rule of thumb manuals expounding Fisher's own test prescriptions. The explicit admission that we reckon the probabilities on the basis of the truth of *the hypothesis that the phenomenon to be demonstrated is in fact absent* implies that the only type of statement justified by the outcome of the test is conditional on the assumption that the hypothesis is true. In its own context the assertion that we need a *reliable method of procedure* in contradistinction to an isolated record seems to be consistent with the framework of indefinitely protracted repetition postulated by Neyman, von Mises, and Wald; but it is the final statement which provokes our attention more especially.

As it stands, its literal form might legitimately convey the author's intention of introducing a new definition of *demonstrability*; but we do not in fact repeat experiments to see whether they fail to give a significant result in Fisher's sense of the term, or in any other. Accordingly, we must interpret it in its own context *vis-à-vis* the sentence italicised in the preceding paragraph, i.e.: (*a*) the so-called *null* hypothesis we select as a basis of test procedure is the *negation of the hypothesis which asserts the reality of the phenomenon*; (*b*) we decline to abandon our suspicion that the alternative hypothesis is true, i.e. that the phenomenon exists, if the null hypothesis assigns an *arbitrarily* low enough probability of occurrence to a specified class of score values including the unique score our single record yields. In any case, we here presume, though in contrariety to the author's apparent intention cited above, a conceptual repetitive framework in which the test procedure operates. Each of the assertions (*a*) and (*b*) above thus invites closer scrutiny, to which the author's failure to distinguish a phenomenon from a hypothesis about a phenomenon and his catholic use of the term *variation* adds a special difficulty arising out of the important distinction to which Kendall directs attention, viz: (*a*) the rarity of the prescribed event in virtue of the hypothetical random distribution implicit in the chosen null hypothesis, *if true*; (*b*) the rarity of the prescribed event in virtue of departures from randomisation in the assembly of data; (*c*) the rarity of the prescribed event in virtue of the inapplicability of the null hypothesis to the occurrence.

As regards the author's view of the proper criterion of choice for a unique null hypothesis, several difficulties arise when we ask what form the denial must take, i.e. what precisely do we mean if we assert that the phenomenon to be demonstrated *is in fact absent*? This is indeed the theme of a lengthy and entertaining analysis by Neyman of the lady and the tea-cup parable referred to for illustrative use in the passage under discussion. The two major difficulties are these: (*a*) many situations in which the author's expositors could certainly advocate a Fisher test procedure admit of no singular statement of the denial appropriate to the situation; (*b*) the object of an experiment may be to assess the validity of equally commendable alternatives each being the denial of the other.

A single example will suffice to elucidate (*a*). We suppose that rival schools, as indeed in the days of the controversy between Weldon and Bateson, assert of a particular class of experiments; (i) a ratio of the most elementary type prescribed by Mendel's hypothesis (3 : 1) holds good; (ii) such a ratio does not hold good with regard to the phenomenon under dispute. Proponents of (ii) may adopt (i) as the denial of their assertion. Proponents of (i) can formulate no such unique denial satisfactory to proponents of (ii). To be sure, they can place themselves in the position of their contestants; but this then imposes the same impracticable obligation on the latter. Few who follow Fisher's test prescriptions appear to realise that there never can be a unique denial of a hypothesis itself formulated, like the theory of the gene, in statistical terms, nor that a 20 to 1 convention can have any relevance to one's assessment of its truth.

In any case, the form of the denial must necessarily satisfy one criterion of adequacy and should rightly satisfy a second. Since a stochastic null hypothesis must prescribe a sample distribution of score values, mathematical tractability of the implicit assumptions in terms of sampling theory will in practice dictate the form that the denial will take. It is therefore of interest to note that the several components of the impressive battery of significance tests associated with Fisher's name consistently imply that the universe of choice is homogeneous with regard to the relevant dimension of classification, i.e. that any manifest discrepancy associated with the relevant taxonomical criterion (or criteria) is such as would arise in random sampling from one and the same infinite hypothetical population. Admittedly, this makes the prescription of a test more tractable from the viewpoint of the mathematician; but if there is any other intelligible reason for adopting such a postulate, it is difficult to find an equally intelligible reason for the silence of Fisher's expositors with respect to the operational intention of the denial. Contemporary literature of therapeutic and prophylactic trials is an uninterrupted record of Chi square tests for 2×2 tables to test the null hypothesis that there is no treatment difference, when the question of real interest is the putative existence of a difference sufficient to justify replacement of one treatment by another.

No doubt Fisher would concede that this is an issue of estimation; but this scarcely explains the object in view when performing the so-called exact test cited in his own words above. In practice, the overwhelming majority of biologists and sociologists who employ Fisher's battery of tests have learned the routine as an army drill from the many manuals following the same method of presentation as his *Statistical Methods*. As stated, this gives no derivation of the sampling distributions invoked and no adequate survey of what principles of statistical inference accredit their use. As also stated, the prescribed null hypothesis conforms to a set pattern in which the idea of randomisation is paramount; but Anscombe (1948) has rightly pointed out that this concept admits of interpretation at no less than three semantic levels when conceived in terms of the assertion that *the phenomenon to be demonstrated is not present*.

In the *Design*, the last passage cited is the only one in which the author expounds his fundamental attitude to significance explicitly or at length. As we have seen, it endows the null hypothesis with a unique status which would be less difficult to interpret if the test battery itself supplied the investigator with a range of choice adequate to what denials may be appropriate to the experimental set-up. We therefore turn with bewilderment to what the author has to say on this topic in a seemingly oblique reference to the Neyman–Pearson theory of alternative test procedure:

> The notion that different tests of significance are appropriate to test different features of the same null hypothesis presents no difficulty to workers engaged in practical experimentation, but has been the occasion of much theoretical discussion among statisticians. The reason for this diversity of viewpoint is perhaps that the experimenter is thinking in terms of observational values, and is aware of what observational discrepancy it is which interests him, and which he thinks may be statistically significant, before he enquires what test of significance, if any, is available appropriate to his needs. He is, therefore, not usually concerned with the question: To what observational feature should a test of significance be applied? This question, when the answer to it is not already known, can be fruitfully discussed only when the experimenter has in view, not a single null hypothesis, but a class of such hypotheses, in the significance of deviations from each of which he is equally interested. We shall, later, discuss in more detail the logical situation created when this is the case. It should not, however, be thought that such an elaborate theoretical background is a normal condition of experimentation, or that it is needed for the competent and effective use of tests of significance (1935a, 5th ed.: 188).

Lack of a clear-cut distinction between observational discrepancy and the hypothesis in seeming contrariety to the recorded observation is at this stage a familiar idiom and need not detain us. The passage is quotable since it implies that the laboratory or field worker has no doubts about what is the appropriate null hypothesis to select and has at his or her disposal no lack of appropriate test prescriptions presumably equipped with suitable tables in one of the good books. We have seen that this is not so, and there is no need to add anything to foregoing comments bearing on this charitable interpretation of the consumer's choice and custom. It is not without interest that Fisher's assertion (1935a, 5th ed.: 13–14) of the unique form the null hypothesis must take is not easy to reconcile with the above and shortly retires from the field:

> Our examination of the possible results of the experiment has therefore led us to a statistical test of significance, by which these results are divided into two

classes with opposed interpretations. Tests of significance are of many different kinds, which need not be considered here. Here we are only concerned with the fact that the easy calculation in permutations which we encountered, and which gave us our test of significance, stands for something present in every possible experimental arrangement; or, at least, for something required in its interpretation. The two classes of results which are distinguished by our test of significance are, on the one hand, those which show a significant discrepancy from a certain hypothesis; namely, in this case, the hypothesis that the judgments given are in no way influenced by the order in which the ingredients have been added; and on the other hand, results which show no significant discrepancy from this hypothesis. This hypothesis, which may or may not be impugned by the result of an experiment, is again characteristic of all experimentation. Much confusion would often be avoided if it were explicitly formulated when the experiment is designed. In relation to any experiment we may speak of this hypothesis as the "null hypothesis," and it should be noted that the *null hypothesis is never proved or established, but is possibly disproved, in the course of experimentation.* Every experiment may be said to exist only in order to give the facts a chance of disproving the null hypothesis (1935a, 5th ed.: 15–16 [Hogben's italics]).

In the light of these remarks and more especially with due regard to the content of the citation from p. 188 of the *Design*, it is permissible to entertain the possibility that Fisher did not initially intend to limit (as on pp. 13–14) the null hypothesis appropriate to every situation by a form of words implying that *the phenomenon to be demonstrated is not present.* Since we cannot explicitly formulate any unique hypothesis conceived as the negation of a particular genetic hypothesis of the Mendelian type, let us therefore explore the consequences of regarding the proper null hypothesis as the Mendelian interpretation itself. Following the prescription of Fisher's school, we should then say that a deviation of $\pm 2\sigma$ from expectation would be a rare event, but a deviation of $\pm 3\sigma$ would be more so. Indeed, any deviation numerically greater than 0.675σ would be less common than the residual score class. Thus the more exacting we make our significance criterion, the greater will be the chance that a true discrepancy between hypothesis and fact will escape detection. All we can say is: (*a*) if the hypothesis is true, score deviations within the range $\pm 0.675\sigma$ will occur as often as score deviations outside this range; (*b*) if our rejection score criterion x_0 is numerically greater than 0.675σ our procedure will more often lead us to correct than to wrong decisions in the long run, *if the hypothesis is* true. Surely it is clear that the conditional risk so specified by the procedure is not necessarily the risk against which the investigator wishes to protect his judgment. Nor is it clear that investigators with opposite views about the *phenomenon to be demonstrated* will assign priority to one and the same alternative risks of recording an erroneous verdict.

This interpretation of an isolated test turns the spotlight on the last words italicised in the citation from pp. 15–16. The assertion that the test cannot demonstrate the truth of the hypothesis, an assertion difficult to dovetail into the intention of the last sentence of the citation from pp. 13–14, deprives it of any usefulness unless conceived as a disciplinary device; but what precisely entitles us to hope that we have *possibly disproved* it? Within the strait-jacket of the unique null hypothesis, we have merely made a reason for rejecting it dictated by considerations relevant to the penalties of doing so when it happens to be true. In any event, the considerations advanced at the end of the last paragraph suggest that the use of the probable error by a generation with no prescience of the 5 percent feeling scarcely merits the stricture:

> The value of the deviation beyond which half the observations lie is called the *quartile* distance, and bears to the standard deviation the ratio 0.67449. It was formerly a common practice to calculate the standard error and then, multiplying it by this factor, to obtain the *probable error*. The probable error is thus about two-thirds of the standard error, and as a test of significance a deviation of three times the probable error is effectively equivalent to one of twice the standard error. The common use of the probable error is its only recommendation; when any critical test is required the deviation must be expressed in terms of the standard error in using the tables of normal deviates (Fisher, 1925a, 10th ed.: 45).

The insertion of italics in this citation is not relevant at this stage. In striking contrast to the pervasive assumption of homogeneity within the framework of test decision is the form of words Fisher uses when he introduces readers of the *Design* to the *fiducial concept of probability that the unknown parameters should be within specified limits*. For the real universe of estimation is in such situations the homogeneous universe of what we elsewhere refer to as Model I; and the probability assignable to a parameter in connexion with anything we say about the interval in which it lies can have only two values, *viz.* zero and unity.

> It is the circumstance that statistics sufficient for the estimation of these two quantities are obtained merely from the sum and the sum of squares of the observations, that gives a peculiar simplicity to problems for which the theory of errors is appropriate. This simplicity appears in an alternative form of statement, which is legitimate in these cases, namely, statements of the *probability that the unknown parameters*, such as μ and σ, *should lie within specified limits*. Such statements are termed statements of *fiducial* probability, to distinguish them from the statements of *inverse* probability, by which mathematicians formerly attempted to express the results of inductive inference. Statements of inverse probability have a different logical content from statements of fiducial probability, in spite of their similarity of form, and require for their truth the postulation of knowledge beyond that obtained by direct observation (1935a, 10th ed.: 195–196 [Hogben's italics]).

In this context it is tempting to cite the curious reasons Fisher gives for restricting the fiducial argument to situations in which we may conceptually invoke a continuous, in contradistinction to a discrete, distribution:

> ... With discontinuous data, however, the fiducial argument only leads to the result that this probability does not exceed 0.01. We have a statement of inequality, and not one of equality. It is not obvious, in such cases, that, of the two forms of statement possible, the one explicitly framed in terms of probability has any practical advantage. The reason why the fiducial statement loses its precision with discontinuous data is that the frequencies in our table make no distinction between a case in which the 2 dizygotic convicts were only just convicted, perhaps on venial charges, or as first offenders, while the remaining 15 had characters above suspicion, and an equally possible case in which the 2 convicts were hardened offenders, and some at least of the remaining 15 had barely escaped conviction (1935b : 50–51).

Some of the enigmas and seeming inconsistencies exhibited in previous citations take on a new aspect when we recall: (*a*) how largely R. A. Fisher by his own admission relies on intuition; (*b*) how much his later views owe to his early experience, assigned as Statistician to an Agricultural Research Institute with the task of extracting any grain reclaimable from a long-standing accumulation of inexpertly designed field trials. This suffices to explain his preoccupation with *sufficiency*, a concept which has so much less prominence in theories advanced by the opposing school. Insistent concern for what his own school refer to as *amount of information* and disdain for excessive consistency go hand in hand in the following from "The Logic of Inductive Inference":

> ... One could, therefore, develop a mathematical theory of quantity of information from these properties as postulates, and this would be the normal mathematical procedure. It is, perhaps, only a personal preference that I am more inclined to examine the quantity as it emerges from mathematical investigations, and to judge of its utility by the free use of common sense, rather than to impose it by a formal definition. As a mathematical quantity information is strikingly similar to *entropy* in the mathematical theory of thermodynamics. You will notice especially that reversible processes, changes of notation, mathematical transformations if single-valued, translation of the data into foreign languages, or rewriting them in code, cannot be accompanied by loss of information; but that the irreversible processes involved in statistical estimation, where we cannot reconstruct the original data from the estimate we calculate from it, may be accompanied by a loss, but never by a gain (1935b : 47).

The importance of this preoccupation with *amount of information* lies in a basically different orientation of Fisher's sect when we set his views in juxtaposition to those of the alternative school. One mode of thought proposes the question: What rules must we impose on our reasoning before we have permitted the data to influence our views? That of Fisher and his followers asks: What course shall we pursue when we have weighed up all the relevant evidence inherent in the data? Thus we find the following in "Uncertain Inference":

> . . . There is one peculiarity of uncertain inference which often presents a difficulty to mathematicians trained only in the technique of rigorous deductive argument, namely, that our conclusions are arbitrary, and therefore invalid, unless all the data, exhaustively, are taken into account. In rigorous deductive reasoning we may make any selection from the data, and any certain conclusions which may be deduced from this selection will be valid, whatever additional data we may have at our disposal. . . . This consideration is vital to the fiducial type of argument, which purports to infer exact statements of the probabilities that unknown hypothetical quantities, or that future observations, shall lie within assigned limits, on the basis of a body of observational experience. No such process could be justified unless the relevant information latent in this experience were exhaustively mobilized and incorporated in our inference (1936 : 254–255).

The special difficulty that arises from Fisher's robust and seemingly contagious confidence in his own intuitions appears in the following citations, the first two of which are from "The Foundations of Theoretical Statistics" (1921) reprinted in the collected works (10.323):

> . . . For the solution of problems of estimation we require a method which for each particular problem will lead us automatically to the statistic by which the criterion of sufficiency is satisfied. Such a method is, I believe, provided by the Method of Maximum Likelihood, although I an not satisfied as to the mathematical rigour of any proof which I can put forward to that effect. Readers of the ensuing pages are invited to form their own opinion as to the possibility of the method of the maximum likelihood leading in any case to an insufficient statistic. For my own part I should gladly have withheld publication until a rigorously complete proof could have been formulated; but the number and variety of the new results which the method discloses press for publication, and at the same time I am not insensible of the advantage which accrues to Applied Mathematics from the co-operation of the Pure Mathematician, and this co-operation is not infrequently called forth by the very imperfections of writers on Applied Mathematics (1921 : 323).

This intrepid belief in what he disarmingly calls common sense, as a substitute for a system of communicably acceptable rules of procedure, has led Fisher, in a source elsewhere cited, to advance a battery of concepts for the semantic credentials of which neither he nor his disciples offer any justification *en rapport* with generally accepted tenets of the classical theory of probability. Thus an operation which appears in the derivation of Behrens' test as a simple error in terms of the classical theory reappears as a novel and *ad hoc* rule of thought (see Yates below) described as fiducial inference in Fisher's own treatment of the same issue. Again and again, we seem to sidestep the notion of inverse probability or the invocation of the highly exceptionable Bayes's scholium either by using a new name for the same reasoning process or by ignoring the issue involved. Thus we come on the following:

> There would be no need to emphasize the baseless character of the assumptions made under the titles of inverse probability and Bayes' theorem in view of the decisive criticism to which they have been exposed at the hands of Boole, Venn and Chrystal, were it not for the fact that the older writers, such as Laplace and Poisson, who accepted these assumptions, also laid the foundations of the modern theory of statistics, and have introduced into their discussion of this subject ideas of a similar character. I must indeed plead guilty in my original statement of the Method of the Maximum Likelihood to having based my argument upon the principle of inverse probability; in the same paper, it is true, I emphasized the fact that such inverse probabilities were relative only. That is to say, that while we might speak of one value of p as having an inverse probability three times that of another value of p, we might on no account introduce the differential element dp, so as to be able to say that it was three times as probable that p should lie in one rather than the other of two equal elements. Upon consideration, therefore, I perceive that the word probability is wrongly used in such a connection; probability is a ratio of frequencies, and about the frequencies of such values we can know nothing whatever. We must return to the actual fact that one value of p, of the frequency of which we know nothing, would yield the observed result three times as frequently as would another value of p. If we need a word to characterize this relative property of different values of p, I suggest that we may speak without confusion of the *likelihood* of one value of p being thrice the likelihood of another, bearing always in mind that likelihood is not here used loosely as a synonym of probability, but simply to express the relative frequencies with which such values of the hypothetical quantity p would in fact yield the observed sample (1921 : 326).

We thus build up a battery of concepts which constitute the exclusive *mystique* of the sect; and a secret language which excludes intercommunication with heretics or schismatics, as in the following from "Inverse Probability":

> ... The process of maximizing π (ϕ) or S (log ϕ) is a method of estimation known as the "method of maximum likelihood"; it has in fact no logical connection with inverse probability at all. The fact that it has been accidentally associated with inverse probability, and that when it is examined objectively in respect of the properties in random sampling of the estimates to which it gives rise, it has shown itself to be of supreme value, are perhaps the sole remaining reasons why that theory is still treated with respect. The function of the θ's maximized is not, however, a probability and does not obey the laws of probability; it involves no differential element $d\theta_1 \, d\theta_2 \, d\theta_3 \ldots \ldots$; it does none the less afford a rational basis for preferring some values of θ, or combination of values

of the θ's, to others. It is, just as much as a probability, a numerical measure of rational belief, and for that reason is called the *likelihood* of θ_1 θ_2 θ_3 . . . having given values, to distinguish it from the probability that θ_1 θ_2 θ_3 . . . lie within assigned limits, since in common speech both terms are loosely used to cover both types of logical situation (1930: 532).

The only notably valiant attempt of one of the apostles of the sect to interpret the teaching of the Leader as a system of logic is that of Yates. In his paper on "An Apparent Inconsistency Arising from Tests of Significance Based on Fiducial Distributions of Unknown Parameters" Yates concedes that in Fisher's extension of the Behrens' integral:

We must frankly recognize that we have here introduced a new concept into our methods of inductive inference, which cannot be deduced by the rules of logic from already accepted methods, but which itself requires formal definition (1939).

What follows is less recognisable as a new concept than as a sequence of *ad hoc* assumptions relevant to a particular test prescription. In simple words, the contention is that : (*a*) the fiducial argument would be invalid if we did not introduce the new concept; (*b*) the acceptance of the new principle leads to no "inconsistencies" if the estimates are sufficient in Fisher's sense; (*c*) if we wish to retain the fiducial argument, we may therefore embrace the new concept with a clear conscience. If there is any ulterior reason for doing so, Yates does not explicitly disclose it.

PART TWO
The Controversy in Sociology

INTRODUCTION

T HE DEBATE over significance tests in sociology has taken place in the last 15 years, but the roots of some of the main issues are found in an earlier treatise that links the sociological debate directly to the statistical controversy discussed by Hogben in the previous section. Margaret Jarman Hagood's *Statistics for Sociologists* (1941) was an early and continuing influence on sociological practice in statistics in general and significance tests in particular.[1] In the passages on significance tests quoted here under the title, "The Notion of a Hypothetical Universe" [Chapter 4], she tentatively develops a basically Fisherian viewpoint on the issues of sampling, population, and meaning and implicitly relates these notions to the issues of scope and process.

Hagood introduced the possibility of using significance tests to make inferences from a complete enumeration of cases to an infinite hypothetical population, under the assumption that such a set of cases can be considered a random sample of all possible samples that could have been generated under similar conditions. She also explored the view that such a rationale might have some potential for increasing the population scope of sociological generalization. While Hagood showed considerable ambivalence about these

---

1. A revised edition of *Statistics for Sociologists* (co-authored by Daniel Price) that contained substantially the same notions on significance tests as the earlier edition was published in 1952.

notions, many sociologists striving for rigor, quantification, and generality in their research paid no heed to her cautious tone and readily expanded her ideas to apply in practice to *any* set of cases.[2] Unfortunately, less controversial but nevertheless crucial aspects of Hagood's view of the tests, such as her suggestion that the proper meaning of the term "significance" is simply that a given statistic "signifies" a basis for rejecting a null hypothesis, were quickly forgotten in the rush to apply the tests to any and all findings.

By the mid-1950's the use of significance tests as well as the tendency to avoid making inferences from findings not meeting conventional significance levels were well established practices in sociological research. In an early critical note on "The Significance of Insignificant Differences" [Chapter 5], Zeisel warned that blanket failure to consider statistically insignificant findings was not a fruitful process of inference for a discipline attempting to improve its state of undeveloped theory. Shortly after the publication of Zeisel's note the authors of two major research monographs from Columbia University's Bureau of Applied Social Research found it necessary to defend their deviation from the norm of significance testing by including brief appendices to explain why they did not consider the use of significance tests relevant for their purposes. Lipset, Trow, and Coleman's appendix on "Statistical Problems" in *Union Democracy* [Chapter 6] and Kendall's "Note on Significance Tests" in *The Student Physician* [Chapter 7] touch on a variety of issues surrounding the tests, but these authors' overriding concern, like Zeisel's, is with the bearing of the tests on the process and purpose of scientific inference.

In a portion of a review of *The Student Physician*, reprinted here under the title, "Some Pitfalls in Data Analysis Without a Formal Criterion," James Davis [Chapter 8], aims his remarks generally at the research reports from Columbia's Bureau and argues that the absence of a formal interpretive criterion such as a significance test leads to problems and inconsistencies in data analysis. His question, "Without significance tests, how is one going to tell when a finding is worth interpretation?" is echoed in several of the subsequent papers that defend the tests, particularly those by Gold [Chapter 20] and Winch and Campbell [Chapter 22].

In his "Critique of Tests of Significance in Survey Research," Selvin [Chapter 9] reflects his association with the Columbia critics of the tests. His critique, however, is more systematic and pointed and, in addition, has been more controversial and widely read. Selvin is concerned generally with the issue of causality, but he implicitly relates his discussion of it to the issues of sampling, population, scope, and process. Specifically, he is

---

2. The extent of Hagood's ambivalence is indicated by the fact that in the previous section Hogben quotes her at length (and selectively) to defend his skepticism of the Fisher notion of an infinite hypothetical population!

concerned with the use of the tests for making causal inferences in nonexperimental explanatory studies in the Lazarsfeld (1955) tradition. Selvin also makes it clear, although he provides no supporting rationale, that he thinks such explanatory studies help sociologists generalize beyond particular populations.

Selvin's argument is as follows: Tests of significance are appropriately used in interpreting an observed relationship between an independent and a dependent variable as definitive evidence that the former is an explanation of (cause of) the latter only when all important variables that are correlated with the independent variable (his "correlated biases") are controlled. Clearly Selvin thinks that tests of significance are at best the last *logical* step in explanatory research, after all correlated biases are controlled, but it is also possible to interpret Selvin to mean that the tests are *technically* illegitimate until control is complete. Since in nonexperimental research control by randomization is by definition impossible and control of all known correlated biases in the analysis of data is a practical impossibility, Selvin thinks that it is almost never appropriate for sociologists to use the tests in explanatory studies. Additionally Selvin points out that researchers often interpret the meaning and level of the tests incorrectly and make errors of technique in applying the tests—further arguments against their use.

The brief letters by Gold [Chapter 10] and Beshers [Chapter 12], which take issue with Selvin's position, make primarily a technical interpretation of his argument and point out that it involves a confusion of statistical and causal inference. In his rejoinders to these letters Selvin [Chapters 11 and 13] denies this confusion and seems to indicate he does not think that the validity of his central argument hinges on whether or not the tests are technically applicable when controls are not complete. Rather, he re-emphasizes his belief that a research strategy that gives priority to assessing random sampling error in nonexperimental causal inference is illogical and meaningless when, as is common, another more important source of error, the failure to control, goes unattended. Clearly certain ambiguities, certain errors, as well as certain overstatements in his paper left Selvin open to the attacks he received; but both Gold and Beshers and later McGinnis and Kish talk past Selvin to some extent since they do not fully address this point, much less attempt to deny its validity.

McGinnis's "Randomization and Inference in Sociological Research" [Chapter 14] was the first and the most focused of two longer papers that attacked Selvin's notions. After formally demonstrating what is involved in a correlated bias, McGinnis shows how randomization removes such biases within the limits of random fluctuation. He then develops a useful classification of hypotheses and shows that tests of significance are technically legitimate for random samples regardless of whether all, some, or no correlated biases are removed by randomization or other means of control. He also argues that the tests can meaningfully and logically be applied in any of these control conditions. This point is true, but it is relevant to Selvin's argument only to the extent that Selvin wants to impose explanatory goals on *all* social research, a questionable interpretation of Selvin's meaning. Similarly, McGinnis's point that the scope of inference allowed by the tests is limited to the particular population sampled and that no completely general relationships exist is true, but its relevance to Selvin's argument requires a narrow interpretation of Selvin's intent. It is reasonable to believe that Selvin's basic concern was with the problem of methods to develop theoretical statements of *greater* generality through explanatory studies, not statements of *absolute* generality, as McGinnis implies.

Kish's article on "Statistical Problems in Research Design" [Chapter 15] is less focused on Selvin's notions than is McGinnis's pointed attack. Kish puts his discussion of significance tests and his critique of Selvin into the general context of research design for explanatory studies and contributes a useful classification of the types of variables involved in such research. Further, he clarifies the sense in which generality and control are achieved in research design. In the part of his essay most specifically addressed to Selvin's paper, Kish reiterates and endorses McGinnis's arguments against the technical basis of Selvin's critique. In addition, Kish points out that the difficulty Selvin notes in achieving controls in nonexperimental studies implies not only the inappropriateness of the tests, but also the impossibility of nonexperimental explanatory studies. This point, however, doubtless reveals more about certain overstatements and confusions in Selvin's paper and the attendant tendency of his detractors to focus on technical and literal interpretations of his argument than it does about Selvin's intent. The total thrust of Selvin's article and of his subsequent responses to Gold and Beshers makes it clear that, just like Kish, Selvin does not think nonexperimental explanatory studies are meaningless unless controls are complete. Selvin thinks tests of significance are meaningless unless controls are complete *because* they direct effort away from and are often erroneously substituted for the more crucial consideration of achieving the best control possible to strengthen the weaknesses of nonexperimental research. Since Kish raises the issues of technique, meaning, level, form, and process and gives a clear discussion of associated common sins in using the tests, there is a strong indication that, despite genuine differences on the technical applicability of the tests, Kish and Selvin also share a considerable area of agreement on the contribution of the tests.

After a brief lull, the debate over significance tests in sociology was continued at a different level and in a much less polemic fashion in Camilleri's "Theory, Probability, and Induction in Social Research" [Chapter 16].

His paper examines the role of significance tests in scientific inference by linking the statistical issues of sampling, population, meaning, and level directly to the philosophy of science issues of scope, form, process, and purpose. Camilleri acknowledges the influence of Hogben but refers to none of the previous papers in the sociology debate; however, his points on sampling, population, and scope implicitly address the questions discussed by Selvin, Kish, and McGinnis. Camilleri's reasons are quite different, but the negative tone of his conclusion about the tests is the same as Selvin's. It is clear, moreover, that Camilleri has general sympathy with Selvin's concerns about generality and explanation, despite basic differences in their conception of the nature and methods of these research goals. Camilleri's paper is, however, broader in perspective and more programmatic than Selvin's. The tests are negatively evaluated in the context of scientific inference by Camilleri, but he gives some notion of how scientific theory can provide explanation without depending on either causal or statistical inference.

Camilleri develops a basic distinction between intrinsic probability (probabilistic relationships between variables), auxiliary probability (probabilistic processes in observation), and inductive probability (the probable validity of a hypothesis, i.e. scientific inference). His central criticism of significance tests is that they have evolved in the practice of research as a means for assessing inductive probability when in fact they are not fruitful for this. Instead he implies that significance tests are perhaps more appropriately construed as a method of auxiliary probability: At best they allow estimates of the relative frequency that decisions about hypotheses will be in error over infinitely repeated trials on the basis of the probability of an observation if the null hypothesis were true. As such they do not satisfy the criteria for scientific inference implicitly developed in his essay.

After several years of silence, problems in using significance tests were again discussed by sociologists in a brief and essentially complementary exchange between Skipper, Guenther, and Nass [Chapter 17], Duggan and Dean [Chapter 18], and Labovitz [Chapter 19]. The issue of level is the central concern in these papers, though the issues of power, meaning, and technique are also raised. While these papers are a contribution to more correct and rational use of the tests, the authors ignore the more fundamental questions connected with the usefulness of the tests that are raised by Selvin and Camilleri.

Gold's "Statistical Tests and Substantive Significance" [Chapter 20] and Winch and Campbell's "Proof? No. Evidence? Yes. The Significance of Tests of Significance" [Chapter 22] reflect a growing awareness that tests of significance are incapable of playing a central role as the sole arbiters of the truth value of hypotheses. The authors argue, however, that the tests nevertheless make a necessary contribution to scientific inference in screening from consideration those findings that have a high probability of having been generated by random fluctuations. Moreover, they maintain that findings are meaningfully screened by the tests regardless of whether a random procedure is known to have generated the data.

In our essay on "Significance Tests Reconsidered" [Chapter 21] we take issues with this argument on theoretical grounds but also on the more pragmatic basis that, even if the appropriateness of using the tests on data generated by nonrandom methods were granted,[3] such a screen is no screen at all outside definitive consideration of the issues of level and power— matters not considered at all by the advocates of the tests as a screen [see also Chapter 2:34].

The broader purpose of our paper, however, is to review a variety of statistical errors in using the tests and to relate these errors to the philosophy of science issues. We attempt to do this by synthesizing critical notions from the full range of sociological, psychological, and other literature on the topic. While our paper does not dwell specifically on the sociological controversy, it reflects and calls attention to the general viewpoint of this Introduction: The focus on technical and literal interpretations of the statistical issue raised by Selvin tended to direct sociological concern away from the more fundamental philosophy of science implications of this and other statistical issues. For the most part even the authors of the more recent sociological papers do not exhibit awareness of critiques of the tests that goes far beyond consideration of statistical issues.[4] As a result, sociological research is still characterized by a substantially ritualistic and naïve use of the tests.

3. The mathematics of random variables involves two main models. The classical statistical inference model on which significance tests are based assumes that there is only *random* variability by virtue of the random sampling or assignment procedure through which sample data are generated from their population. The random measurement errors and other "chance" errors that are the subject of inquiry in the other main use of random variable mathematics, random process models, are thus not considered in significance testing. At the very least, then, Gold and Winch and Campbell add to terminological confusion when they speak of the result of the random process model they employ as a "test of significance." As indicated here and in our paper we think they add to methodological and theoretical confusion generally, since many of the features of the tests *other* than the requirement of random generation make use of the tests for inferring (minimal) substantive importance with a random process model no less beset with difficulties than is the case with their traditional use in statistical inference.

4. With the exception of our own paper and one reference to Bakan in Gold's (1969) paper, none of the sociological papers published since Camilleri's exhibit awareness of Camilleri's or Hogben's contributions or of the philosophy of science critiques by psychologists reprinted in the next section.

4. The Notion of a Hypothetical Universe

MARGARET JARMAN HAGOOD

Situations requiring descriptive statistics. Let us consider the situations in sociology requiring the different methods of statistics. It seems that there is a rough parallel between the differentiation between social surveys and social research made by Pauline V. Young (1939:61) and the differentiation between descriptive and inductive statistics (in its second use described below). If a survey has as its purpose the description of a set of conditions for a certain area at a certain time, usually with the implied purpose of measuring them in order to plan some program of reform to alter them, then descriptive statistics may be adequate for the analysis of such quantitative data as are collected in this type of project.

Practical sampling situations requiring inductive statistics. If, on the other hand, either of the following situations is the case, inductive statistics is indicated. The first situation is the same as that described above for the survey, except that the area is too great for every unit to be counted or measured, and only a sample of units is observed. Here one uses inductive statistics to form estimates of various summarizing measures for the whole area (universe) with accompanying measures of unreliability of the estimates. This function of inductive statistics is simply a substitute for descriptive statistics when all units cannot be surveyed.

Hypothetical sampling situations requiring inductive statistics. The other use of inductive statistics is not so easy to state or explain. In this second situation, from observations made either from a sample or from complete survey of some limited universe, we generalize to a hypothetical universe which is difficult to define. It is the universe of all possible samples (which may be limited universes) which could have been produced under similar conditions of time, place, culture, and other relevant factors. Generalizing to such a hypothetical universe provides the sociologist with hypothetical "universals," which are not "universal" in an absolutist sense, however. They differ from the universals similarly arrived at by chemists, physicists, and others in a greater degree of complexity in the specification of the "similar conditions." A chemist can state as a "universal" that under specified conditions of temperature, humidity, pressure, etc., such and such a chemical reaction will take place in such and such a fashion when certain elements are brought together. His statement will be true "universally," which means, in ordinary thought, regardless of time, location, or the culture or civilization of that location. Since, however, the phenomena which interest sociologists are affected by time, location, and the nature and stage of a culture, any sample taken at one time in one area and observed as to phenomena associated with culturally conditioned human beings cannot be considered as representative of all time, all locations, all cultures. Without considering further the existence or meaning of universals in physical sciences, we can say that no one knows whether or not real—in contradistinction to hypothetical—sociological universals exist. Certain societal processes such as competition have been observed in many different time-place-culture combinations, and nonstatistical generalizations have been made as to their universality. But the practical limitations of securing data representative for all societies at all stages in all times is beyond the scope of any immediate possibility in the present stage of sociological research. For the present, at least, we shall have to content ourselves with "universals" whose universality is greatly diminished. For certain phenomena on which we have observations extending over a period of years, we can broaden the time span for which a generalization holds. But for most research projects which cut a cross section in time our generalizing is to the hypothetical universe of such universes as could have been produced within the stated conditions. It may help to imagine an expanded or prolonged present, where the dynamic forces continue to operate as of the specified time, producing phenomena which show only "chance" variation.

To an absolute determinist this conception has no validity. To him the observed results are the only ones that could have been caused and therefore could have been observed. To those who are attempting to integrate into the theory of causation of sociological phenomena the newer ideas of indeterminacy expressed in probabilities, the procedure of making a statistical induction in such a research situation seems to be a promising method. The matter is somewhat controversial, partly because the idea as well as the utility of the concept of such a hypothetical universe is relatively new and not too well defined.

Barring for the present the possibility of establishing and describing truly universal universals in sociology (they are at least barred from quantitative research, although the organismic, philosophy of history, or other types of generalizing may arrive at them by means of nonstatistical inductions), it seems that we have to accept this concept of generalization to a hypothetical universe of possibilities if we wish to proceed beyond the stage of mere description of a unique set of data toward the unraveling of the order or regularity underlying varied manifestations observed. The path is not very clear. We shall explore it further after the procedures for making statistical inductions have been explained. Perhaps more tools are available than we can use meaningfully. Yet careful and critical use of the ones which appear to be appropriate may lead to a clearer understanding of their full meaning. The eventual test of the usefulness of these statistical tools will, of course, be the fruitfulness of the results they yield [from Hagood, 1941: 302–304] . . . .

The universe of possibilities. Let us examine what possible meanings standard errors and statistical tests of hypotheses could have in such cases where all units in a finite universe have been measured or enumerated. First, there must be some superuniverse to which the hypotheses, and such terms as "true" or "real" value refer. This superuniverse must be a universe from which our finite universe can be considered a random sample. It has been defined as an unlimited or infinite hypothetical universe of possibilities—the universe of all the possible finite universes that could have been produced at the instant of observation under the conditions obtaining.[1] It is therefore only an imagined possibility, and whether or not one wishes to utilize the concept is still at the discretion of the individual research worker.

Meaning of the universe of possibilities in experimental work. Since the construct of a hypothetical universe of possibilities has proved useful and fruitful in other fields of research, let us look at its application in one of them, that of agricultural experimentation. The ideal of any experimental set-up is to control all conditions save the factors being studied and then in terms of correlation coefficients and other measures developed in the statistics of relationship to describe the relationship existing between the one or more "independent" or "causal" factors and the "dependent" or "effect" characteristic. With modern methods of analysis of data, the effect of several

1. The clearest exposition of the concept of the universe of possibilities is found in Samuel A. Stouffer (1934).

"independent" facts such as fertilizer, variety of plant, spacing, and other controllable factors on the "dependent" characteristic, or yield, may be studied simultaneously. We wish to consider the simplest case, however, where the relationship between only one independent fact such as amount of fertilizer and the dependent factor, yield, is being investigated. Let us suppose that for a certain range in amount of fertilizer used, there is found to be a linear correlation between amount of fertilizer and amount of yield described by the coefficient r based upon an experiment involving N observations, one on each of N plots. As far as the methods of descriptive statistics go, the value of r precisely describes the degree of association between amount of fertilizer and amount of yield for this particular group of N plots for one season. But no experimentalist trained in statistical methods would stop at that point; he would test his observed r to see if it were significantly different from zero. More exactly, he would test the hypothesis that his sample of N observations with an observed correlation of r could have been a randomly drawn sample from a universe where fertilizer and yield were uncorrelated, or where $\rho = 0$. If the test showed that the probability of observing a sample of N cases with a correlation as great as r from such a universe was very small, the experimenter would reject the hypothesis $\rho = 0$ and implicitly or explicitly affirm the hypothesis that $\rho > 0$. He would interpret his results to mean that he had observed a *significant* coefficient of correlation in his sample of N observation —significant in the sense that the observed r *signified* a ρ different from zero in the universe from which the sample had been drawn.

So far the procedures are straightforward, but the point we are interested in is more elusive. It is in answering this question: What is the precise definition of the universe concerning which the hypothesis was tested and from which the set of N observations may be considered a random sample? The experimentalist defines the universe as all the possible results that would be obtained by repeating the experiment under identical conditions an infinite number of times. One may ask how he knows he is justified in assuming that his one set of experimental data may be expected to yield values of parameter estimates which obey the laws of a sampling distribution theoretically derived for the mathematical model of random sampling. The answer is that a great mass of empirical proof has shown this to be the case. In fact, experimental techniques are often submitted to testing in what is known as uniformity trials where the criterion is conformity of their results to such sampling distributions (Snedecor, 1937, 2nd ed.: 46).

Two alternative interpretations of chance factors. Granted that experimental technique has not been perfected in any science to the extent where absolute identity of conditions can be obtained in repeated experiments, empirical evidence shows that in some fields it can be secured to the degree that only such variations as would be expected in random sampling are uncontrollable.

The next question to be answered is: To what are these residual variations due? Their distributions resemble the distributions we would expect if there were actually random sampling from an infinite universe, and therefore they are often said to be due to "chance" or to "random sampling variation." Again we reach the matter of defining "chance," and any attempts lead immediately into theories of causation. It may be that there is rigid and specific determinism of each event which occurs, and that our ascribing the residual variations we cannot explain to "chance" is merely a way of placing the limits of present day scientific knowledge. In such an interpretation the level at which knowledge stops may be different for different fields of science—it certainly reaches to subatomic levels in modern physics, whereas in certain aspects of agricultural biology it may stop with factors external to the living organism. If such an interpretation be accepted, we can expect to see advances in science reduce the unexplained variation we now ascribe to "chance" toward the limiting value of zero when knowledge is perfect, whether or not the limit ever be achieved. Just what interpretation is to be attached to the fact that the variations we cannot at any one stage explain are distributed according to the expectation of random sampling is not completely clear. The usual explanation for the close correspondence between observation and theory is that the factors labeled "chance" are numerous, each relatively small in importance, and independent of each other; for it can be theoretically deduced that this sort of situation would lead to such variation as is actually observed.

Certain developments in modern physics have implications for a less rigid determinism of the occurrence of events. Such interpretations of causation admit the impossibility of predicting unique events, because of the fact that only group averages are determined, with dispersion of individual events around the average in distributions similar to those described by random sampling theory. Indeterminacy according to such an interpretation of causation is not due to a mere limitation of the state of knowledge, but is an inherent property of the behavior of events, which are determined only in terms of group averages. By this interpretation, probability is a central concept of all treatments of causation, and the distributions expected from random sampling are a basic and fundamental aspect of the description of ultimate facts of causation. The implication of this interpretation is that no matter how advanced the state of knowledge becomes, scientific prediction will always have to be done in terms of probabilities less than unity.

The experimentalist's practical interpretation. The meaning and implications of these two interpretations of causation should be faced, but one does not have to "believe" in one or the other to make adequate interpretations of statistical analysis. To get back to the case of the agricultural biologist, his interpretation of the "significance" of his observed correlation coefficient is usually oriented to a practical situation. On the practical level, the fact that his *r* is "significantly" different from zero and (let us assume) positive means that he has grounds for confidence that a greater amount of fertilizer on this particular variety of plant will cause a higher yield from it, not only in a rigidly controlled experimental situation, but under actual farming conditions also. His confidence is based upon the approximate correspondence between the variation observed under experimental conditions controlled to the extent that identical conditions are approximated, and the variation of a distribution theoretically expected from random sampling of an infinite universe. Since the results of the experimental and mathematical models correspond, he can express his expectation of variation in his experimental situation in terms of the probabilities deduced from the mathematical situation, realizing that the approximate nature of the correspondence limits to some extent the literalness with which he must interpret precise probabilities. The validity of his practical predictions is not determined by what he interprets the "chance" forces to be, so long as he can describe the results of their operation. Thus, the experimentalist utilizes the statistical analysis of his results to explain phenomena (insofar as measuring degree of association where causation is imputed may be called explanation), to predict future phenomena (for instance, more yield from more fertilizer), and thereby to provide a scientific basis for the prediction and control of phenomena. This is an exaggeratedly simple case, of course, for purposes of illustration to contrast the situation of the experimentalist with that of the mere observer who has little or no control over the phenomena he is studying.

Interpretation in the nonexperimental situation. Sociologists, especially those in the field of social psychology, are developing experimental techniques for attack on certain of their problems where they have the advantage of more or less control over relevant factors. In general, however, this is not the case, and we are concerned here with the situation where the sociologist can only observe at one time measures or attributes of a series of units. We are concerned especially with the case where the observations are made for all of a series of units as of a certain time, as, for instance, on all of the rural counties in the United States for 1930. Let us suppose that for 1930 one observes a negative correlation coefficient of value *r* between fertility ratio and level of living for all rural counties in the United States. With the same formulas and procedures used by the experimentalists, the sociologist tests the significance of his observed *r* and finds it "significantly" different from zero. What does this test mean to the sociologist? In any test of significance there is a testing of some hypothesis about a universe from which the set of observations (a limited universe itself in this case) may be considered a random sample. That is, the logical structure of a superuniverse and the variation expected in random samples from it is the same for the observer sociologist as for the experimentalist.[2] Imagining any experiential counterpart of the logical model is a more difficult matter, however. It is easy enough for the experimentalist to imagine repeated experiments under identical condititions, whether or not he can actually perfect his technique to the degree that he can reproduce conditions identically. His universe of possibilities can therefore be put into meaningful terms; it can at least be imagined, even if it cannot actually be produced. It is not so easy for the sociologist to imagine a set of observations repeated under conditions identical with those of one date. The fact of change in social and cultural phenomena renders unrealistic any conception of identical repetition of the complex of factors conditioning characteristics such as fertility and level of living. The concept of the universe of possibilities —that is all possible sets of measures on fertility and level of living that could possibly be produced in the thousand rural counties of the United States under conditions exactly similar to those of 1930—the concept has neither a realistic counterpart nor a readily imaginable counterpart. To what, then, does the variation expected from random sampling from such a universe of possibilities correspond? Only a feat of imagination involving an infinite prolongation of a present moment, where conditioning factors remain the same but "chance" factors continue to produce random variation can supply the answer. With this done, the observer sociologist along with the experimentalist still faces the problem of interpretation of the chance variation— with the alternatives of ascribing it to the present limitations in knowledge or to the statistical nature of the occurrence of events.

Other more realistic models may occur to the reader, but so far none has been proposed which is satisfactory. It has been suggested that the limited universe of measures on all of a series of demographic units as of a certain date be considered a sample in time; that the random variations of sampling from a superuniverse have their counterpart in the fluctuations which would be observed if we made observations on successive days, or for successive years, while the general influencing conditions would not have altered appreciably. It is evident, however, that such successive fluctuations would

2. This sentence deserves certain qualification. The observational situation often differs from the experimental in that other relevant factors are not constant for all of the observed units. The concept of the dynamic universe as explained by Thomas C. McCormick involves the requirement that each observed unit in the sample have the same probability of possessing an attribute or a given value of a variable. Such homogeneity can sometimes be approximately obtained in the observational situation by successive subclassification with respect to all known influencing factors as is illustrated by McCormick (1937). Or a statistical substitute for homogeneity can sometimes be obtained by partial correlation techniques. But in neither the experimental nor the observational situation can absolute conformity to the criterion of homogeneity be obtained, and the essential difference between the two situations seems to be one of degree of approximation to the criterion. How gross this approximation may become before the validity of generalizing to the dynamic universe of possibilities is completely vitiated is an unsolved question.

not be independent, nor could they be thought of as being produced by forces independent of each other, and therefore they would not be expected to have the same distribution as those produced by chance factors in random sampling (as in fact can be shown to be the case). This approaches the problem of the economist in the interpretation of fluctuation in a time series.[3]

Another suggestion is that the measures on demographic units may be conceived of as one of an infinite set of such measures secured by dividing the total area surveyed into different series of areal units by shifting of boundaries under certain conditions of contiguity and uniformity of size. The matter of the arbitrary nature of the "lumps" in which our demographic information is secured, and the possible variations to be expected by recombining the information into different lumps have not been explored adequately. While the matter needs attention, it is probably not the answer to the search for a realistic counterpart of the universe of possibilities and to the random variation expected in samples from such a universe.

Reasons for using the construct of a hypothetical universe. At present the sociologist must face the fact that the postulated, hypothetical, infinite universe of possibilities, concerning which he tests hypotheses to establish the "significance" of his results, is merely a logical structure, for which he can offer no real counterpart in his research situation. Then what is the utility of such a construct and of the tests of significance based upon it? The answer to this question is not perfectly clear at the present stage of the application of statistical methods to sociological research. A case for the use of such a construct may be made on the basis of the following considerations, however. The amount of variation due to chance factors expected in statistics of random samples can be used as a standard, against which we evaluate variation observed in two different samples or hypothesized between a universe of possibilities and an observed sample. This affords a criterion for differentiating between variations which may be regarded as accidental or fortuitous, and those which cannot be so regarded. Another consideration in favor of the use of such a construct is that it has proved fruitful in other fields of research and

3. Excerpts from a noted economist's work on certain time series will illustrate this:
    Now time series, especially those relating to social and economic phenomena, are likely to violate in a marked degree the fundamental assumption which underlies the use of the methods sketched above, namely, that not only the successive items in the series but also the successive parts into which the series may be divided must be random selections from the *same* universe. Time series are, in fact, a group of successive items with a characteristic conformation. Such series . . . cannot be considered as a random sample of any definable universe except in a very unreal sense. Nor are the successive items in the series independent of one another. . . . The fact is that the "universe" of our time does not "stay put," and the "relevant conditions" under which the sampling must be carried out cannot be re-created. . . . It is clear, then, that standard errors derived from time series relating to social and economic phenomena do not have the same heuristic properties that they have, or are supposed to have, in the natural sciences (Schultz, 1938: 214–215).

therefore deserves a fair trial in sociological research, although interpretations of other fields cannot be slavishly imitated since the situations are so different. Finally, there is another reason for trying to generalize to the superuniverse of possibilities, which can only be suggested, since the concept has not been well clarified by those engaged in sociological research. It is based on the premise that there is a stability, a regularity, an orderliness in the occurrence of sociological phenomena, even though it is dynamic and ever-changing, and that one task of developing a scientific sociology embraces the description and formulation of the stable and regular, though dynamic, relationships underlying two or more series of phenomena. We have stated previously that the fact of differences in geographic location, culture, and time seems to preclude the possibility of developing any truly universal laws, or descriptions of relationships, among series of social phenomena which would be valid for all times, places, and cultures. Therefore, our goal in developing a scientific sociology is necessarily limited in the description of relationships. Yet somehow, there seems to be a place for the sifting from sets of observed measures of relationships the irrelevant variations which particularize them as unique, in a search for meaningful relations, impermanent and varied with location though they be. This goal is so far short of those of the physical sciences that it may be misleading to use the term scientific in our field. And yet at the present stage, the goal seems to be the only realistic one. It seems also that the transition from analysis of observed data on a finite universe by the methods of descriptive statistics to the use of the methods of inductive statistics in inferring information about the universe of possibilities—be it only a logical construct—is one approach to this limited goal [from Hagood, 1941: 425–432] . . . .

An example. The observations of the measures of the 48 states on migration and natural increase for the decade 1930–1940 may be considered as comprising an entire limited universe. There is no sampling situation here in the practical sense of the word where one measures only a fraction of the phenomena in which he is interested and generalizes to all the phenomena. Nevertheless, in research articles which include correlation coefficients based upon measures of the 48 states as varying units, one usually finds tests of significance, or standard errors, or other devices which have been developed from sampling theory. It is again the situation where one imagines a hypothetical infinite universe of all the possible limited universes of 48 pairs of observations which might have been observed. This hypothetical universe is the universe from which the 48 pairs of observations analyzed may be considered a random sample. Two people might discuss the problem thus.

> *A:* I have finished the description of the association between percentage net change due to migration and the average annual rate of

natural increase during the decade 1930–1940 as summarized above. The description is complete and exact for the 48 states of the United States for the decade treated. There is no need to make any tests of significance of the coefficients determined.

B: But are you not interested in the relationship between the two characteristics in general, rather than just in the way they were associated in this particular situation?

A: I would be interested in their association "in general" if I knew precisely what "in general" means for these characteristics. I certainly couldn't generalize these results to other decades where the factors might have been behaving differently; and certainly not to South America or Europe or some other place where the factors might have been behaving differently. My information is only on the 48 states of the United States for the period 1930–1940, and I cannot project the description of an association of characteristics beyond the time and place to which the basic information refers.

B: I grant that the description would not be valid for other times or other places except insofar as other relevant conditions in them are the same as in this actual situation. But for the time and place to which these data relate, can you not from this analysis and description of the association between the two characteristics generalize beyond the historical account of the unique facts recorded?

A: What is there beyond a historical account? What manifestations of association between the two characteristics could there possibly have been other than these which actually happened and were recorded as the set of observations I analyzed?

B: That is just the question. Given the same conditions which produced these migration and natural increase measures, operating repeatedly in the same place and time, would they always produce identically the same results?

A: But your question refers to a hypothetical situation, for we cannot imagine the 1930–1940 decade repeated with all factors which influence migration and natural increase remaining the same.

B: If, for the moment, you do imagine such a hypothetical situation, what would you expect as the resulting association between migration and natural increase?

A: I would expect minor variation in the coefficients describing association, but essentially the same sort of association that I have described.

B: Do you have any idea of the magnitude of the variation which might be expected to occur?

A: Only this idea, drawn from analogy, that it would probably be the type of variation observed in physical experiments when all the factors known to influence a result are held constant, but still the results show a residual variation.

B: The type of variation you are describing, which is produced by a very great number of very small factors, is called "random" variation, and a great deal of theoretical work has been done in describing the form and amount of random variation to be expected under various assumptions.

A: But since I *know* what happened in the 48 states in the 1930–1940 decade, what difference would the existence of random variation in a hypothetical situation make?

B: Suppose that in the imaginary universe of all the possible sets of observations on migration and natural increase which might have occurred during the decade 1930–1940 there is really no correlation between the two characteristics. Then if correlation coefficients were computed for every set of 48 observations, their mean would be zero. Most of these coefficients would cluster around zero in value, with a small proportion of them having moderately great positive or negative values. Suppose in such a case you have drawn in your particular set of 48 observations a value of the coefficient which is no greater than one would expect fairly frequently just from random variation. Would not that fact have a bearing on your interpretation, in terms of your problem, of the description of association you have made?

A: Why yes, if the association I have described could be regarded as "accidental" in the sense of being fairly common in a sample of 48 drawn from a universe with no association, then I would not attach much importance to it. Because doing so would give the false impression that the two characteristics were meaningfully associated in some way— whether by an intermediate factor or by some common antecedent factor or factors I do not know—when the association is possibly "accidental," if "accidental" can be defined in this way.

B: Then wouldn't it make you feel more secure in using your results for interpreting the situation, or for planning further research, if in some way you found that an association of the degree you observed would *not* be expected to occur one time out of 100 in samples of 48 from a universe with no correlation?

A: I suppose so, for in spite of my carefully qualified statements of interpretation, a correlation coefficient does make me suspect an actual relationship between the two characteristics, be it ever so indirect. And I should want to know that I was not wasting my time looking for the meaning of something which might be regarded as 'accidental." Nor would I want to suggest to those who read my report that there is a meaningful relation between the two characteristics if the only evidence I have may be regarded as "accidental."

B: Well, that is exactly what tests of significance of correlation coefficients do. They tell you what is the probability that a coefficient of the size you find for the number of observations you have might occur "accidentally."

A: But does "accidentally" as you have defined it actually apply to the situation we are discussing?

B: That is the crucial point of the whole argument as to whether or not you should use tests of significance in such a situation as this. One might justify the use of the concept of "accidental" in either of two ways. The first justification is based upon the assumption that demographic characteristics, like others that can be tested empirically, are subject to a residual fluctuation of the "random" variety when all the influencing factors remain the same, and that any magnitude of fluctuation, which is expected fairly frequently from these unknown but probably innumerable and minor factors, is called "accidental." Since the assumption can never be tested empirically—we cannot, for instance, turn time back and repeat the 1930–1940 decade—there can be no proof as to whether this assumption is justified.

The second justification is more abstract. Without any analogy to physical experiments or to any real situation, "accidental" is simply defined as encompassing the range of fairly frequently observed fluctuations which can be theoretically deduced for a random sampling situation. Then a hypothetical universe is postulated, a universe from which the actual set of observations may be considered a random sample, and the information obtained from the actual observations is used to infer information about this hypothetical universe. The universe and the tests of significance relating to it are simply logical structures. They do not have any objective counterparts; they have no everyday interpretation. There is no one convincing argument which justifies their use; but many statisticians feel that they may be useful in getting at a concept of the regularity or order underlying social phenomena.

A: Neither of your arguments convinces me of the utility of tests of significance where there is no practical sampling situation. Since I am merely beginning the study of statistics as applied to sociological research, however, I am willing to examine carefully the practice of persons with more experience in the field. What do other sociologists do about tests of significance in such situations?

B: They do everything imaginable. Some ignore tests of significance altogether, implying the same point of view which you hold. Others attach to their coefficients of correlation either their probable[4] or standard errors, assuming that their readers will know what to do with them. Others tabulate "critical ratios" (the ratio of an observed coefficient to the standard error of samples of the same size in a universe which has the same value of its coefficient as the observed), also assuming that their readers can make the correct interpretation of them. Others merely state whether or not an observed coefficient is "statistically significant" without specifying what tests they use—and incorrect tests as well as correct tests are in common use. Still others, though fewer, give the results of specified correct tests of significance. But even in the case of these last, there is seldom an explanation of what they mean by labeling a particular coefficient of correlation as "significant" or "statistically significant." Therefore, you can find good precedent among sociologists for whatever you choose to do—except for explaining exactly what you mean when you use tests of significance.

A: Then if after more thought on the matter I decide to test the significance of my coefficients, in my report of a quantitative sociological research project should I go through all of the explanation you have given of the logical structure, the universe of possibilities, the nature of random variation, etc., every time I give the results of the application of a test of significance to an observed correlation coefficient?

B: Probably not. Such a procedure would be repetitious and boring. One should, however, himself understand the meaning of his tests; he should choose correct tests; and at this point of development of the application of statistical methods to sociological research, he should specify the test he has used. But what is of equal importance is that the reader should know what the test means, since otherwise he may be unduly impressed with the magic phrase, "statistically significant." It is an example of the common situation where the research worker should use correctly and carefully the appropriate statistical methods, but because of the necessity of economy of words in reporting his results, he has to assume the statistical literacy of his readers.

Since tests of significance of correlation coefficients are found increasingly in sociological literature, regardless of whether those who take the point of view of B in the discussion above can *prove* one should use them (when there is no practical sampling situation), one should learn how they can be made correctly. For developments in statistical theory of the last decade or two

---

4. The probable error, which assumes a symmetrical sampling distribution, cannot be applied correctly to any coefficient of correlation different from zero, for only in universes where there is no correlation is the sampling distribution of *r*'s symmetrical. Therefore, the probable error should *never* be attached to an observed coefficient.

have shown the tests formerly used to be incorrect, and those who are using as guides texts published 10 years or more ago are likely to be using unacceptable tests of significance for their correlation coefficients. The most common test of significance answers for the universe the question as to whether or not association *exists* in the universe—that is, it investigates for the universe the first aspect of association [from Hagood, 1941: 612–616].

5. The Significance of Insignificant Differences

HANS ZEISEL

IT HAS BECOME fashionable to chide social scientists for drawing inferences from data which, on analysis, prove to be not significant by customary statistical standards.

This criticism typically arises with respect to fringe data from survey or experimental statistics, when through consecutive breakdowns the sample begins to wear thin. Eventually, from a fourfold table a relationship may emerge, the confidence limits of which are very low. Nevertheless, the analyst is often tempted to explicitly note the relationship and draw an inference from it.

It is easy enough to avoid misrepresentation. "If, in a larger sample, this relationship should hold . . ." or "Although this difference is not statistically significant . . ." or "There are indications that . . ." or simply "It would seem . . ." are household safeguards for this situation.

But such exculpating clauses merely conceal the real issue. The question is whether or not to encourage theoretical inferences from such statistical fringe data, at the considerable risk of their being proved wrong later on.

There are good reasons, both theoretical and practical, for such encouragement. The theoretical reasons are these: First, even a high level of statistical significance entails by definition the possibility of error. Secondly, unless the

Reprinted from *Public Opinion Quarterly*, 17 (Fall, 1955), 319–321, by permission of the author and the publisher. Copyright © 1955 by the Columbia University Press.

data are derived from controlled experiments, there is the ever present possibility that hidden variables account for the spuriously high significance of the result. But even if the data are derived from controlled experiments, extrapolating from the universe from which the sample was actually drawn always creates the possibility of error.

Whether, in view of all these threats, to draw or not to draw an inference is a practical question transcending the statistical data proper. It is a commonplace to point this out, but one should realize that a theoretical inference in a piece of social science research is much less of a risk than making a wrong decision, for instance, in a vaccine test.

Until very recently, social science theorems were derived largely unencumbered by statistical data. Even now, probably the greater part of social science theorizing is kept apart from any attempts of adducing statistical evidence. On the other hand, the researchers who follow the statistical way of life often distinguish themselves by a certain aridity of theoretical insights.

Both these situations suggest that the social analyst should not be discouraged from theorizing, guided by his data, even if these data be far from conclusive. Considering the sociologists of the old school, he may draw the consolation that to be lead by *any* data is better than to be led by none; and considering the often discouragingly pedestrian analysis of statistical studies, he may draw encouragement for somewhat more imaginative if sometimes unwarranted inferences.

There is, now, in the social sciences no greater need than the development of theoretical insights guided by empirical data. At such times, to provide this guidance and serve as a stimulant is the significance of statistically insignificant data. Even if the probability is great that an inference will have to be rejected later, the practical risk of airing it is small. Subsequent and more elaborate studies may disprove some of these inferences; but for those that survive social science will be the richer.

To be sure, a physicist would frown on such recommendations. But his is a world of generalizations on a high level. By comparison, the social sciences are at a stage where for decades to come the formation of even tentative theoretical structures will be at a premium.

6. Statistical Problems

SEYMOUR MARTIN LIPSET, MARTIN A. TROW, and JAMES S. COLEMAN

IN THIS BOOK, no statistical tests of significance have been used. This may seem unaccountable, particularly in view of the numerous quantitative comparisons which constitute much of the analysis. Can it be defended, and if so how? It can be defended, and we shall defend it at length because there seems to be no good statement of our position in print.

Statistical tests are used for a number of different purposes. One use is to indicate the precision of a descriptive statement about a population. When a random sample of a population is measured in terms of some attribute, then the sample distribution is used to make a statement about the population distribution. For example, on the basis of the ages of men in a random sample, a statement about the mean age of a population may be made. But because only a sample of the population was measured, such a statement is subject to sampling error. The statement may be made in terms of confidence intervals (e.g. "One can say [with a confidence that 95 percent of statements like this will be true] that the average age of printers is between 42 and 48 years") or in terms of a mean value with certain limits (e.g. "The average age of printers is 45 years, plus or minus 3 years, at the .05 level").

This, then, is one way in which statistical inference is used in survey

Reprinted with permission of the Macmillan Company from *Union Democracy* by Seymour Martin Lipset, Martin A. Trow, and James S. Coleman. © by The Free Press, a Corporation 1956.

analysis. However, such a use is confined almost entirely to descriptive studies, whose primary aim is to describe a population in terms of the attributes by which the sample is measured. If the aim of this study were to accurately describe the New York Typographical Union in terms of the proportion of members with this or that attribute, then this type of statistics would be necessary to indicate the precision of the descriptive statements. But such description is not the aim in a study like this, in which the aim is rather to establish the existence or nonexistence of relationships between attributes. For this a different kind of statistical inference is necessary.

To determine the existence of relationships in survey analysis, the usual procedure is to present contingency tables.[1] Sometimes such a relationship appears to be strong; in other cases it is quite weak. How does one decide whether there really is a relationship in the population? The usual method is through Chi-square tests of independence in contingency tables. The familiar Chi-square values presented in most published psychological work and increasingly in sociological analysis are such tests. If the Chi-square value is low, the null hypothesis (that the two variables are independent) is accepted; if it is above a certain critical value, the null hypotheis is rejected at the .05 or .01 level of confidence.[2] In the latter case, the alternative hypothesis—that a relationship does exist—is accepted.

The Chi-square test for independence is thus the test applicable to the kind of analysis carried out here, an analysis which attempts to determine the existence of relationships. But however applicable the test is in the ideal case, it is not so practically, except in special circumstances. Some factors in an analysis like the present one tend to make the test too weak, for they violate the assumptions on which the test is based. Thus a supposed relationship which is "significantly different from chance" as judged by a Chi-square test, may not be significantly different from chance at all. On the other hand, there are certain factors which tend to make usual Chi-square tests too strong; acceptance of the null hypothesis may occur even though a strong relationship does exist. Finally, there are serious questions about the relevance of such tests for analyses like this, even when they are neither too weak nor too strong. These three classes of factors will be considered in turn.[3]

1. If the variables are continuous, the usual method is to present an estimate of the correlation between the two variables. This is technically a different procedure, but the logic of the statement to be made here applies to such cases.
2. That is, only five times (or one time) out of a hundred will such a rejection be made when the hypothesis—that no relationship exists—is true.
3. This is not to say that no empirical analyses in social science need statistical inference. Many do, and among these are the studies which are primarily descriptive. For an excellent review of statistical problems in such studies, see William C. Cochran, Frederick Mosteller, and John W. Tukey (1953).

1. Factors Which Tend to Make Significance Tests Too Weak

In *a posteriori* analyses, the investigator can do all of the following things, any of which violates the assumptions behind the test of significance described above:

(a) He can modify in the light of the data whatever *a priori* hypothesis he had; if he develops an hypothesis *on the basis of* relations found in the data, then it is clearly foolish to turn around and test this hypothesis by testing the "statistical independence" of those same relations.

(b) If he is blessed with an abundance of data, some of which are substitutable for others, he can select those data which confirm his hypothesis that a relationship exists. It is easy to see how this might occur: Tests of independence in contingency tables allow some such statement as "Less than five times out of a hundred would such a relationship have occurred by chance." If the investigator looks at a hundred tables, by chance alone he will find about five which show "significance at the 5 percent level." We have admittedly selected and discarded from many tables, just as have almost all analysts of interview material; and the rationale for such action is by no means naïve. Social researchers usually begin with a plethora of hypotheses and half-formed ideas. The selection and discarding of tables is at the same time a selection and discarding from among this wealth of vague ideas. It is in this way that a consistent analysis develops from a mass of mutually contradictory or confused notions. At the same time some data must be discarded because measurement is bad. Errors of measurement tend to obscure relationships; and partly because of this, investigators often accept the table which shows a relationship and discard another bearing on the same hypothesis if it fails to show one.

(c) An added freedom is allowed the investigator in using data which must be "collapsed" in the contingency tables (such as the liberal-conservative scale or the ideological-sensitivity scale in this analysis). These scales may be collapsed at points advantageous to the hypothesis, rather than at others. If the collapsing of such scales is not done independently of the relationship being investigated, significance may be found spuriously. In both these scales we have collapsed independently of the hypothesized relations (on the basis of the numbers in each category after collapsing), but at other places we may have collapsed answer categories so as to make differences appear as large as possible.

2. Factors Which Tend to Make the Tests Too Strong

On the other side of the fence, ordinary tests of significance of a single contingency table would be too strong on the following grounds.

(a) Tests of significance of a single contingency table assume isolated hypotheses, each to be confirmed or disconfirmed by the single table, and each independent of the others. But it is one of the essential characteristics of the present study that the hypotheses *are* related, and that a given table acts not to confirm a single hypothesis, but a whole network. Conversely, a given hypothesis is confirmed not by a single table, but by many. Clearly, then, what should be tested is the significance of the total set of interlocking tables (and, by inference, the interlocking causal relations). This would be simple if the data for each table were independently gathered: the *product* of the probabilities that each table had occurred by chance would be the probability that the set had occurred by chance. This would mean that less restrictive restrictions could be placed on each table. For example, a set of four such interlocking tables would show a statistically significant relation at the 1 percent level if each were significant at about the 30 percent level, which is a remarkable lack of restriction.

But things are not so simple. In the present analysis, as in most like it which are reasonably complex, the data were not collected independently, but are from the same sample, and there is no tight interlocking of hypothesized relations. Though there is an attempt at such interlocking, there is still much looseness in the set of hypothesized relations. The system of relations would not collapse if one empirical result were taken away. There is some interlocking between tables, and some independence between measures used in different tables, but it is impossible to say just how much.

3. Why the Tests May Be Irrelevant

Finally, it is not clear that tests such as this are even relevant to a study such as the present one, for the following reasons.

(a) This study, like many in social science, is an exploratory study, not a confirmatory one, while statistical tests of hypotheses are designed for confirmation; Chi-square tests of independence are designed to confirm and consolidate what is already believed to be true. A study like the present one is designed to find out what was not even guessed at before. That this new knowledge is not fully confirmed is no great cause for concern. Further studies upon different organizations will constitute more reliable confirmation, for they test the hypotheses in a different population, which a Chi-square test used on this data could never do. It is probably better to place one's faith in further studies to confirm hypothesized relationships than to place it in Chi-square tests. Even if all the assumptions for such tests are fulfilled, the population to which the result is to be generalized is not the population from which the sample was taken. It is a theoretical population, of all men in certain kinds of organizations. To replicate the study in another organization would confirm the result under quite different conditions, and this would seem of more value than the assurance offered by the usual Chi-square test. This is the method through which the natural sciences have made most of their remarkable advances, and there is no indication that many of these advances would have come earlier if modern statistical inference had been used instead.[4]

(b) It is useful to ask again just what is being tested by a Chi-square test. This test shows whether a relationship between two measurements could have been due to chance. Often, however, these measures are simply crude indicators of the variables which it is desired to test. For example, to show that ideologically sensitive men are more active politically, we use as an indicator the amount of talk about politics reported by the men. This is only a crude indicator of political activity and might easily fail to show the relationship if one actually exists. Thus, because the variables being related hypothetically are seldom the same as the measures being related empirically, a test of independence appears relevant.

(c) It is important to ask whether we really want to test the existence or nonexistence of a relation. Suppose a relation is extremely weak: Is such a relation of interest? Probably not, in most cases; yet a large enough sample would find such a relation to be significantly different from chance. On the other hand, an extremely strong relationship would be found not significantly different from chance if the sample were very small.

It is probably true that we are ordinarily interested in knowing two things: whether a relation could have been produced by chance, and how strong the relation is. But probably the latter is more important to most investigators. To know this, tests of significance are not necessary; correlation coefficients or some other measures of association are.[5] Thus there seems more reason to compute the latter than the former.

(d) Finally, it is important to question whether it is the *statistical* relationship one should be concerned with testing. In the development of a science, it is causal relations which are sought after, and statistical relations are only a bad reflection of this. If two variables are both affected by a third, this can produce a statistical relation between the two where no causal relation exists at all. Thus it appears that a statistical test of the hypothesis might not be as

---

4. For example, a careful reading of W. P. D. Wightman's *The Growth of Scientific Ideas* (1953), which charts the course of scientific discoveries from early times, failed to show more than one or two possible places where statistical inference might have contributed significantly to the development of any science in its early years.

5. Oftentimes Chi–square-type measures have been used to measure the size of a relationship; the size of a Chi-square value in a contingency table is thus some measure of the degree of association. But it is only a crude one and would be better replaced by a measure designed for the purpose. For a review of tests of association in contingency tables, see Leo Goodman and William H. Kruskal (1954).

useful as other means of testing whether a statistical relation represents a causal one. Such means are ordinarily nonstatistical, and rather of this sort: "If it is true that a variation in X causes a variation in Y (as the existence of a statistical relation between X and Y might suggest), then it should also be true that a variation in X causes a variation in Z." Through such means, followed by tests of the X–Z relation, the existence of a causal relation between X and Y may be more and more confirmed, without bothering to see if the statistical relation is significantly different from chance.

There is no intention here of suggesting that statistical inference is generally irrelevant in social research. It is certainly relevant for descriptive studies, to attach some measure of precision to statements made about a population on the basis of measuring a sample.

Statistical tests of hypotheses, however, seem to be of quite limited aid in building theoretical social science. As tests in certain experimental situations, to determine the effect or noneffect of an experimentally introduced condition, they are of aid. But in testing hypotheses in field researches, they appear to be of questionable value, for the reasons given above. Since we are not really interested in statistical relationships anyway, but in causal relationships, it appears to be of much more value to test whether a statistical relationship ("significant" or not) represents a causal relation, or merely covariation between two variables. Such tests are made not by statistical tests of hypotheses, but by examining further hypotheses which would be true if the relation is a causal one, false if it is not. To devise and examine empirically such hypotheses seems a much more reasonable way of building a theoretical structure than to test the original hypothesis with a Chi-square test, however well the assumptions for such a test may be fulfilled.

7. Note on Significance Tests

PATRICIA KENDALL*

THE READER will find that no traditional significance tests have been reported in connection with the statistical results in this volume. This is intentional policy rather than accidental oversight. It is a policy, furthermore, which the Bureau of Applied Social Research has always adhered to in reporting the results of exploratory studies such as are presented in this volume. Some of the reasons for this decision may be described as follows.

To begin with, traditional tests of significance have been developed to study the probable correctness or incorrectness of *single, isolated* statements, for example, that Drug A is more likely than Drug B to cause nausea. But it is an essential feature of these exploratory studies in medical education, and of many other empirical investigations in the social sciences, that there is no single hypothesis which can be viewed independently of other hypotheses. Instead, there is a series of loosely interrelated hypotheses which must be looked at in combination. For example, we may hypothesize that medical

Reprinted from Appendix C of Robert K. Merton, George G. Reader, M.D., and Patricia Kendall, *The Student Physician: Introductory Studies in the Sociology of Medical Education*, Cambridge, Mass.: Harvard University Press (for the Commonwealth Fund), 301–315, by permission of the authors and the Commonwealth Fund. Copyright © 1957 by the Commonwealth Fund.

* Dr. James Coleman and Dr. Hanan C. Selvin contributed to the formulation of some of these points while they were members of the Columbia University Bureau of Applied Social Research staff.

students will respond more favorably to some features of an educational innovation than they do to others. But a corollary hypothesis, which must be borne in mind when this first one is stated, is that those aspects of the innovation which are reacted to negatively may have as much—or more—impact on student attitudes and self-images as those which are better liked. In other words, it can be misleading to focus attention on one hypothesis without taking into consideration other related hypotheses. In a situation such as this it does not help to apply significance tests as standard procedure, for this assumes that each result is isolated from and independent of every other. Perhaps the solution is to consider the probability of the joint occurrence of the several results; but this has the shortcoming of assuming that we know, quite explicitly, which of the hypotheses are related, and in what manner.

Another reason for feeling that traditional significance tests are not appropriate relates to the purposes of our studies. These tests are designed to keep one from making statements about percentage differences (or other differences) when there is too little evidence to justify the statement. In order to do this, in order to avoid making unjustified claims about the magnitude and importance of an observed difference, one sets a rather high level of significance; one says, in other words, that, if there is more likelihood than 1 in 100 (or 5 in 100) that the observed difference could have come about by chance, one must conclude that it is not a "real" difference. But in the effort to avoid mistaken decisions that an observed difference is significant, one runs the danger of making another kind of error. By insisting on a high level of significance (a low probability that there is no real difference) one increases the possibility of rejecting as insignificant differences which actually do exist. And this is a serious error in exploratory studies of the kind which we have been conducting. At this early stage of thinking about the processes of medical education it would seem desirable to assemble a wide array of evidence, even if some of it is not conclusive. We want to gather together as much information as we can about the relations between experiences in medical school and the development of particular attitudes and values. Later on, some of the hypotheses which we set up before starting the research, and others which have emerged during the course of our investigation, can be submitted to more definitive and rigorous tests. Until that time, however, we do not want to cut short possibly productive lines of investigation by insisting now that our preliminary results prove themselves significant. We do not want to say now that, because some of our early results do not meet stringent criteria of significance, they should be disregarded, or discounted, or not reported.

There are still other reasons why we believe that the routine application of significance tests to our data is not fully justified. For one, these tests presuppose that the units being studied were sampled randomly from the populations to which they belong. We have administered our questionnaires to the total student bodies of three medical schools; in no sense can these groups be considered random samples, or even, for that matter, samples. There is a real question, therefore, whether there would be any justification at all for making use of traditional tests of significance.

Moreover, there is a distinction to be made between the importance of a result and its statistical significance. Reliance on tests of significance generally carries with it an obligation to report only those results which meet the criteria of the tests; where a significant difference is not found the result is not presented. But this can often be quite misleading in social research. In this field some of the most important results are those in which an expected relationship is not revealed by the data. Statistically, such results are not significant; conceptually, they have great significance. For example, we should expect students in the lower half of their class to be more likely than top students to favor a system of anonymous examinations in medical school. We would consider it highly significant if they did not, if, in other words, there were no relationship between class standing and preferences for an examination system in which the students' anonymity is preserved.

These are the most important reasons why we have intentionally refrained from applying traditional tests of significance to our findings. This is not to say, of course, that we have been indiscriminate in our presentation of statistical results. We have established criteria of significance for ourselves, even though these do not involve the usual and more familiar tests. Essentially, these criteria involve two related types of *consistency*.

The first criterion is that, to be considered significant, results must be *internally* consistent. That is, a finding with regard to one question is held to be valid only if it also holds true in connection with a closely related question. For example, we may find that, as they advance through medical school, students become increasingly likely to say that they want to know their patients only on a doctor–patient basis, and not also as friend to friend. We attach significance to this finding only when we uncover other evidence that preferences for such detached relations are progressively learned; if we find, for instance, that advanced students are more likely than their junior colleagues to say that it is important for the physician to control his affective reactions in the doctor–patient relationship.

The second criterion is that results must be *replicatively* consistent if they are to be considered significant. That is, a finding in one group must also hold true in a second independent group, if the same general conditions prevail in both. The groups in which such consistency is studied may be comparable classes in two or more medical schools, different classes at comparable stages of their training within one medical school, or particular subgroups of the classes. The only requirement is that they be independent of each other. In these instances, according to this criterion of consistency, results found in one group should be repeated in others. For example, we may find in one medical school that, as the students move from one stage to another in their training, they become progressively more likely to say that they view their contact with patients as an opportunity both to learn medicine and to help patients, rather than just to learn medicine. We consider this a valid result only if the same pattern is observed in a second medical school, or in the same medical school at different times.

There is, of course, one type of situation where the criterion of replicative consistency must be modified. If two medical schools differ markedly in their organization or curriculum, then we should not expect to find that attitudes or preferences affected by these organizational features will be replicated from one school to the other. For example, in a school where students do not have substantial contact with patients until their third or fourth years, we may find that first-year students worry more about their performance in their medical studies than they do about their competence to deal with patients. In a second school, where first-year students have considerable contact with patients, we may find exactly the reverse—namely, that first-year students in this school worry more about their adequacy in dealing with patients than they do about their performance in their studies. Because this discrepancy can be related to significant differences in the organization of medical studies in the two schools, we consider both findings valid. Or, if two classes within the same school are very differently composed, we do not expect that attitudes or behavior affected by these differences will be replicated from one class to the other. For example, one entering class may contain a relatively large number of married students, while the class entering the next year contains relatively few. Under these conditions, we expect very different findings with regard to such matters as the way in which friendships are formed, the extent to which friends function as significant points of reference, and patterns of leisure-time activity.

8. Some Pitfalls of Data Analysis Without a Formal Criterion

JAMES A. DAVIS

AT FIRST GLANCE, we note the trademark of this school*—the absence of significance tests—and, as in other recent Bureau publications, an appendix purporting to justify their exclusion. While some of the argument is trenchant (the cases are, as they say, a universe, not a sample) and some of it plain foolish (e.g. If you use significance tests, you don't report non-significant findings, and sometimes the failure of a hypothesis is interesting), it is my opinion that it is not the absence of significance tests but the absence of *any formal criterion* for arriving at a conclusion which typifies this approach.

The essence of the Columbian exposition of research data seems to consist of intuitive interpretations of heavily partialed tables. While almost anyone is in favor of the partials, the intuitive part raises some problems.

Consider, for a moment, "percentage differences." Without significance tests, how is one going to tell when a finding is worth interpretation? On page 201, a 13 percent difference between the extremes of three groups is

Excerpted from James A. Davis, "Review of Robert K. Merton, George Reader, and Patricia Kendall, *The Student Physician*," *American Journal of Sociology*, 63 (January, 1958), 445–446, by permission of The University of Chicago Press. Copyright 1958 by The University of Chicago.

* [The "school" Davis refers to is the Bureau of Applied Social Research at Columbia University, which conducted the research reported in *The Student Physician* and *Union Democracy* (see Chapters 6 and 7). Davis' page references, below, are to Merton *et al.*, *The Student Physician* (1957)—The Editors.]

considered a fairly firm finding. On page 135, however, another 13 percent difference among three groups is shrugged off as relatively minor. On page 128 a 19 percent difference among five groups (which, in addition, shows an internally inconsistent trend) is presented as a flat finding, with no qualification at all. The reader is left with no idea of how the researchers determine which tables do or do not show relationships.

Again, on page 191, one of the authors wants to sum scores on five items to get an index. He says that the normality of the distribution of the sums (which, of course, is not tested) "suggests that the questions differentiate students according to their preferential attitudes" and proceeds to use the sum as an index. Six pages later he notes that another such distribution "is markedly skewed" (it is also bi-modal) and then proceeds to use the sum as an index. We don't know whether normal or non-normal distributions make better indexes, but we are entitled to doubt that the author is using any criterion at all in deciding whether it is justifiable to use a sum as an index.

And again, on page 143, it is claimed that medical students perceive more undergraduate scholastic competition on the basis of 49 percent reporting "a great deal of competition" as compared with 18 percent of the law students. However, seven pages later the same author, in another context, says: "It is natural that students, perceiving the competitive situations in which they are engaged, might well react with feelings of concern," and then he examines data on "concern with school standing." The table here shows 13 percent of the medical students and 14 percent of the law students reporting such concern. Now in the methodological appendix to the book it is claimed that "internal consistency" can replace significance tests as a criterion. We suggest that either it is *not* natural that feelings of concern arise from competition or the criterion of internal consistency has been applied in a somewhat relaxed form.

Finally, on page 181, the finding that there is a relationship between students' perception of themselves as "doctors" and their being treated as such by others is considered as "lending support" to the proposition that "there is a consonance between the status that students believe others in their role set assign them and the status to which students assign themselves." The author says that this proposition has long been recognized but little investigated. However, on page 203, another author reports that there is no relationship between the students' self-confidence and their grades, and "it is therefore not the case that self-evaluations correspond to faculty judgment." Is this a "dis-confirmation" of the proposition? No, it is suggested that the two measures tap different dimensions! The reader wonders exactly what sort of evidence it would take to "dis-confirm" hypotheses in the new methodology.

The point is not that there are internal inconsistencies in the materials. Rather we would suggest that, without formal criteria, tests of significance, or other numerical criteria, it is almost impossible for the research worker to be consistent, regardless of the elegance of his arguments in a methodological appendix. Furthermore, without some communicable decision process, it is even less possible for different investigators to arrive at the same conclusions from data. The net result is "art," not "science."

In a very literal sense, the new survey methodology comes close to an art form in which a verbal argument is presented, accompanied by complex and fascinating counterpoint on the IBM machine. Since, however, none but aesthetic grounds can be given for evaluating the harmony between the two, it is the artistic skill of the performer which occupies the attention of the reader. That the performers in this volume are talented, sophisticated, and probably correct in most of their interpretations only makes their method, as method, more dangerous.

9. A Critique of Tests
of Significance in Survey
Research

HANAN C. SELVIN

DIFFERENCES OR CORRELATIONS are now routinely tested to ensure the statistical soundness of sociological inferences. The popularity of these tests has even reached the point where their absence occasions criticism (see, for example, Furfrey, 1955; and Katz, 1955). Considering their increasing acceptance in the twenty-odd years since they first appeared in sociological writings, one might ask why their use has not become universal, why every empirical study based on standardized data does not test its hypotheses statistically.

Perhaps this is nothing more than cultural lag, which will eventually disappear with more widespread knowledge of statistics. However, the failure of many sociologists to "test for significance" may well be deliberate, even though they may not give their reasons. With only slight exaggeration it is possible to divide empirical researchers into two groups: (1) those who test each conclusion for significance but seldom cross-tabulate extensively to discern causal or explanatory factors; and (2) those whose substantive analyses are based on extensive cross-tabulations, with no tests of significance. Although the members of the first group are by no means of one mind, the few critics within it have generally concluded that the tests do perform a valid function in sociological research (Hagood and Price, 1952: 286–294, 419–423; Chapin, 1955: 176–186, 261–283). Exactly the opposite point of view has been argued in four recent discussions, three by sociologists and one by a statistician (Lipset *et al.*, 1956 [Chapter 6]; Patricia Kendall, 1957 [Chapter 7]; Selvin, 1956; and Wold, 1956).[1] Each of these discussions, however, is somewhat specialized. It seems useful, therefore, to consider tests of significance in sociological research more generally, to inquire into the conditions under which they may properly be used and the kinds of inferences that may justifiably be drawn. The conclusions are essentially negative: that the conditions under which tests of significance may validly be used are almost impossible of fulfillment in sociological research (save perhaps in one or two limited areas, to be discussed briefly) and that, even when these conditions are met, the nature of the research situations faced by sociologists is such that correct inferences from the tests are equally difficult to reach.

It should be made clear that we are not concerned with technical errors in statistical procedure—for example, the improper use of the normal distribution in place of Student's distribution or the use of formulas based on strictly random sampling for data obtained by stratified, cluster, or systematic sampling, each of which requires a different approach.[2] Such errors are easy enough to discover and remedy. Nor are we concerned with describing a particular population on the basis of a sample. This is an important practical problem in such fields as governmental policy and public-opinion polling, but in most sociological research and in most other fields of "pure" research as well the population sampled is chosen because it is convenient; the emphasis is on explanation rather than description, on uncovering general relationships rather than precisely depicting a unique situation.[3]

Our task is to analyze the problems raised by the application of significance tests to explanatory empirical studies. These problems fall naturally into two main groups: (1) problems stemming from the difficulty of designing appropriate procedures for testing hypotheses, and (2) problems raised by the interpretation of the results of these tests.

1. Coleman, Kendall, and I independently came to these overlapping and complementary points of view while working on different empirical studies at Columbia University's Bureau of Applied Social Research. The imbalance of three sociologists and one statistician in this citation bears out Wold's indictment of the "stepchild treatment" that these problems have received from professional statisticians.
2. Some elementary texts do recommend clustering and stratification, but the computational formulas usually given are for simple random sampling. See, for example, Lillian Cohen (1954: Chapters 5–7) and Hagood and Price (1952: 289). Others, such as Allen Edwards (1954), do not consider these other types of sampling at all, yet simple random sampling is almost never used in empirical research. For a thorough discussion of this topic see Leslie Kish (1957).
3. An extensive discussion of the differences in logic and procedure between these two types of studies is contained in Herbert H. Hyman (1955b: 66–89). See also Wold (1956).

Problems of Design

The principal difficulty in applying tests of significance to sociological research centers about the concept of *experimental control*. What sociologist has not envied the physicist's ability to hold the temperature of an enclosed gas constant while studying the relationship between pressure and volume? We are not alone in our envy; for many years biological scientists—especially agriculturalists and animal husbandrymen—were confronted with the problem of controlling irregular variations in the characteristics of their experimental animals or in the fertility of their experimental plots. Observed differences were a combination of "treatment effects" and of uncontrolled variations in the experimental subjects. If a brood of chicks raised on a new kind of poultry mash gained more weight than a control brood fed the standard mash, one could not be sure that the difference resulted from the mash, rather than from variations in the characteristics of the chicks.

The statisticians came to the rescue of the biologists. Guided by the pioneering work of R. A. Fisher, they developed a statistical approach to problems of experimental inference where "all relevant factors" cannot be controlled. The essential idea is to control what can be controlled and to *randomize the uncontrollable*. Since it is impossible to control all idiosyncratic differences between chicks, the experimenter forms an experimental group and a control group from a single brood[4] by some random process. Once this is done, there are statistical techniques for measuring the probability that the difference in weight gained between the two groups could have been produced by the accidental assignment to the experimental group of a greater proportion of chicks idiosyncratically prone to gain weight. If this probability is small enough, then one can be confident that the difference in weight gained really results from the difference in feeding.

The theory of experimental inference has developed far beyond such simple procedures, but every explanatory experimental design, no matter how complex, still rests on the principle of randomization.[5] The difficulty in applying such designs in sociology is that sociologists are seldom able to randomize. Sociologists cannot randomize except in laboratory studies of small groups and in the few policy-oriented experiments in which it is possible and meaningful to assign subjects randomly to control and experimental groups.[6]

Moreover, sociologists are often interested in situations where the classical division into control and experimental groups is not applicable, since there is

4. The use of a single brood makes the genetic differences between chicks less than if different broods had been used.

5. See Wold (1956: 30), for a discussion of the general prerequisites of controlled experimentation.

6. The difficulties of randomizing the subjects between experimental and control groups are presented in Chapin (1955: 167–169, 194–195).

no stimulus or set of treatments to be administered to one group and withheld from the other. To exemplify the problems of nonexperimental[7] research in sociology, let us examine the hypothesis that urban residents have a higher level of political interest than do rural residents. Suppose that appropriate rural and urban populations have been chosen and that a probability sample has been drawn from each.[8] Suppose further that when these two samples are compared, the urban residents do turn out to be more likely than the rural residents to express an interest in politics. Our question then is: Under which conditions, if any, can this difference be subjected to a test of statistical significance?

Since there is no random assignment of subjects to experimental and control groups, it might seem that the test is beside the point. However, there is another source of random differences between the two groups, the accidental variations produced in sampling from the two populations. Although random sampling does not have the same effect as randomization on the uncontrolled variables, it does introduce random differences between the two groups.[9] Therefore, we can ask whether the accidents of sampling could have produced an urban sample more interested than the rural sample, even though the average level of interest in the two populations is the same. In other words, what is the probability that two populations with equal proportions of interested people could have given rise to samples that differ by as much as our samples do, simply because the process of random selection happened to yield a greater proportion of interested people from one sample than the other? It would seem that this question is closely comparable to the one asked in the chick-feeding experiment and that the same logic should apply.

On closer examination, however, a fundamental difference appears. Both processes produce random differences between the two samples. But randomization removes the systematic effect of uncontrolled variables; in effect, it converts systematic differences between the experimental and control groups into random differences, thus allowing the statistical measurement of the possibility that the observed differences could have been produced by the randomization. Where two groups are sampled without randomization there is no statistical procedure for assessing the possible effects of the uncontrolled variables. The two groups in our example may differ in many ways in addition

7. Since sociologists use "observation" to denote a particular set of techniques for gathering data, we have used the terms "survey" and "nonexperimental" to describe those situations where randomization is impossible.

8. The logic of our analysis would be unchanged if there had been only one sample whose respondents were subsequently classified as rural or urban. We have chosen a two-sample illustration to parallel the preceding discussion of experimental inference.

9. "... It seems to be a frequent misconception that the random selection of sampling units can serve as a surrogate for the *ex ante* randomization of controlled factors in a genuine experiment" (Wold, 1956: 60). In the section on problems of meaning, we shall analyze the sources of random differences in more detail.

to the differences produced by sampling and by the rural-urban difference whose effect we are trying to measure. Thus urban residents are, on the average, more educated, have higher incomes, smaller families, more memberships in formal organizations, and so on. That is, these characteristics are correlated with the variable of urban-rural residence, which is why they are called "correlated biases" (Cochran, 1953: 305).[10]

Where so many factors are uncontrolled, it is obviously impossible to say that political interest depends only on place of residence and on random variables. To apply statistical tests to our data on political interest would therefore be misleading, since the effects of the correlated biases may well be greater than the random errors. At the very least, without evidence that the correlated biases are much smaller than the random errors (which seems unlikely in this and similar cases), one should not "test for significance," even though elegant statistical techniques are available for dealing with random errors, and ignore the effect of the correlated biases, which cannot be measured so easily.

Furthermore, the effects of these uncontrolled variables are not necessarily in the same direction; some may act to increase the urban-rural difference in political interest, others to decrease it. The errors made by ignoring the correlated biases may thus be in either direction and cover an unspecifiable range of magnitudes. Under these conditions, to ask whether the observed difference could have been produced by random errors is like wondering whether one's car has stopped because of random misbehavior of the electrons in the ignition system without first making sure that there is gasoline in the tank.

To many sociologists these last few paragraphs may seem beside the point. Most sociological variables can be controlled by cross-tabulation after the data are gathered, if they were not controlled beforehand in the design of the study. That is, the correlated biases could be eliminated by comparing only rural and urban residents with the same income, education, size of family, and so on. Is it not legitimate, therefore, to control these correlated biases first and then test for significance? Although this argument is fundamentally correct, there are important practical difficulties in applying it to such an extent that significance tests would be justified. We shall consider these difficulties shortly. However, most users of significance tests do not even attempt to deal with the correlated biases; instead, they move directly from the observed difference to a test of significance. *But only when all important*

---

10. Cochran also refers to "systematic biases" that affect all members of the population equally. Instrument errors are an example of systematic biases in the natural sciences. Systematic biases in the social sciences are far less common. Moreover, since sociologists are usually interested in subgroup comparisons, in which systematic biases cancel out, there is no need to consider this type of error further.

correlated biases have been controlled is it legitimate to measure the possible influence of random errors by statistical tests of significance. These tests must be the last step in statistical analysis, not the first.

In principle, then, tests of significance have a place in nonexperimental research. But in practice conditions are rarely suitable for the tests. We can distinguish three factors that work against the complete removal of the correlated biases. (1) In few if any surveys is the number of cases large enough to control simultaneously all the important uncontrolled variables. Even with upwards of 5,000 cases Kinsey was not able to control more than seven variables in any one analysis; some of the uncontrolled variables had previously been shown to affect the dependent variable under consideration (Kinsey *et al.*, 1948: 681–735). Our illustrative example of political interest could easily embody four or five additional variables whose importance has been demonstrated in other studies (Berelson *et al.*, 1954: 337–338). Few studies of political behavior have enough cases to study combinations of six or more independent variables. (2) All variables known to be relevant may not have been included in the design of the study, perhaps for reasons of economy, or because they were not part of the investigator's field of interest. Sociologists and psychologists working with the same set of dependent variables—prejudice, for example—usually consider different sets of independent variables.[11] (3) Some variables are so "confounded" with the variables whose relationship is being analyzed that they *cannot* be controlled. Consider the problem of studying attitudes toward desegregation among Southern whites and Negroes: What technique of cross-tabulation could remove the differential effects of the interviewer's race on the two groups of respondents? While one can conceive of experimental designs in which both races are interviewed by Negro and white interviewers, the practical difficulties are obvious.[12]

Designing nonexperimental studies so that tests of significance can be used validly is at best difficult. The tests are applicable only when all relevant variables have been controlled, either by prior design (randomization) or subsequent cross-tabulation, and these requirements are difficult to fulfill.

11. A sociologist might study the relation between status and prejudice, while a psychologist might consider the effects of personality variables. In either case the other set of variables may act as correlated biases unless it can be shown that status and personality variables are independent (and there is abundant evidence that they are highly dependent). Of course, we are not advocating here the blanket inclusion of psychological variables in sociological studies and vice-versa, but these considerations do have a bearing on the interpretation of significance tests.

12. As Hyman points out, survey conditions typically require each interviewer to work in a small and relatively homogeneous area. "If the number of interviews is increased, . . . subgroup comparisons will tend to a greater extent to be comparisons between respondents of *different* interviewers and will be affected to a correspondingly greater degree by interviewer variability" (Herbert H. Hyman *et al.*, 1955a: 374).

Moreover, even if all correlated biases could be eliminated there would remain the task of correctly interpreting the results of the tests.

Problems of Interpretation

The problems of interpreting significance tests fall into three groups— problems of *meaning*, of *random process*, and of *selection*. Of these three, the problems of meaning are the most easily recognized and the most easily dealt with. The difficulty here is in the misinterpretation of "significance" and "level of significance." These concepts can, of course, be defined precisely. The level of significance of a difference between two groups[13] is the frequency with which a difference as large as, or larger than, the observed value would occur if (1) there were actually no difference between the two populations from which the groups were drawn, and (2) the only factors operating to differentiate the two groups are random. Or, as statisticians prefer to put it, the level of significance is the probability of rejecting the null hypothesis that there is no difference between the two populations when it is actually true but appears false because of random accidents. The smaller the level of significance, the greater the "insurance" against taking accidental differences in the sample as indicative of true differences in the populations.

Insofar as the level of significance is interpreted in this essentially negative way, there is no problem of meaning. But one frequently finds the level of significance given an erroneous positive interpretation—as a virtual "seal of approval" or a measure of substantive importance. In many research reports the authors point with pride to differences significant at the 1 percent level and apologize for those significant at only the 5 percent level. But to cite the level of significance without *first* being certain that the observed difference could not have been produced by correlated biases is to convey the impression that the results are highly precise when, in fact, they may not be. The level of significance is only one link in a chain of methodological evidence that the results are substantially as claimed; to offer it as the only piece of evidence is misleading.

A related difficulty is the confusion of statistical significance with substantive importance. Some researchers give more prominence to the level of significance than to the size of the difference; often only the level of significance and the direction of the finding are reported, and occasionally the level of significance appears alone. High levels of predictability, explanation, and association are legitimate goals for social scientists; they are not the same as

---

13. The same logic applies to differences between means, proportions, correlation coefficients, and other parameters; we have adopted a more general formulation in referring to "the difference" between groups. Similarly, we have used "groups" here instead of "samples" because the logic of testing hypotheses is not restricted to samples.

a high level of significance, nor is statistical significance a substitute for them. A 1 percent difference may be significant at the .001 level if the sample is large enough, yet such a small difference is essentially meaningless for sociology at present. Correspondingly, a large difference that is not significant at the 5 percent level, simply because it was based on a small sample, may be of major theoretical importance.

Once they are pointed out, these problems of meaning are easy to guard against. Less easily eliminated are problems concerned with the sources of the differences between groups. There are three sets of factors which may operate to make the observed differences between groups depart from a true value of zero: (a) the experimental variable or independent variable whose effects on the dependent variable are being studied; (b) other extraneous but known variables, which presumably would have a measurable effect on the dependent variable but which cannot be controlled in the design of the study (the correlated biases); and (c) the "random" variables, which we shall now consider.

"Randomness" is a difficult concept, and philosophers of science find it hard to define satisfactorily (Nagel, 1955: 374). Instead of grappling with this problem we shall examine the "random processes" or sources of those discrepancies between true and observed values called "random errors." Some knowledge of these processes or mechanisms is essential if sound inferences are to be made from samples. In Nagel's words,

> If nothing is known concerning the mechanism of a situation under investigation, the relative frequencies obtained from samples may be poor guides to the character of the indefinitely large population from which they are drawn (Nagel, 1955 : 401).

It is generally assumed that random processes are those in which many small, independently operating factors are simultaneously at work. Thus the usual explanation for the observed randomness in tosses of a coin is that the outcome is affected by a large number of independent mechanical factors, varying slightly from one toss to the next. In survey research such complexes of small, independent factors may occur at three points: in the selection of a sample, in the act of responding to a question, and in the processing of the responses.[14] Sociologists have paid most attention to sampling error, the accidental selection of a sample whose characteristics differ markedly from those of the population from which it was drawn. For example, it is possible, though hardly likely, to draw a random sample of 1,000 cases from the population of New York City and have it be 99 percent Republican. A

---

14. For another classification of random errors, see M. B. Wilk and O. Kempthorne (1955).

second source of random error is located in the respondent—unpredictable fluctuations in mood, attitude toward the interviewer, perceptions, and so on. These phenomena are more often studied by psychologists as "unreliability" than by sociologists.[15] Finally, there are random errors of recording and tabulating: Interviewers may accidentally check the wrong response, coders may misclassify, key-punch operators may hit the wrong key, and even machines may miscount. These "errors of measurement" have been most systematically studied in the physical sciences, where the other two sources of random errors are relatively less important.

One purpose served by this catalog of random errors is to clarify what is meant by attributing a relationship to "chance" factors.[16] Obviously, a test of significance cannot discriminate one source of random errors from the others. It merely answers the question: How likely is it that the totality of random processes could have produced the observed result? Recognition of this fact should clarify some ambiguous interpretations of significance tests, for example, where total populations are being compared. If the populations of two census tracts or of two school classes are being compared, there is obviously no "sampling error" producing the observed differences. To cope with this situation, one textbook suggests that such total populations be considered as "samples from still larger hypothetical universes of possibilities" (Hagood and Price, 1952). This concept is difficult to grasp intuitively, and it is largely unnecessary if one recognizes that sampling is not the only source of random errors. Even where there is no sampling in the usual sense, discrepancies between the true situation and the observed results may be produced by random errors of response or processing.

It might seem, therefore, that tests of significance could be used to compare total populations, if the tests are interpreted as dealing with random errors of response or processing. However, new difficulties arise in connection with the errors of response. Underlying all statistical tests is the assumption that the random errors of each respondent, from whatever source, are independent

15. Note that psychologists pay little heed to sampling error; compare Hagood and Price (1952) with Edwards (1954). The former have almost nothing on unreliability; the latter book, by a psychologist, omits any mention of sampling error. As long as psychologists are seeking to generalize, rather than merely to measure the characteristics of unique individuals, they should *consider* sampling error also, although they may not compute tests of significance, for the reasons discussed in this paper.

16. It is probably an example of the "fallacy of misplaced concreteness" to speak of "chance factors" or to say that some phenomenon *is* random. At least for units larger than atomic particles, nothing is *inherently* random; it is more or less satisfactorily *treated* as random in some discipline. And what is random for one discipline may be the object of study for another. Thus the tosses of a coin are treated as random in assigning chicks to control and experimental groups; in principle, however, a physicist would hold that the outcome is almost completely determinate, if he could measure all the factors involved. Similarly, a sociologist can treat momentary variations in perception as random, but a psychologist might be able to predict these variations.

of the random errors of every other respondent.[17] This assumption would seem to be satisfied with regard to the errors of response, as long as the respondents are physically and socially separated from each other, for example, in public-opinion polling. But when people interact frequently, one cannot assume that their response errors are independent. The factors leading one person to distort his answers in a certain way tend to operate similarly in the people he associates with. Since frequent interaction is more likely among members of a total population than among the scattered members of a sample, this is a second reason why tests of significance should not be used in comparing total populations.[18]

This list of sources of random errors also helps to determine when *not* to be concerned with the possible effects. Tests of significance measure the possibility that the observed differences *could* have resulted from random errors; they do not, of course, indicate whether random errors were present. Now, if through one means or another all three sources of error are inoperative in some situation, there is clearly no point in computing the probability that random errors could have produced the observed difference. Such situations are not impossible: Comparing total populations rules out the possibility of sampling errors; coding and punching checks minimize errors of processing; and random fluctuations in response may be reduced by "motivating" the respondent to answer carefully.[19] This "drying up" of the sources of random errors makes them less important as possible explanations of observed differences. If these sources of random error cannot operate, then it is meaningless to measure the probability that they could have operated.

The last of the three major problems of interpretation is *selection*, the ways in which one chooses hypotheses and draws valid inferences from data. Again we are assuming that all other necessary conditions for using tests of significance have been met. Here we shall consider the implications of the ways in which survey analysts reach conclusions about statistical significance. The principal difficulty in selection is the very richness of survey data. Unlike the experimenter who designs a study to test at most a half-dozen hypotheses simultaneously because he can control only a few variables in a single

17. Comparisons of subgroups may be much more seriously affected by errors of nonresponse than are the results for total populations or samples if, as usually happens, the nonrespondents are more concentrated in certain subgroups than in others. This is another way of saying that errors of nonresponse are not independent; they tend to cluster.

18. If this were the only difficulty in the application of significance tests—if the other problems of design and interpretation were not also present—the solution would not be too difficult. Econometricians have already dealt with the same problem in the form of "autocorrelated time series"; for example, the level of prices in a given year is not independent of the previous year's prices. But, since the other problems still remain, it does not seem useful to use these more powerful techniques on the correlated response errors.

19. For an analysis of the effects of a respondent's motivations on his response errors, see Patricia L. Kendall (1954).

experiment, the survey analyst has scores or even hundreds of variables in one questionnaire. Frequently, therefore, he discovers relationships that he had not thought of in constructing the questionnaire. Since hypotheses precisely formulated in advance of gathering data are less common in present-day sociology than conceptual schemes or sets of important variables, it is probably safe to assume that most hypotheses in survey research are formulated *after* examining the data.

There is nothing intrinsically wrong with this procedure, as long as the hypotheses are subsequently tested on *other* data. But when the hypotheses are tested on the same data that suggested them and when tests of significance are based on such data, then a spurious impression of validity may result. The computed level of significance may have almost no relation to the true level. We can best demonstrate this point by an example, which, although extreme, is not too far from what often happens in survey research. Suppose that twenty sets of differences have been examined, that one difference seems large enough to test and that this difference turns out to be "significant at the 5 percent level." Does this mean that differences as large as the one tested would occur by chance only 5 percent of the time when the true difference is zero? The answer is *no*, because the difference tested has been *selected* from the twenty differences that were examined. The actual level of significance is not 5 percent, but 64 percent![20] Almost two times out of three there will be at least one apparently "significant" difference in examining sets of twenty differences drawn from populations where the true differences are zero. Thus tests of hypotheses formulated *after* examining the data may give a false impression of statistical significance when, in fact, one has found only results that might easily have been produced by random errors. This is not an argument against such *a posteriori* hypotheses, for sociologists have as yet few *a priori* hypotheses that are precise enough to test empirically. It is, however, an argument against improperly using tests of significance to gain an unwarranted security.

Curiously enough, this analysis of *a posteriori* hypotheses, which began as an indictment of statistical tests, can be extended to show how the tests might legitimately be used on such hypotheses. Consider once more the twenty differences drawn from populations where the true differences are zero. We

20. The probability of finding at least one difference significant at the 5 percent level when twenty independent samples are drawn from populations with true differences of zero is: $1 - Pr$ (no such differences). That is, either there is at least one such difference, or there is no such difference. The term in parentheses is the probability of not getting a significant difference on one trial (0.95), raised to the 20th power: $(0.95)^{20} = 0.36$. Then the required probability is: $1 - 0.36 = 0.64$. We have assumed independent samples here for ease of computation. With real data the various tables formed from the same population are certainly not independent, and the computation of the true level of significance would be extremely difficult.

have seen that the probability of *at least one* difference "significant" at the 5 percent level is 0.64. By similar calculation it can be shown that the probability of at least *two* "significant" differences is 0.26, that the probability of at least *three* is 0.07, and that the probability of at least *four* is 0.01.[21] In other words, if one examines twenty differences and finds four or more "significant" at the 5 percent level, then the *set* of differences is significant at the 1 percent level, since this combined result would have happened only one time in a hundred if the true differences were zero. A single result may be misleading, but the consistency of a set of results may be a valid indication of statistical significance. But even this argument depends on a careful consideration of the correlated biases and of the lack of independence of the twenty tables.

Another problem of selection stems from the analyst's freedom to "collapse" his tables in different ways. An attribute or a scale with several values, such as a nine-point scale of ideological sensitivity (Lipset *et al.*, 1956: 434) can be dichotomized or trichotomized by using several different cutting points, some of which may increase the correlation between the scale and another variable, while others may reduce it. In fact, one occasionally finds a table where the *direction* of the association can be changed by shifting the cutting points. In such cases, where the analyst can drastically alter the apparent relationship between the variables by manipulating the cutting points, the test of significance would seem to be pointless. Why test for statistical significance when even the very direction of the results is in question?

In this situation, however, the test of significance is blameless. The real problem is: Under what conditions can one legitimately manipulate his data by shifting the cutting points? A partial answer to this question has been available for over forty years, yet few sociologists seem to be aware of it. In his classic discussion of the statistics of attributes, Yule points out that the sign of association of an "isotropic" contingency table is unaffected by the choice of cutting points for the two variables (Yule and Kendall, 1937, 14th ed.: 57–59).[22] That is, combining the rows or columns of a contingency table is legitimate only when the table is isotropic or can be made isotropic by re-arranging rows or columns. Since tests of significance are not too seriously affected by the choice of cutting points in an isotropic table, the survey analyst's freedom to manipulate the cutting points is not *per se* a strong argument against tests of significance. The essential idea is that cutting

21. The probability of exactly r differences "significant" at the 5 percent level in twenty independent tables is: $_{20}C_r (0.05)^r(0.95)^{20-r}$. The probability of k or more such differences is: $1-[Pr$ (exactly one difference) $+ Pr$ (exactly two differences) $+ \ldots + Pr$ (exactly k−1 differences)].
22. The three chapters on the statistics of attributes contain many valuable insights for sociologists, who, as Paul F. Lazarsfeld has long urged, need the simple statistics of attributes far more than the complicated statistics of variables.

points should be manipulated only in isotropic tables, regardless of whether one tests for significance or not.

Summary

Statistical tests are unsatisfactory in nonexperimental research for two fundamental reasons: It is almost impossible to design studies that meet the conditions for using the tests, and the situations in which the tests are employed make it difficult to draw correct inferences. The basic difficulty in design is that sociologists are unable to randomize their uncontrolled variables, so that the differences between "experimental" and "control" groups (or their analogs in nonexperimental situations) are a mixture of the effects of the variable being studied and the uncontrolled variables or correlated biases. Since there is no way of knowing, in general, the sizes of these correlated biases and their directions, there is no point in asking for the probability that the observed differences could have been produced by random errors. The place for significance tests is after all relevant correlated biases have been controlled.

Even if studies could be designed so that the correlated biases were controlled, there would remain the problem of correctly interpreting the tests. Many users of tests confuse statistical significance with substantive importance or with size of association. Others, depending on their discipline, restrict themselves to considering a single source of random errors, although other sources may be present in addition to, or instead of, this one. Perhaps the most common problem of interpretation results from sociologists' usually having to formulate their hypotheses after examining the data and then "testing" these hypotheses on the same data. This produces apparent levels of significance that may have no relation to the true levels.

In design and in interpretation, in principle and in practice, tests of statistical significance are inapplicable in nonexperimental research. Sociologists would do better to re-examine their purposes in using the tests and to try to devise better methods of achieving these purposes than to continue to resort to techniques that are at best misleading for the kinds of empirical research in which they are principally engaged.

10. Comment on "A Critique of Tests of Significance"

DAVID GOLD

HANAN SELVIN (1957 [Chapter 9]) has performed a long overdue service for sociologists in calling attention to the uncritical use and abuse of statistical tests of significance. At the very least, it is to be hoped that the article will serve to remove tests of significance from the realm of ritual behavior. There are undoubtedly a number of matters in Selvin's treatment which call for further elaboration and discussion. This note is addressed to two of them.

(1) One argument offered for not using tests of significance appears to be particularly weak (and, incidentally, detracts from some of the other considerably more important arguments) as a result of ignoring the descriptive features which are necessarily a part of any explanatory survey. If the analyst is concerned with testing Selvin's illustrative hypothesis that urban residents have a higher level of political interest than rural residents, the question of whether there is in fact a difference between rural and urban residents in the defined populations with respect to political interest is relevant. The analysis and interpretation that will be developed is dependent upon the *observed fact* that such a difference does or does not exist. And this observed fact, in a probability sample drawn from appropriate rural and urban populations, within probability limits, can be legitimately established with a test of

Reprinted from the *American Sociological Review*, 23 (February, 1958), 85–86, by permission of the author and the publisher. Copyright © 1958 by the American Sociological Society.

significance. The uncontrolled variables or "correlated biases," such as education or income, which may "explain" an observed rural-urban difference or lack of difference in political interest do not vitiate the legitimacy of a test of significance *for the existence of a difference*. What produced the difference or obscured the difference does not alter the gross fact that rural and urban residents in the given populations do or do not differ. Thus, the injunction cannot hold that one cannot legitimately test for significance when there is lack of randomization or correlated biases have not been controlled. If the ultimate goal is "explanation" or "interpretation" (in the Hyman-Lazarsfeld-Kendall sense), there are indeed "problems of interpretation" of tests of significance, but these are not the "problems of design" in the manner that Selvin conceives them to be. A reasonable case can be made that tests of significance of zero-order relationships in explanatory surveys are not particularly important, but this is definitely not the same argument that the tests are not legitimate.

(2) An important weakness of much analysis in current social research is the failure of the analyst to consider the distinction between statistical significance and substantive importance. As Selvin correctly points out, the former does not provide a substitute for the latter.

> A 1 percent difference may be significant at the .001 level if the sample is large enough, yet such a small difference is essentially meaningless for sociology at present. Correspondingly, a large difference that is not significant at the 5 percent level, simply because it was based on a small sample, may be of major theoretical importance (1957: 524 [Chapter 9: 101]).

Since, at a given level of significance, statistical significance demands a greater *degree* of relationship from a small sample than a large sample, it might appear that the researcher can more easily treat substantively important differences by selecting small samples rather than large samples. This, of course, is not true. It is simply that smaller samples produce statistics more frequently which deviate widely from parameters than do larger samples. Thus the large difference in a small sample must always be replicated in large samples to assess substantive importance.

These comments should not be construed as expressing disagreement with the general tenor of Selvin's critique.

11. Reply to Gold's Comment on "A Critique of Tests of Significance"

HANAN C. SELVIN

DAVID GOLD (1958 [Chapter 10]) is correct in calling attention to my virtual neglect of tests of significance in descriptive surveys. If one is concerned simply with establishing the existence of a difference in political interest between urban and rural residents, then there may indeed be some value in testing statistically the null hypothesis that there is no difference. But this is not usually the end toward which sociologists are working; rather, we seek to make general inferences from specific data, and it seems to me that tests of significance provide a false reassurance in such cases. Moreover, the tests may not even be altogether applicable to descriptive surveys; even with probability sampling as currently practiced, the difference that is tested may stem from nonrandom factors other than the differences between urban and rural residents.

When Gold says that "The uncontrolled variables . . . do not vitiate the test of significance for the existence of a difference," he is correct *insofar as these uncontrolled variables refer to characteristics of the respondents*, such as income or education. But there are other uncontrolled variables that are characteristics of the interview or of the survey procedures. For example, there is the set of nonresponse biases: It may be that the true difference in political interest between urban and rural residents is zero, but that the

Reprinted from the *American Sociological Review*, 23 (February, 1958), 86, by permission of the author and publisher. Copyright © 1958 by the American Sociological Society.

nonresponses are more likely to occur among interested rural residents and dis-interested urban residents, thus partly accounting for the observed difference. Why raise the question of sampling error and other kinds of random error when this large and frequent source of nonrandom error is not controlled, measured, or even considered? Again, there is the problem of interviewer bias: It may be that the reported difference between urban and rural residents stems from the behavior of a few interviewers. It is possible to control this source of error by randomizing the assignment of interviewers to respondents, so that the possible biases of a few interviewers are not concentrated in one group of respondents. This, of course, is impractically expensive in most studies, but the possibility of such biases is real nonetheless, particularly when the number of interviewers is small.

These two examples suggest that "problems of design" in the application of significance tests are encountered in descriptive surveys, as well as in explanatory surveys. Certainly these kinds of errors are not unknown to those who routinely use significance tests, yet at the moment I cannot recall a single instance in which they were taken into account in applying the tests. As Gold says, the tests have become a form of ritual behavior: Reciting the magic phrase "significant at the .01 level" is often a substitute for hard thinking about the quality of one's data.

I am grateful to Gold for noting the possibility that the passage quoted in his letter might be misinterpreted as an argument for small samples in place of large ones. One of the dangers of small samples is the discarding of valid results simply because of the relatively high probability that they *might* have occurred by chance. The larger the sample, the less the probability of such an error (the oft-neglected "Type II error," as against the "Type I error," the probability of which is measured in significance tests).

12. On "A Critique of Tests of Significance in Survey Research"

JAMES M. BESHERS

HANAN SELVIN (1957 [Chapter 9]) has confused statistical inference with causal inference. All his statements to the effect that sociologists need not employ significance tests in survey analysis are based upon this confusion.

Statistical inference may be employed in (1) establishing descriptive generalizations, and (2) unravelling causal relationships. An example of the former use would be: Urban residents have a significantly higher level of political interest than rural residents. The latter use would entail such statements as: The level of political interest is determined by a combination of factors. These factors and their relationships would then be indicated. Clearly these two usages are in no way equivalent. Moreover, the use and interpretation of statistical methods differ considerably with these two applications.

In general, the unravelling of causal relationships includes three steps: (1) the identification of important factors; (2) the estimation of relationships between these factors; and (3) the testing of hypotheses about these factors. Cross-tabulation is an important phase of the second step, but it should not be the terminal stage in analysis.

Selvin argues that statistical inference may be substituted for causal inference in experimental research, but may not be substituted in survey research. In fact, statistical inference may not be substituted in either case.

Reprinted from the *American Sociological Review*, 23 (April, 1958), 199, by permission of the author and the publisher. Copyright © 1958 by the American Sociological Society.

Survey research can only uncover causal relationships if the survey design encompasses all pertinent variables. Experimental research can only uncover causal relationships if the experimental design encompasses all pertinent variables. Randomization in an experiment may permit accurate estimation of the relationships among a set of variables (step 2), but randomization does not guarantee that this is an important set of variables (step 1), nor does randomization necessarily eliminate correlated biases. Research technique is no substitute for theory; experimental research is no exception.

Selvin claims to find support for his views in an article by Herman Wold (1956). No statement in Wold's article suggests that significance tests should be abandoned in survey research.

Significance tests are of little value for surveys which (1) ignore the principle of sampling, and (2) are not guided by theory. Perhaps such exploratory surveys may generate hypotheses to be verified by a subsequent well-designed survey utilizing significance tests.

13. Reply to Beshers

HANAN C. SELVIN

IN LIMITING MY PAPER to ". . . the application of significance tests to explanatory empirical studies" (1957: 520 [Chapter 9: 95]) I tried to focus on the role of statistical significance in assessing causal relationships. As Hyman says, explanatory surveys seek to establish reliably *the nature of the relationship between one or more phenomena, or dependent variables, and one or more causes or independent variables* (1955b: 66 [italics in original]). It is surprising, therefore, that James Beshers (1958 [Chapter 12]) accuses me of confusing causal inference and statistical inference.

Perhaps the confusion that Beshers finds may come from our different conceptions of causal inference. In his third paragraph Beshers lists three steps in studying causal relationships. I completely agree with his list, but not with what I take to be his interpretation of the third step, "the testing of hypotheses," in which he seems to argue for tests of statistical significance and against cross-tabulation. How, after all, does one "test a hypothesis?" In general, to determine whether two variables are causally related, one must show that the observed relationship is not, or could not reasonably be, the result of "extraneous" variables. These extraneous variables can, in turn, be divided into two classes: those variables that can be identified by the researcher and those that cannot. The question whether the identifiable variables

Reprinted from the *American Sociological Review*, 23 (April, 1958), 199–200, by permission of the author and the publisher. Copyright © 1958 by the American Sociological Society.

have produced the observed relationship is handled by studying the original relationship with the possibly extraneous variables held constant. As Hyman points out, "The analysis of such complex relationships provides a solution to the problem we shall call *spuriousness* and permits the analyst to infer that the original relationship involves cause and effect" (1955b: 242). Cross-tabulation is central in studying causality, insofar as identifiable extraneous variables are concerned.

Now what of the complex of factors that cannot be identified? Here, as I tried to show in my paper, is the legitimate place for tests of significance, to determine the probability that this complex of unknown and presumably independent factors could have produced the original relationship. But this should be done only after the known and identifiable factors have been analyzed. On this point Wold's comments are in order: "With observational data the modern approach in hypothesis testing is more or less hampered by the scanty knowledge about the properties of the residual variation, and this is particularly true in the treatment of counter-hypotheses. The need for testing the statistical inference is no less than when dealing with experimental data, but with observational data other approaches come to the foreground. . . . various sources of evidence are taken into account by way of checks and cross-checks" (1956: 39). At another point Wold says that, in explanatory studies without randomization ". . . there is no clearcut distinction between explanatory factors which are explicitly accounted for in the hypothetical model and disturbance factors which are summed up in the residual variation" (1956: 35). That is, when the effects of the independent variable have been taken into account, some of the residual variation in the dependent variable may well involve other "explanatory factors." Is it unreasonable to infer from this that the residual variation in observational data may be largely nonrandom? Or, to put the question somewhat differently, should one not try to account for as much of the residual variation as possible by means of identifiable explanatory factors, rather than assume (without any empirical justification) that what one has not troubled to look at is "random"? The question whether random variables *could* have produced the observed relationship should not be raised without determining whether identifiable nonrandom factors actually *did* produce the relationship. (In a letter received after this reply was written, Wold states that "I am delighted to see that your arguments and conclusions are on the same wave length as mine in the JRSS paper to which you refer. . . .")

Beshers' fourth paragraph is, in my opinion, a mixture of the acceptable, the unacceptable, and the irrelevant. First he attributes to me the view that "statistical inference may be substituted for causal inference in experimental research. . . ." I do not hold this view, nor is it so stated in my paper. Beshers then goes on to advocate the inclusion of "all relevant variables" in explanatory surveys. That this is almost a platitude does not make it less valid, but I do not see its relevance here. However, when Beshers says that an experiment "can only uncover causal relationships if the experimental design encompasses all relevant variables" [sic]* and that "randomization does not eliminate correlated biases" [sic], he is simply wrong. What is essential is ". . . some means of insuring that a treatment will not be continually favored or handicapped in successive replications by some extraneous source of variation, known or unknown. This is done by the device of randomization, due to Fisher . . . who has shown how tests of significance and confidence intervals can be constructed, using only the fact that randomization has been properly applied in the experiment" (Cochran and Cox, 1950: 6–7).

I am glad that Beshers is in favor of theory in research. So am I.

* [Beshers' actual statements were: ". . . all *pertinent* variables," and ". . . nor does randomization *necessarily* eliminate correlated biases" (italics ours).—The Editors.]

14. Randomization and Inference in Sociological Research

ROBERT McGINNIS

AN ACCUSING FINGER has been leveled at those of us who test survey research data for statistical significance. In a recent paper, Professor Hanan C. Selvin concludes that ". . . conditions under which tests of significance may validly be used are almost impossible of fulfillment in sociological research . . ." (1957: 520 [Chapter 9: 95]). If his critical appraisal causes sociologists to re-examine their research procedures, Selvin will be due a great deal of credit. But, if some are convinced that his conclusion is correct, and if they act accordingly, the results could be disastrous for quantitative sociology. Selvin arrived at this conclusion by reasoning that ". . . tests of statistical significance are inapplicable in nonexperimental research" (1957: 527 [Chapter 9: 106]). Now most sociological research is nonexperimental, and more often than not it involves probability sampling from some population. Thus, the door is opened to the possibility of making statistical inferences. In this light, Selvin's charges certainly merit the most careful consideration.

This paper contains an analysis of Selvin's position and presents some of the arguments against it.[1] It is not meant to be a philippic, but is concerned

Reprinted from the *American Sociological Review*, 23 (August, 1958), 408–414, by permission of the author and the publisher. Copyright © 1958 by the American Sociological Society.

1. For additional critical comments, see David Gold (1958 [Chapter 10]) and James Beshers (1958 [Chapter 12]).

almost exclusively with an evaluation of his claims. Specifically, it is contended that the statements quoted above are false and that the stated reasons for making them have nothing to do with statistical tests of significance.

The Reasons Against

Professor Selvin marshals several reasons against the use of significance tests in sociological research. These are divided into two categories which he calls problems of design and problems of interpretation. The principal problems of design are, first, that randomization usually is impossible in the assignment of subjects to experimental and control groups in social research and, second, that sociologists often have no "... stimulus or set of treatments to be administered to one group and withheld from the other" (Selvin, 1957: 521 [Chapter 9: 97]). The second of these two reasons for disallowing statistical tests in social research is actually very little different from the first. It refers to the fact that subjects are assigned by "nature" to classes of most variables which interest sociologists. Its importance evidently rests in the fact that this natural assignment prevents the investigator from randomizing. The critical problem for sociologists, then, in Selvin's view, rests in the fact that the experimental procedure of randomization is impossible in most of their research.

Problems of interpretation which Selvin cites include the confusion of statistical significance with substantive importance, problems of interpreting random errors, and the sin of *a posteriori* selection of hypotheses to be tested. Such problems of interpretation, or more properly of misinterpretation, are real and important, and Selvin's discussion of them is a genuine contribution. But they have nothing whatsoever to do with the admissibility of significance tests in sociological research. Misinterpretation of the results of statistical analyses is a mark of naïveté, but not necessarily of improper statistical procedures. Surely Professor Selvin would not have us abandon a valuable set of research tools because some of those who apply them make a very poor job of it. These commonplace abuses are not considered in this paper.

It is clear that the act of randomization plays a crucial part in Selvin's argument. Thus, to understand his position it is necessary to have some knowledge of the effects of randomization and of the possible consequences of failure to randomize. These effects are illustrated in the following section, after which it is contended that Professor Selvin's main conclusion, namely that tests of significance are inadmissible in social research, is incorrect.

The Mathematics of Randomization

The process and consequences of randomization are considered here only with relation to the simplest possible situation, that of a continuous dependent variable, a dichotomous independent variable, and one dichotomous confounding factor.[2] We consider only the problem of testing a simple hypothesis about the difference between two population mean values, but the discussion applies quite generally to tests of hypotheses about parameters of almost any form. Although Selvin does not mention estimation procedures, the following remarks apply to statistical estimation as well as to tests of hypotheses.

In the illustration below it is supposed that there exists one variable only, the so-called confounding factor, which is related jointly to the independent and to the dependent variables. With this assumption in mind it is supposed that mean values of the dependent variable are found to differ between the two classes of the independent variable, but that these differences vanish within the two sub-classes of the confounding variable. This is the situation that troubles Selvin. Under these conditions he would assert that the obtained difference between means in the two independent classes is meaningless or spurious. For this reason we ask what it is that gives rise to the difference.

In the following paragraphs it is shown that the difference between means of the two independent variable classes is a function of two parameters, labeled α and β. The first of these is found to be a measure of association between the confounding factor and the independent variable and the second is found to be a measure of association between the confounding factor and the dependent variable. These parameters are such that whenever either takes the value zero, the difference between means of the independent classes also becomes zero. Finally, it is shown that the process of randomization has the effect of setting the first parameter equal to zero. These two parameters are the crux of the matter. Should they be viewed as real and meaningful components to be investigated, or merely as nuisance effects to be eliminated in the research design? Evidently Selvin would take the latter view. It is contended later that such a position is entirely unsatisfactory in social research.

To establish the necessary structure, let X represent an independent variable which is divided into classes X_1 and X_2. Let Y be a continuous variable and let Z be a variable which is divided into Z_1 and Z_2. Nine different population mean values of Y, to be noted μ, can be defined if the population is classified with respect to X and Z. These are displayed in Table 14.1. Let N_{1j} (i, j = 1, 2) be the number of subjects in the population which are in both classes Z_1 and X_j. Let $N_{1.}$ be the (marginal) number of cases in class Z_1 and $N_{.j}$ be the number in class X_j. Also let $N = N_{1.} + N_{2.} = N_{.1} + N_{.2}$.

---

2. The terms dependent and independent are meant to connote nothing about cause and effect. They reflect only the direction of analysis which interests the researcher. By a confounding factor is meant any variable whose distribution is statistically independent of neither the dependent nor the independent variable. For a discussion and bibliography on randomization, see Oscar Kempthorne (1955).

*Table 14.1. Nine mean values
of Variable Y*

	X_1	X_2	
Z_1	μ_{11}	μ_{12}	$\mu_{1.}$
Z_2	μ_{21}	μ_{22}	$\mu_{2.}$
	$\mu_{.1}$	$\mu_{.2}$	μ_y

Since X is the independent variable of the problem, the hypothesis to be tested, in null form and in the notation of Table 14.1, is the following:

$$H: \mu_{.1}=\mu_{.2}(=\mu_y) \tag{1}$$

Note that the variable Z plays no explicit part in the hypothesis. It might seem that the following is an equivalent hypothesis:

$$H: \mu_{11}=\mu_{12}(=\mu_{1.}) \ and \ \mu_{21}=\mu_{22}(=\mu_{2.}) \tag{2}$$

It is not true, however, that (1) always implies or is implied by (2). In fact whenever (2) is true (1) is also true if and only if at least one of two conditions is satisfied.[3]

These are:

Condition (A) $\mu_{1.}=\mu_{2.}$. $\qquad\qquad\qquad\qquad\qquad\qquad\qquad$ (3)
Condition (B) $N_{11}N_{22}=N_{12}N_{21}$ $\qquad\qquad\qquad\qquad\qquad$ (4)

Condition (A) is implied by statistical independence of variables Y and Z. Condition (B) both implies and is implied by statistical independence of variables X and Z. Whenever neither condition is satisfied and (2) is true, then (1) must be false. In such a situation the following important question is raised: If variables X and Y are unrelated within classes of variable Z, is there any "true" difference between $\mu_{.1}$ and $\mu_{.2}$, or is the apparent difference wholly spurious? It is Selvin's implicit negative answer to this question which underlies his entire argument.

The relations described above can be summarized with the aid of two definitions. Let

$$\alpha = \left| \frac{N}{N_{.1}N_{.2}} [N_{1j} - \frac{1}{N}(N_{1.}N_{.j})] \right| \ (i, j=1, 2) \tag{5}$$

which is known to be unique over all i,j in any four-celled table. Let

$$\beta = |\mu_{1.}-\mu_{2.}| \tag{6}$$

---

3. Proofs are omitted throughout this paper, but are available upon request.

With these definitions it can be demonstrated that, if (2) is true, then

$$|\mu_{.1}-\mu_{.2}|=\alpha\beta \tag{7}$$

and that if the row-wise variances, say $\sigma_1{}^2$ and $\sigma_2{}^2$ are such that

$$\sigma_1{}^2=\sigma_2{}^2=\sigma_0{}^2 \tag{8}$$

then the overall variance, say $\sigma_y{}^2$ is given by

$$\sigma_y{}^2=\sigma_0{}^2+\frac{N_1.N_2.\beta^2}{N^2} \tag{9}$$

Clearly, whenever (3) is satisfied, $\beta=0$ and whenever (4) is satisfied, $\alpha=0$. Thus, by (7), whenever either condition is satisfied, and (2) is true, the major hypothesis, $H: \mu_{.1}=\mu_{.2}$ also is true. Moreover, whenever these conditions obtain, the overall variance is equal to either of the homogeneous row-wise variances, as is seen by (9). Now we are in a position to assess the influence of randomization.

The physical process of randomization consists of assigning subjects to classes of the independent variable (X_1 and X_2 in the illustration) by some random process. This act therefore generates a random variable so that we can consider expected or average values of the two terms in (4).[4] Randomization has the effect of setting the two expectations equal, that is, $E(N_{11}N_{22})=E(N_{12}N_{21})$. It follows from this that (4) is always satisfied after randomization except for random fluctuations so that the value of α becomes approximately 0. In this way, whenever (2) is true, (1) is also true within the limits of random fluctuation. This process then has the effect of eliminating the need to consider the question raised above about spurious differences.

In a situation where condition (A) is not satisfied with respect to X and all variables Z, the association between X and Z is sometimes called correlated bias. Failure to eliminate this effect is at the heart of Selvin's argument, illustrated by his claim that "... *only when all important correlated biases have been controlled is it legitimate to measure the possible influence of random errors by statistical tests of significance* (Selvin, 1957: 522 [Chapter 9: 98–99, italics in original]). The force of this assertion may be illustrated by the following situation.

Suppose that in some population there are three variables, X, Y, Z, which conform to the conditions of Table 14.1 and for which neither of the conditions of (3) and (4) is satisfied.[5] Suppose also that (2) is true for these three

---

4. For discussion of random variables and expectations, see William Feller (1950: Chapter 9).

5. For simplicity, suppose that for all other variables, U, condition (B) is satisfied for X and U. In this way statistical control of variable Z would be completely equivalent to randomization.

variables. In this population then, variables X and Y are independent among the subgroup Z_1 and the subgroup Z_2, but, when the two groups are combined this independence is lost.

Suppose that two investigators are unaware of these facts and that each independently sets out to test the hypothesis that there is no difference in the two means $\mu_{.1}$ and $\mu_{.2}$. Imagine that both draw random samples from the population and that they happen to select exactly the same set of subjects. Suppose that the first investigator elects to control the correlation of variables X and Z, which is the equivalent of randomization with respect to this one confounding effect, but that the second fails to do this. Then despite the fact that both observed the same subjects, they drew random samples from two quite different populations. In the population of the first investigator the hypothesis $H: \mu_{.1} = \mu_{.2}$ is true and the variable Y has variance $\sigma_Y^2 = \sigma_0^2$. In the second investigator's population the hypothesis is false, as is seen by (7), and the variable Y has variance $\sigma_Y^2 = \sigma_0^2 + \dfrac{N_1 . N_2 . \beta^2}{N^2}$. The first researcher would probably be led to accept the null hypothesis while the second probably would be led to reject it. All of these differences are attributable solely to the fact that the first investigator controlled the correlated bias of variables X and Z while the second did not.

Now Selvin would have us ask which of the two procedures is correct, implying that one is not. He has already given us his answer. He would evidently contend that the procedure of the first investigator is correct while that of the second is in serious error. The claim here is that *neither* procedure is incorrect, that the basic error is made by those who ask the question in the first place.

Selvin's confusion stems mainly from the failure to recognize that there are different classes of hypotheses. Classifications based on the number of parameters under consideration, directionality and the like are ignored in this discussion. Rather, a three-level classification based on the *conditionality* of empirical hypotheses is used. To establish this classification one definition is added to the notation already developed. Let γ represent any parameter of a distribution function. In the preceding sections, γ would be the difference between two means, $\mu_{.1} - \mu_{.2}$. Then three classes of hypotheses about γ can be considered. In null form these are:

Type I $\quad H: \gamma = 0$
Type II $\quad H: \gamma = 0 \mid (\alpha_i \beta_i) = 0$ for X, Y, and certain Z_1 $(i = 1, 2, \ldots, N)$
Type III $\quad H: \gamma = 0 \mid (\alpha_i \beta_i) = 0$ for X, Y, and all Z_1 $(i = 1, 2, \ldots)$

The first type might be called an *absolute hypothesis* in that no conditions are established regarding the relation of the variables under consideration to any others. A Type II hypothesis might be termed *finitely conditional* since a condition of statistical independence is required between variables X, Y, and a finite number of other variables, Z_1, if the hypothesis is to be true. Similarly, hypotheses of the third type might be called *infinitely conditional* in that conditions are specified for the variables X, Y, and an infinite number of variables Z_1. With these definitions, we can proceed to an appraisal of Selvin's claims.

Reasons Against the Reasons Against

At the outset, note should be made of what amounts to a grossly misleading passage in Professor Selvin's discussion. In an attempt to formulate a statement of research objectives, he writes the following: "... In most sociological research ... the population sampled is chosen because it is convenient; the emphasis is on explanation rather than description, on uncovering general relationships rather than precisely depicting a unique situation" (1957: 520 [Chapter 9:95]). It is doubtful that he believes this generalization to be literally true, so we must view the passage as an unfortunate choice of terms. But, the statement as it stands implies that if a relation is found to obtain after randomization in one population, then it must obtain in all populations. This, of course, does not hold. Any population can be viewed as an element in the field of sets generated by the set of all possible elements.[6] The fact that a relation obtains in one set implies that it obtains in all others only if XYZ correlations are identical for X, Y, and all variables Z in all sets. Since this is hardly ever the case, it is not true that one can uncover "general" relationships by examining some arbitrarily selected population. Undoubtedly Selvin recognizes these facts, but his use of the terms "explanation," "description," "general," and "unique" could easily mislead some readers. There is no such thing as a completely general relationship which is independent of population, time, and space. The extent to which a relationship is constant among different populations is an empirical question which can be resolved only by examining different populations at different times in different places.

Selvin's statement of research goals bears careful examination for another reason. Although the terms are cloudy in the extreme, it is fairly evident that by "description" is meant the process of testing a Type I hypothesis, while "explanation" refers to a test of a Type III hypothesis. The former asserts only that a set of parameters, γ, has equal values. The latter asserts that this equality holds even after all correlated biases have been removed. In Selvin's view, "explanation" rather than mere "description" is our cup of tea. From the lack of restrictions placed on these terms, one could almost claim that the task he sees for social scientists is to uncover the eternal verities.

6. For example, the set of all persons now living in the United States is an element in the field induced by the set of all creatures ever having lived on earth. For a discussion of fields and sample spaces, see Harold Cramér (1946: Chapter 1, esp. § 1.6).

Selvin may not have such a large order in mind, to be sure, but the job which he prescribes for social researchers nonetheless is one of overwhelming proportions. For it seems to be his belief that social scientists should be concerned immediately and exclusively with Type III hypotheses and their tests. This conviction leads Selvin to the conclusion that *all* hypotheses of sociology are automatically of Type III.

This mistaken impression is illustrated by Selvin's consideration of the hypothesis that ". . . urban residents have a higher level of political interest than do rural residents" (1957: 521 [Chapter 9: 97]). Note that this is a Type I hypothesis since no conditions are imposed on the relationship of the two variables to any others. Selvin supposes that adequate measures of these two variables are at hand and that appropriate samples have been drawn at random from the two defined populations. Should an investigator then attempt to test the hypothesis by statistical inference, Selvin would object because of failure to control the correlated bias of such variables as education and income. "Where so many factors are uncontrolled it is obviously impossible to say that political interest *depends only* on place of residence and random variables" (1957: 522 [Chapter 9: 98, McGinnis's italics]). Obviously impossible, yes, but this is not what the investigator sets out to do in the first place. At the outset a Type I hypothesis is to be tested. As the illustration proceeds, however, Selvin transforms it into one of Type III and then takes the researcher to task for failing to satisfy the necessary conditions. Here is clear evidence of failure to recognize the distinctions between these hypothetical types.

It is this failure which generates the major fallacy of Professor Selvin's paper. We are told that it is "legitimate" to use tests of significance only after all correlated bias has been removed. If by legitimate is meant conformity to the assumptions of a statistical test, this statement is incorrect. Moreover, the argument at this point does not, in fact, apply to statistical test procedures.

All statistical tests require that certain conditions be satisfied if they are to be used legitimately. Some of these conditions concern the form of frequency distributions and the manner in which observations are made; others are imposed by the statement of the hypothesis under test. Whatever the hypothesis, some statistical models require that distribution functions be continuous and normally distributed. Other models make different demands. Regardless of the test statistic, some hypotheses demand that certain correlated biases be controlled, others do not. Only Type III hypotheses impose the condition that *all* correlated biases be controlled, which can be accomplished only by randomization. Type II hypotheses require that some finite number of related effects be eliminated, presumably in the statistical model itself. Type I hypotheses make no demands whatsoever of this sort.

No test of significance requires *of itself* that all correlated biases be removed, that is, that randomization be effected. It is true that this process assures that some requisite conditions of certain test statistics will be reasonably well met. Failure to randomize, however, in no way assures that such conditions will be violated. In general, then, the claim that all statistical tests of all hypotheses require the experimental procedure of randomization is unwarranted.

These considerations show that Selvin's assertion about legitimate applications of tests of significance is incorrect. They also show the reasons for his claim. Selvin insists that Type III interpretations be made of *all* verified hypotheses; and his statement is concerned exclusively with interpretations. Thus, the innocent test of significance becomes a false villain.

This situation becomes even more clear when it is recognized that there is only one difference between testing a hypothesis by observing an entire population and by drawing a sample, observing it, and making inferences to the population. In either case, acceptance or rejection of the hypothesis is made according to some prearranged decision procedure. In the former instance a correct decision is made with certainty, in the latter with probability less than one.[7] A statistical test of significance tells nothing more than the probability of making errors in the decision given that certain conditions obtain. If correlated biases are present in the population, they influence the outcome of the test whether or not the population is sampled or enumerated. If Selvin's researcher had enumerated his rural and urban populations rather than sampling them, or if he had used no tests of significance, the problem would not have been altered. So long as he did not randomize, it would be impossible in either case for him to conclude that political interest depended solely on place of residence and random variables.[8]

Apart from this unwarranted assertion, the tone of his paper suggests that Selvin would still refuse or be reluctant to allow the use of statistical test procedures in most sociological research. This reluctance seems to stem from the view that the only worthwhile hypotheses are those of the third type. There is little doubt that Selvin advocates a sociology of cause and effect. Few would dispute that this is a desirable goal. The latter does not justify the insistence, however, that all researchers restrict their attention to testable Type III hypotheses. What would be the consequences of such a restriction?

One consequence is clear: Most social researchers would soon be out of business. Since randomization is required to test a Type III hypothesis and since this is usually impossible in sociology, the business slump would follow swiftly and inevitably. A second consequence, assuming that some hypotheses

7. This assumes that no errors are made in observation, recording, computation, and so on.

8. This is not to say that randomization is sufficient to permit legitimate imputations of cause and effect. Problems of temporal paths of cause as well as epistemological questions would have to be settled first.

of this sort could be tested, would be a major increase in interpretive power. Rather than concentrating on correlations, we could at least begin to interpret relations, with all due caution, of course, in terms of cause and effect. Such a result would be valuable, of course, but this does not deny the considerable value of "descriptive" knowledge.

This less potent form of knowledge is useful for at least two reasons. First, the more exhaustive the descriptive knowledge which theorists have at hand, the sounder the basis for the construction of general theory—theory which eventually might yield testable hypotheses of the third type. A second utility of testing nonconditional hypotheses is that the practice yields at least partially satisfactory bases for prediction. In fact, a test of a Type I hypothesis may sometimes contribute considerably more to accurate short-range prediction than would a test of a Type III hypothesis. For example, from an explanatory point of view, it is imperative to understand that racial differences in measured intelligence tend to diminish or to disappear altogether as one increases the number of variables whose correlated effects are controlled. We know, however, that because of the effects of these many variables, there are sometimes considerable differences among races in measured intelligence. For purposes of prediction and policy formation, who would maintain that such knowledge is trivial?[9]

We contend, therefore, that a sociologist performs a valuable service whenever he investigates relationships among sociologically important variables, even though he is able to establish only the most paltry number of controls. So long as his design is as carefully constructed as possible, his measurements as accurate as instruments permit, and his interpretations no broader than the data and test procedures warrant, he is performing a worthwhile service.

If Selvin had said no more than that randomization is desirable in sociological research, we should have concurred enthusiastically. Had he made fresh, useful suggestions toward this end, these would have been a most welcome contribution. Since none was made, we may ask: What would Selvin have us do? The answer is not to be found in his paper. The less stringent position advocated here, on the other hand, seems to have some suggestions. The elementary facts noted above have at least a few implications for action.

The first of these may be phrased as follows: Where randomization is impossible, seek to maximize statistical controls. We have depicted three classes of hypotheses, including those of the second type which require control of a finite number of variables. By employing appropriate statistical devices it is

9. In this connection it should be noted that Type II hypotheses have a value which is afforded by those of neither Type I nor Type III. Since tests of Type II hypotheses require that subjects be classified with respect to all variables under consideration, joint frequency distributions result which give information, often extremely important, about interactions at all levels among the variables.

possible to control the correlated bias of a number of related variables. Partial correlation coefficients and the analysis of covariance are examples of such devices.[10] As the number of tested Type II hypotheses is increased and as more and more variables are brought under control in each test, we shall come to an ever more satisfactory approximation of Selvin's "explanatory knowledge."

We recognize with Professor Selvin the difficulty of putting this dictum into effect. He observes that thousands of cases may be required to control the merest handful of related variables. This is undeniably the case, at least when a simple random sampling plan is used to locate the cases. Multiphase sampling, however, in which the population or a large areal sample is prelisted with respect to control variables, can greatly reduce the necessary number of cases. The judicious selection of control variables and the use of more sophisticated sampling plans can add immeasurably to the strength of sociological research.

Even where these suggestions are carried out, there remains the problem of selecting appropriate statistical tests. Many of these models are based upon assumptions which are not tenable with respect to most sociological variables. All too few of our variables are known to be continuous, for example, and to possess workable distribution functions. This remains a serious problem despite the encouraging progress of theoretical statisticians in the area of nonparametric statistics. Perhaps, then, sociologists would be wise to devote more of their attention to problems of measurement and standardization, even if this means that less time can be devoted to testing complicated hypotheses. Only when we have measuring instruments with more powerful scaling properties than ordinality can we make use of the most effective statistical tools to control correlated bias.

10. Control features also are available for use with some noncontinuous variables. See for example M. Ezekiel (1941: Chapter 17).

15. Some Statistical Problems in Research Design

LESLIE KISH

STATISTICAL INFERENCE is an important aspect of scientific inference. The statistical consultant spends much of his time in the borderland between statistics and the other aspects, philosophical and substantive, of the scientific search for explanation. This marginal life is rich both in direct experience and in discussions of fundamentals; these have stimulated my concern with the problems treated here.

I intend to touch on several problems dealing with the interplay of statistics with the more general problems of scientific inference. We can spare elaborate introductions because these problems are well known. Why then discuss them here at all? We do so because, first, they are problems about which there is a great deal of misunderstanding, evident in current research; and, second, they are *statistical* problems on which there is broad agreement among research statisticians—and on which these statisticians generally disagree with much in the current practice of research scientists.[1]

Reprinted from the *American Sociological Review*, 24 (June, 1959), 328–338, by permission of the author and the publisher. Copyright © 1959 by the American Sociological Association.

1. "The statistician cannot evade the responsibility for understanding the processes he applies or recommends. My immediate point is that the questions involved can be disassociated from all that is strictly technical in the statistician's craft, and *when so detached*, are questions only of the right use of human reasoning powers, with which all intelligent people, who hope to be intelligible, are equally concerned, and on which the statistician, as such, speaks with no special authority. The statistician cannot excuse himself from the duty of getting his head clear on the principles of scientific inference, but equally no other thinking man can avoid a like obligation" (Fisher, 1935a, 6th ed.: 1–2).

Several problems will be considered briefly, hence incompletely. The aim of this paper is not a profound analysis, but a clear elementary treatment of several related problems. The literature references contain more thorough treatments. Moreover, these are not *all* the problems in this area, nor even necessarily the most important ones; the reader may find that his favorite, his most annoying problem, has been omitted. The problems selected are a group with a common core, they arise frequently, yet they are widely misunderstood.

Statistical Tests of Survey Data

That correlation does not prove causation is hardly news. Perhaps the wittiest statements on this point are in George Bernard Shaw's preface to *The Doctor's Dilemma*, in the sections on "Statistical Illusions," "The Surprises of Attention and Neglect," "Stealing Credit from Civilization," and "Biometrika." (These attack, alas, the practice of vaccination.) The excellent introductory textbook by Yule and Kendall (1937: Chapters 4, 15, and 16) deals in three separate chapters with the problems of advancing from correlation to causation. Searching for causal factors among survey data is an old, useful sport; and the attempts to separate true explanatory variables from extraneous and "spurious" correlations have taxed scientists since antiquity and will undoubtedly continue to do so. Neyman (1952) and Simon (1954; 1956) show that, beyond common sense, there are some technical skills involved in tracking down spurious correlations. Econometricians and geneticists have developed great interest and skill in the problems of separating the explanatory variables.[2]

The researcher designates the explanatory variables on the basis of substantive scientific theories. He recognizes the evidence of other *sources of variation* and he needs to separate these from the explanatory variables. Sorting all sources of variation into four classes seems to me a useful simplification. Furthermore, no confusion need result from talking about sorting and treating "variables," instead of "sources of variation."

I. The *explanatory* variables, sometimes called the "experimental" variables, are the objects of the research. They are the variables among which the researcher wishes to find and to measure some specified relationships. They include both the "dependent" and the "independent" variables, that is, the

2. See the excellent and readable article by Wold (1956); and the two-part, technical article by M. G. Kendall (1951; 1952). The interesting methods of "path coefficients" in genetics have been developed by Wright for inferring causal factors from regression coefficients. See Wright (1954) and Tukey (1954a). See also Li (1956). I do not know whether these methods can be of wide service in current social science research in the presence of numerous factors, of large, unexplained variances, and of doubtful directions of causation.

"predictand" and "predictor" variables.[3] With respect to the aims of the research all other variables, of which there are three classes, are extraneous.

II. There are extraneous variables which are *controlled*. The control may be exercised in either or both the selection and the estimation procedures.

III. There may exist extraneous uncontrolled variables which are *confounded* with the Class I variables.

IV. There are extraneous uncontrolled variables which are treated as *randomized* errors. In "ideal" experiments (discussed below) they are actually randomized; in surveys and investigations they are only assumed to be randomized. Randomization may be regarded as a substitute for experimental control or as a form of control.

The aim of efficient design both in experiments and in surveys is to place as many of the extraneous variables as is feasible into the second class. The aim of randomization in experiments is to place all of the third class into the fourth class; in the "ideal" experiment there are no variables in the third class. And it is the aim of controls of various kinds in surveys to separate variables of the third class from those of the first class: These controls may involve the use of repeated cross-tabulations, regression, standardization, matching of units, and so on.

The function of statistical "tests of significance" is to test the effects found among the Class I variables against the effects of the variables of Class IV. An "ideal" experiment here denotes an experiment for which this can be done through randomization without any possible confusion with Class III variables. (The difficulties of reaching this "ideal" are discussed below.) In survey results, Class III variables are confounded with those of Class I; the statistical tests actually contrast the effects of the random variables of Class IV against the explanatory variables of Class I confounded with unknown effects of Class III variables. In both the ideal experiment and in surveys the statistical tests serve to separate the effects of the random errors of Class IV from the effects of other variables. These, in surveys, are a mixture of explanatory and confounded variables; their separation poses severe problems for logic and for scientific methods; statistics is only one of the tools in this endeavor. The scientist must make many decisions as to which variables are extraneous to his objectives, which should and can be controlled, and what methods of control he should use. He must decide where and how to introduce statistical tests of hypotheses into the analysis.

As a simple example, suppose that from a probability sample survey of adults of the United States we find that the level of political interest is higher in urban than in rural areas. A test of significance will show whether or not

3. M. G. Kendall points out that these latter terms are preferable. See Kendall (1951; 1952) and Kendall and Buckland (1957). I have also tried to follow in IV below his distinction of "variate" for random variables from "variables" for the usual (nonrandom) variable.

the difference in the "levels" is large enough, compared with the sampling error of the difference, to be considered "significant." Better still, the confidence interval of the difference will disclose the limits within which we can expect the "true" population value of the difference to lie.[4] If families had been sent to urban and rural areas, respectively, after the randomization of a true experiment, then the sampling error would measure the effects of Class IV variables against the effects of urban *versus* rural residence on political interest; the difference in levels beyond sampling errors could be ascribed (with specified probability) to the effects of urban *versus* rural residence.

Actually, however, residences are not assigned at random. Hence, in survey results, Class III variables may account for some of the difference. If the test of significance rejects the null hypothesis of no difference, *several* hypotheses remain in addition to that of a simple relationship between urban *versus* rural residence and political interest. Could differences in income, in occupation, or in family life cycle account for the difference in the levels? The analyst may try to remove (for example, through cross-tabulation, regression, standardization) the effects due to such variables, which are extraneous to his expressed interest; then he computes the difference, between the urban and rural residents, of the levels of interest now free of several confounding variables. This can be followed by a proper test of significance—or, preferably, by some other form of statistical inference, such as a statement of confidence intervals.

Of course, other variables of Class III may remain to confound the measured relationship between residence and political interest. The separation of Class I from Class III variables should be determined in accord with the nature of the hypothesis with which the researcher is concerned; finding and measuring the effects of confounding variables of Class III tax the ingenuity of research scientists. But this separation is beyond the functions and capacities of the statistical tests, the tests of null hypotheses. Their function is not explanation; they cannot point to causation. Their function is to ask: "Is there anything in the data that *needs* explaining?"—and to answer this question with a certain probability.

Agreement on these ideas can eliminate certain confusion, exemplified by Selvin:

> Statistical tests are unsatisfactory in non-experimental research for two
> fundamental reasons: It is almost impossible to design studies that meet the

4. The sampling error measures the chance fluctuation in the difference of levels due to the sampling operations. The computation of the sampling error must take proper account of the actual sample design, and not blindly follow the standard simple random formulas. See Leslie Kish (1957).

conditions for using the tests, and the situations in which the tests are employed make it difficult to draw correct inferences. The basic difficulty in design is that sociologists are unable to randomize their uncontrolled variables, so that the difference between "experimental" and "control" groups (or their analogs in nonexperimental situations) are a mixture of the effects of the variable being studied and the uncontrolled variables or correlated biases. Since there is no way of knowing, in general, the sizes of these correlated biases and their directions, there is no point in asking for the probability that the observed differences could have been produced by random errors. The place for significance tests is after all relevant correlated biases have been controlled. . . . In design and in interpretation, in principle and in practice, tests of statistical significance are inapplicable in nonexperimental research (1957: 527 [Chapter 9: 106]).[5]

Now it is true that in survey results the explanatory variables of Class I are confounded with variables of Class III; but it does not follow that tests of significance should not be used to separate the random variables of Class IV. Insofar as the effects found "are a mixture of the effects of the variable being studied and the uncontrolled variables;" insofar as "there is no way of knowing, in general, the sizes" and directions of these uncontrolled variables, Selvin's logic and advice should lead not only to the rejection of statistical tests; it should lead one to refrain altogether from using survey results for the purposes of finding explanatory variables. *In this sense*, not only tests of significance but any comparisons, any scientific inquiry based on surveys, any scientific inquiry other than an "ideal" experiment, is "inapplicable." That advice is most unrealistic. In the (unlikely) event of its being followed, it would sterilize social research—and other nonexperimental research as well.

Actually, much research—in the social, biological, and physical sciences— must be based on nonexperimental methods. In such cases the rejection of the null hypothesis leads to several alternate hypotheses that may explain the discovered relationships. It is the duty of scientists to search, with painstaking effort and with ingenuity, for bases on which to decide among these hypotheses.

As for Selvin's advice to refrain from making tests of significance until "after all relevant" uncontrolled variables have been controlled—this seems rather far-fetched to scientists engaged in empirical work who consider themselves lucky if they can explain 25 or 50 percent of the total variance. The control of all relevant variables is a goal seldom even approached in practice. To

5. In a criticism of Selvin's article, McGinnis (1958 [Chapter 14]) shows that the separation of explanatory from extraneous variables depends on the type of hypothesis at which the research is aimed.

postpone to that distant goal all statistical tests illustrates that often "the perfect is the enemy of the good."[6]

Experiments, Surveys, and Other Investigations

Until now, the theory of sample surveys has been developed chiefly to provide descriptive statistics—especially estimates of means, proportions, and totals. On the other hand, experimental designs have been used primarily to find explanatory variables in the analytical search of data. In many fields, however, including the social sciences, survey data must be used frequently as the analytical tools in the search for explanatory variables. Furthermore, in some research situations, neither experiments nor sample surveys are practical, and other investigations are utilized.

By "experiments" I mean here "ideal" experiments in which all the extraneous variables have been randomized. By "surveys" (or "sample surveys"), I mean probability samples in which all members of a defined population have a known positive probability of selection into the sample. By "investigations" (or "other investigations"), I mean the collection of data—perhaps with care, and even with considerable control—without either the randomization of experiments or the probability sampling of surveys. The differences among experiments, surveys, and investigations are not the consequences of statistical techniques; they result from different methods for introducing the variables and for selecting the population elements (subjects). These problems are ably treated in recent articles by Wold (1956) and Campbell (1957).

In considering the larger ends of any scientific research, only part of the total means required for inference can be brought under objective and firm control; another part must be left to more or less vague and subjective—

6. Selvin performs a service in pointing to several common mistakes: (a) The mechanical use of "significance tests" can lead to false conclusions. (b) Statistical "significance" should not be confused with substantive importance. (c) The probability levels of the common statistical tests are not appropriate to the practice of "hunting" for a few differences among a mass of results. However, Selvin gives poor advice on what to do about these mistakes, particularly when, in his central thesis, he reiterates that "tests of significance are inapplicable in nonexperimental research," and that "the tests are applicable only when all relevant variables have been controlled." I hope that the benefits of his warnings outweigh the damages of his confusion.

I noticed three misleading references in the article. (a) In the paper which Selvin appears to use as supporting him, Wold (1956: 39) specifically disagrees with Selvin's central thesis, stating that "The need for testing the statistical inference is no less than when dealing with experimental data, but with observational data other approaches come to the foreground." (b) In discussing problems caused by complex sample designs, Selvin writes that "Such errors are easy enough to discover and remedy" (1957: 520 [Chapter 9: 95]) referring to Kish (1957). On the contrary, my article pointed out the seriousness of the problem and the difficulties in dealing with it. (c) "Correlated biases" is a poor term for the confounded uncontrolled variables and it is not true that the term is so used in literature. Specifically, the reference to Cochran is misleading, since he is dealing there only with errors of measurement which may be correlated with the "true" value (1953: 305).

however skillful—judgment. The scientist seeks to maximize the first part, and thus to minimize the second. In assessing the ends, the costs, and the feasible means, he makes a strategic choice of methods. He is faced with the three basic problems of scientific research: measurement, representation, and control. We ignore here the important but vast problems of measurement and deal with representation and control.

Experiments are strong on control through randomization; but they are weak on representation (and sometimes on the "naturalism" of measurement). Surveys are strong on representation, but they are often weak on control. Investigations are weak on control and often on representation; their use is due frequently to convenience or low cost and sometimes to the need for measurements in "natural settings."

Experiments have three chief advantages: (1) Through randomization of extraneous variables the confounding variables (Class III) are eliminated. (2) Control over the introduction and variation of the "predictor" variables clarifies the *direction* of causation from "predictor" to "predictand" variables. In contrast, in the correlations of many surveys this direction is not clear—for example, between some behaviors and correlated attitudes. (3) The modern design of experiments allows for great flexibility, efficiency, and powerful statistical manipulation, whereas the analytical use of survey data presents special statistical problems (Kish, 1957).

The advantages of the experimental method are so well known that we need not dwell on them here. It is the scientific method *par excellence*—when feasible. In many situations experiments are not feasible and this is often the case in the social sciences; but it is a mistake to use this situation to separate the social from the physical and biological sciences. Such situations also occur frequently in the physical sciences (in meteorology, astronomy, geology), the biological sciences, medicine, and elsewhere.

The experimental method also has some shortcomings. First, it is often difficult to choose the "control" variables so as to exclude *all* the confounding extraneous variables; that is, it may be difficult or impossible to design an "ideal" experiment. Consider the following examples: The problem of finding a proper control for testing the effects of the Salk polio vaccine led to the use of an adequate "placebo." The Hawthorne experiment demonstrated that the design of a proposed "treatment *versus* control" may turn out to be largely a test of *any* treatment *versus lack* of treatment (Roethlisberger and Dixon, 1939).[7]

7. Troubles with experimental controls misled even the great Pavlov into believing *temporarily* that he had proof of the inheritance of an acquired ability to learn: "In an informal statement made at the time of the Thirteenth International Physiological Congress, Boston, August, 1929, Pavlov explained that in checking up these experiments it was found that the apparent improvement in the ability to learn, on the part of successive generations of mice, was really due to an improvement in the ability to teach, on the part of the experimenter" (Greenberg, 1929: 327).

Many of the initial successes reported about mental therapy, which later turn into vain hopes, may be due to the hopeful effects of *any* new treatment in contrast with the background of neglect. Shaw, in "The Surprises of Attention and Neglect" writes: "Not until attention has been effectually substituted for neglect as a general rule, will the statistics begin to show the merits of the particular methods of attention adopted."

There is an old joke about the man who drank too much on four different occasions, respectively, of scotch and soda, bourbon and soda, rum and soda, and wine and soda. Because he suffered painful effects on all four occasions, he ascribed, with scientific logic, the common effect to the common cause: "I'll never touch soda again!" Now, to a man (say, from Outer Space) ignorant of the common alcoholic content of the four "treatments" and of the relative physiological effects of alcohol and carbonated water, the subject is not fit for joking, but for further scientific investigation.

Thus, the advantages of experiments over surveys in permitting better control are only relative, not absolute (Cornfield, 1954). The design of proper experimental controls is not automatic; it is an art requiring scientific knowledge, foresight in planning the experiment, and hindsight in interpreting the results. Nevertheless, the distinction in control between experiments and surveys is real and considerable; and to emphasize this distinction we refer here to "ideal" experiments in which the control of the random variables is complete.

Second, it is generally difficult to design experiments so as to represent a specified important population. In fact, the questions of sampling, of making the experimental results representative of a specified population, have been largely ignored in experimental design until recently. Both in theory and in practice experimental research has often neglected the basic truth that in causal systems the distributions of relations—like the distributions of characteristics —exists only within specified universes. The distributions of relationships, as of characteristics, exist only within the framework of specific populations. Probability distributions, like all mathematical models, are abstract systems; their application to the physical world must include the specification of the populations. For example, it is generally accepted that the statement of a value for mean income has meaning only with reference to a specified population; but this is not generally and clearly recognized in the case of regression of assets on income and occupation. Similarly, the *statistical* inferences derived from the experimental testing of several treatments are restricted to the population(s) included in the experimental design.[8] The clarification of the

---

8. McGinnis points out that usually "it is not true that one can uncover 'general' relationships by examining some arbitrarily selected population. . . . There is no such thing as a completely general relationship which is independent of population, time, and space. The extent to which a relationship is constant among different populations is an empirical question which can be resolved only by examining different populations at different times in different places" (1958: 412 [Chapter 14: 122]).

population sampling aspects of experiments is now being tackled vigorously by Wilk and Kempthorne (1955; 1956) and by Cornfield and Tukey (1956).

Third, for many research aims, especially in the social sciences, contriving the desired "natural setting" for the measurements is not feasible in experimental design. Hence, what social experiments give sometimes are clear answers to questions the meanings of which are vague. That is, the artificially contrived experimental variables *may* have but a tenuous relationship to the variables the researcher would like to investigate.

The second and third weaknesses of experiments point to the advantages of surveys. Not only do probability samples permit clear statistical inferences to defined populations, but the measurements can often be made in the "natural settings" of actual populations. Thus in practical research situations the experimental method, like the survey method, has its distinct problems and drawbacks as well as its advantages. In practice one generally cannot solve simultaneously all of the problems of measurement, representation, and control; rather, one must choose and compromise. In any specific situation one method may be better or more practical than the other; but there is no overall superiority in all situations for either method. Understanding the advantages and weaknesses of both methods should lead to better choices.

In social research, in preference to both surveys and experiments, frequently some design of controlled investigation is chosen—for reasons of cost or of feasibility or to preserve the "natural setting" of the measurements. Ingenious adaptations of experimental designs have been contrived for these controlled investigations. The statistical framework and analysis of experimental designs are used, but not the randomization of true experiments. These designs are aimed to provide flexibility, efficiency, and, especially, some control over the extraneous variables. They have often been used to improve considerably research with controlled investigations.

These designs are sometimes called "natural experiments." For the sake of clarity, however, it is important to keep clear the distinctions among the methods and to reserve the word "experiment" for designs in which the uncontrolled variables are randomized. This principle is stated clearly by Fisher (1935a, 6th ed.: 17–20)[9], and is accepted often in scientific research. Confusion is caused by the use of terms like "*ex post facto* experiments" to describe surveys or designs of controlled investigations. Sample surveys and controlled investigations have their own justifications, their own virtues; they are not just second-class experiments. I deplore the borrowing of the prestige word "experiment," when it cloaks the use of other methods.

Experiments, surveys, and investigations can all be improved by efforts to overcome their weaknesses. Because the chief weakness of surveys is their low

9. "Controlled investigation" may not be the best term for these designs. "Controlled observations" might do, but "observation" has more fundamental meanings.

degree of control, researchers should be alert to the collection and use of auxiliary information as controls against confounding variables. They also should take greater advantage of changes introduced into their world by measuring the effects of such changes. They should utilize more often efficient and useful statistics instead of making tabular presentation their only tool.

On the other hand, experiments and controlled investigations can often be improved by efforts to specify their populations more clearly and to make the results more representative of the population. Often more should be done to broaden the area of inference to more important populations. Thus, in many situations the deliberate attempts of the researcher to make his sample more "homogeneous" are misplaced; and if common sense will not dispel the error, reading Fisher may.[10] When he understands this, the researcher can view the population base of his research in terms of efficiency—in terms of costs and variances. He can often avoid basing his research on a comparison of one sampling unit for each "treatment." If he cannot obtain a proper sample of the entire population, frequently he can secure, say, four units for each treatment, or a score for each.[11]

Suppose, for example, that thorough research on one city and one rural county discloses higher levels of political interest in the former. It is presumptuous (although common practice) to present this result as evidence that urban people in *general* show a higher level. (Unfortunately, I am not beating a dead horse; this nag is pawing daily in the garden of social science.) However, very likely there is a great deal of variation in political interest among different cities, as well as among rural counties; the results of the research will depend

10. Fisher says: "We have seen that the factorial arrangement possesses two advantages over experiments involving only single factors: (i) Greater *efficiency*, in that these factors are evaluated with the same precision by means of only a quarter of the number of observations that would otherwise be necessary; and (ii) Greater *comprehensiveness* in that, in addition to the four effects of single factors, their 11 possible interactions are evaluated. There is a third advantage which, while less obvious than the former two, has an important bearing upon the utility of the experimental results in their practical application. This is that any conclusion, such as that it is advantageous to increase the quantity of a given ingredient, has a wider inductive basis when inferred from an experiment in which the quantities of other ingredients have been varied, than it would have from any amount of experimentation, in which these had been kept strictly constant. The exact standardisation of experimental conditions, which is often thoughtlessly advocated as a panacea, always carries with it the real disadvantage that a highly standardized experiment supplies direct information only in respect of the narrow range of conditions achieved by standardisation. Standardisation, therefore, weakens rather than strengthens our ground for inferring a like result, when, as is invariably the case in practice, these conditions are somewhat varied" (1935a, 6th ed.: 99–100).

11. For simplicity the following illustration is a simple contrast between two values of the "explanatory" variable, but the point is more general; and this aspect is similar whether for true experiments or controlled observations. Incidentally, it is poor strategy to "solve" the problem of representation by obtaining a good sample, or complete census, of some small or artificial population. A poor sample of the United States or of Chicago *usually* has more over-all value than the best sample of freshman English classes at X University.

heavily on which city and which county the researcher picked as "typical." The research would have a broader base if a city and a rural county would have been chosen in each of, say, four different situations—as different as possible (as to region, income, industry, for example); or better still in twenty different situations. A further improvement would result if the stratification and selection of sampling units followed a scientific sample design.

Using more sampling units and spreading them over the breadth of variation in the population has several advantages. First, some measure of the variability of the observed effect may be obtained. From a probability sample, statistical inference to the population can be made. Second, the base of the inference is broadened, as the effect is observed over a variety of situations. Beyond this lies the combination of results from researches over several distinct cultures and periods. Finally, with proper design, the effects of several potentially confounding factors can be tested.

These points are brought out by Keyfitz in an excellent example of controlled investigation (which also uses sampling effectively): "Census enumeration data were used to answer for French farm families of the Province of Quebec the question: Are farm families smaller near cities than far from cities, other things being equal? The sample of 1,056 families was arranged in a 2^6 factorial design which not only controlled 15 extraneous variables (income, education, etc.) but incidentally measured the effect of 5 of these on family size. A significant effect of distance from cities was found, from which is inferred a geographical dimension for the currents of social change (1953: 470). The mean numbers of children per family were found to be 9.5 near and 10.8 far from cities; the difference of 1.3 children has a standard error of 0.28.

Some Misuses of Statistical Tests

Of the many kinds of current misuses this discussion is confined to a few of the most common. There is irony in the circumstance that these are committed usually by the more statistically inclined investigators; they are avoided in research presented in terms of qualitative statements or of simple descriptions.

First, there is "hunting with a shot-gun" for significant differences. Statistical tests are designed for distinguishing results at a predetermined level of improbability (say at $P = .05$) under a specified null hypothesis of random events. A rigorous theory for dealing with individual experiments has been developed by Fisher, the Pearsons, Neyman, Wold, and others. However, the researcher often faces more complicated situations, especially in the analysis of survey results; he is often searching for interesting relationships among a vast number of data. The keen-eyed researcher hunting through the results of one thousand random tosses of perfect coins would discover and display about fifty "significant" results (at the $P = .05$ level).[12] Perhaps the problem has become more acute now that high-speed computers allow hundreds of significance tests to be made. There is no easy answer to this problem. We must be constantly aware of the nature of tests of null hypotheses in searching survey data for interesting results. After finding a result improbable under the null hypothesis the researcher must not accept blindly the hypothesis of "significance" due to a presumed cause. Among the several alternative hypotheses is that of having discovered an improbable random event through sheer diligence. Remedy can be found sometimes by a reformulation of the statistical aims of the research so as to fit the available tests. Unfortunately, the classic statistical tests give clear answers only to some simple decision problems; often these bear but faint resemblance to the complex problems faced by the scientist. In response to these needs the mathematical statisticians are beginning to provide some new statistical tests. Among the most useful are the new "multiple comparison" and "multiple range" tests of Tukey (1949), Duncan (1955), Scheffé (1953), and others. With a greater variety of statistical statements available, it will become easier to choose one without doing great violence either to them or to the research aims.

Second, statistical "significance" is often confused with and substituted for substantive significance. There are instances of research results presented in terms of probability values of "statistical significance" alone, without noting the magnitude and importance of the relationships found. These attempts to use the probability levels of significance tests as measures of the strengths of relationships are very common and very mistaken. The function of statistical tests is merely to answer: Is the variation great enough for us to place some confidence in the result; or, contrarily, may the latter be merely a happenstance of the specific sample on which the test was made? This question is interesting, but it is surely *secondary*, auxiliary, to the main question: Does the result show a relationship which is of substantive interest because of its nature and its magnitude? Better still: Is the result consistent with an assumed relationship of substantive interest?

The results of statistical "tests of significance" are functions not only of the magnitude of the relationships studied but also of the numbers of sampling

12. Sewell points to an interesting example: "On the basis of the results of this study, the general null hypothesis that the personality adjustments and traits of children who have undergone varying training experiences do not differ significantly cannot be rejected. Of the 460 Chi-square tests, only 18 were significant at or beyond the 5 percent level. Of these, 11 were in the expected direction and 7 were in the opposite direction from that expected on the basis of psychoanalytic writings. . . . Certainly, the results of this study cast serious doubts on the validity of the psychoanalytic claims regarding the importance of the infant disciplines and on the efficacy of prescriptions based on them" (1952: 158–159). Note that by chance alone one would expect 23 "significant" differences at the 5 percent level. A "hunter" would report either the 11 or the 18 and not the hundreds of "misses."

units used (and the efficiency of design). In small samples significant, that is, meaningful, results may fail to appear "statistically significant." But if the sample is large enough the most insignificant relationships will appear "statistically significant."

Significance should stand for meaning and refer to substantive matter. The statistical tests merely anwer the question: Is there a big enough relationship here which *needs* explanation (and is not merely chance fluctuation)? The word *significance* should be attached to another question, a substantive question: Is there a relationship here *worth* explaining (because it is important and meaningful)? As a remedial step I would recommend that statisticians discard the phrase "test of significance," perhaps in favor of the somewhat longer but proper phrase "test against the null hypothesis" or the abbreviation "TANH."

Yates, after praising Fisher's classic *Statistical Methods* (1925a), makes the following observations on the use of "tests of significance":

> Second, and more important, it has caused scientific research workers to pay undue attention to the results of the tests of significance they perform on their data, particularly data derived from experiments, and too little to the estimates of the magnitude of the effects they are investigating.
>
> Nevertheless the occasions, even in research work, in which quantitative data are collected solely with the object of proving or disproving a given hypothesis are relatively rare. Usually quantitative estimates and fiducial limits are required. Tests of significance are preliminary or ancillary.
>
> The emphasis on tests of significance, and the consideration of the results of each experiment in isolation, have had the unfortunate consequence that scientific workers have often regarded the execution of a test of significance on an experiment as the ultimate objective. Results are significant or not significant and this is the end of it (1951).

For presenting research results statistical estimation is more frequently appropriate than tests of significance. The estimates should be provided with some measure of sampling variability. For this purpose confidence intervals are used most widely. In large samples, statements of the standard errors provide useful guides to action. These problems need further development by theoretical statisticians (Cox, 1958).

The responsibility for the current fashions should be shared by the authors of statistical textbooks and ultimately by the mathematical statisticians. As Tukey puts it (see also Cox, 1958; and Duncan, 1955):

> *Statistical methods should be tailored to the real needs of the user.* In a number of cases, statisticians have led themselves astray by choosing a problem which they could solve exactly but which was far from the needs of their clients. . . . The broadest class of such cases comes from the choice of significance procedures rather than confidence procedures. It is often much easier to be "exact" about significance procedures than about confidence procedures. By considering only

the most null "null hypothesis" many inconvenient possibilities can be avoided (1954b: 710 [italics in original]).

Third, the tests of null hypotheses of *zero* differences, of no relationships, are frequently weak, perhaps trivial statements of the researcher's aims. In place of the test of zero difference (the nullest of null hypotheses), the researcher should often substitute, say, a test for a difference of a specific size based on some specified model. Better still, in many cases, instead of the tests of significance it would be more to the point to measure the magnitudes of the relationships, attaching proper statements of their sampling variation. The magnitudes of relationships cannot be measured in terms of levels of significance; they can be measured in terms of the difference of two means, or of the proportion of the total variance "explained," of coefficients of correlations and of regressions, of measures of association, and so on. These views are shared by many, perhaps most, consulting statisticians—although they have not published full statements of their philosophy. Savage expresses himself forcefully: "Null hypotheses of no difference are usually known to be false before the data are collected; when they are, their rejection or acceptance simply reflects the size of the sample and the power of the test, and is not a contribution to science" (1957).

Too much of social research is planned and presented in terms of the mere existence of some relationship, such as: Individuals high on variate x are also high on variate y. The *exploratory* stage of research may be well served by statements of this order. But these statements are relatively weak and can serve *only* in the primitive stages of research. Contrary to a common misconception, the more advanced stages of research should be phrased in terms of the quantitative aspects of the relationships. Again, to quote Tukey:

> There are normal sequences of growth in immediate ends. One natural sequence of immediate ends follows the sequence: (1) Description, (2) Significance statements, (3) Estimation, (4) Confidence statement, (5) Evaluation. . . . There are, of course, other normal sequences of immediate ends, leading mainly through various decision procedures, which are appropriate to development research and to operations research, just as the sequence we have just discussed is appropriate to basic research (1954b: 712–713 [italics in original]).

At one extreme, then, we may find that the contrast between two "treatments" of a labor force results in a difference in productivity of 5 percent. This difference may appear "statistically significant" in a sample of, say, 1000 cases. It may also mean a difference of millions of dollars to the company. However, it "explains" only about one percent of the total variance in productivity. At the other extreme is the far-away land of completely determinate behavior, where every action and attitude is explainable, with nothing left to chance for explanation.

The aims of most basic research in the social sciences, it seems to me, should be somewhere between the two extremes; but too much of it is presented at the first extreme, at the primitive level. This is a matter of over-all strategy for an entire area of any science. It is difficult to make this judgment off-hand regarding any specific piece of research of this kind: The status of research throughout the entire area should be considered. But the superabundance of research aimed at this primitive level seems to imply that the over-all strategy of research errs in this respect. The construction of scientific theories to cover broader fields—the persistent aim of science—is based on the synthesis of the separate research results in those fields. A coherent synthesis cannot be forged from a collection of relationships of unknown strengths and magnitudes. The necessary conditions for a synthesis include an *evaluation* of the results available in the field, a coherent interrelating of the *magnitudes* found in those results, and the construction of models based on those magnitudes.

16. Theory, Probability, and Induction in Social Research

SANTO F. CAMILLERI

MOST SOCIAL RESEARCHERS employ probabitity concepts in their research either explicitly or implicitly whether or not they are aware of it. The chief notable exceptions are those who are proponents of what has come to be called "analytic induction,"[1] a rationale that explicitly excludes certain probability formulations. But these are small minorities in the sociological society. Most of us, in one way or another, are probability oriented in our research. Yet, for all this collective commitment to the use of probability, there is a great deal of confusion and obscurity about its nature and use in research.

The disagreements connected with the use of probability in social research stem from many sources. There is a general failure to recognize the different purposes probability reasoning has served, with the result that critics and defenders of probability often are talking about different things. A perhaps even more important source of difficulty lies in an inadequate understanding of the nature of science and the proper business of the scientist. Without presuming to be definitive, the present paper undertakes to make a contribution

Reprinted from the *American Sociological Review*, 27 (April, 1962), 170–178, by permission of the author and the publisher. Copyright © 1962 by the American Sociological Association.

1. For a description and analysis of this school of thought see William S. Robinson (1951).

to the clarification of these basic issues. We shall discuss briefly some aspects of scientific theory and the verification of scientific hypotheses. In the course of this we shall distinguish three basic uses of probability: intrinsic, auxiliary, and inductive.[2] In a critical analysis of inductive probability we shall introduce the concept of systematic import as a part of the inductive process.

The Dual Responsibility of the Scientist

Discussions about science usually begin with a statement to the effect that the purpose of science is to make predictions for the purpose of control. We agree with these objectives, but they do not clearly reflect what we believe is the chief function of the scientist: To create and provide empirical support for a particular means of making predictions, *viz.*, systematic theory. Our contention is that the scientist has a direct responsibility to both these tasks, the creation of theory and the providing of empirical support for theory, and that to fail to meet either responsibility is to fail fundamentally.

Some Aspects of Scientific Theory

FORMAL AND EMPIRICAL TRUTH[3]

Scientific theory has a formal aspect. It consists of statements of fact in propositional form which are connected by explicit rules. These rules provide an objective, communicable means of operating on some of these propositions to produce additional "lower level" propositions or theorems. The *formal truth* of a theorem rests essentially on a demonstration that it has been generated by a correct application of the rules (has been correctly inferred).

Scientific theory has an empirical aspect, as well. To make the theory "say something" about reality it has to be interpreted or given empirical content. This is accomplished by means of a set of coordinating definitions that are essentially rules for identifying the empirical referents of the theorems. The *empirical truth* of a theorem refers to the judged correspondence of its interpreted assertion with the observed state of affairs. In their narrow conception, operational definitions serve as an important type of coordinating definition (see Hempel, 1956).

GENERALITY

The propositions embodied in the theory refer to all instances of the phenomena to which they are linked by the coordinating definitions. By *all* we mean

2. The present discussion owes much to Lancelot Hogben, who treated this topic at great length in his important book, *Statistical Theory* (1957 [see Chapters 1, 2, 3]).
3. For a valuable treatment of the logical structure of scientific theories, see Richard B. Braithwaite (1953).

literally those that occurred in the past, those that now exist, and those that may yet occur in the future. That is, the universes to which the propositions refer are always infinite and hence conceptual. Their elements never all exist at one time. Connected with this is the requirement that the coordinating definitions must not contain time or place specifications, except as these factors are variables in the system. For example, a proposition connecting vote with residence must not be interpreted so as to refer only to the voters in a given election who reside at that time in a given community. This conception of the universe and the correlated restriction on the coordinating definitions provide the system with its predictive relevance with regard to future events and its explanatory relevance with regard to events which have already occurred.[4]

DETERMINISTIC AND PROBABILISTIC THEORIES

The propositions of a theory can be arranged in a hierarchy based upon their order in the deductive evolvement of the theory. A deterministic theory is one whose propositions at all levels within the hierarchy express universal relationships of the sort "All A is B." A probabilistic theory is one that contains at least one highest level proposition that expresses a probability relationship between some classes of the elements referred to by the theory. The reason that such a proposition must be a highest level one is that probability statements can be *inferred* only from probability statements, so if they appear at all in a theory, they must appear at a highest level (Braithwaite, 1953: Chapter 6).

When probability propositions are so incorporated in a theory, they express a fundamental property of the phenomena to which they refer. To remove this kind of proposition from a theory literally involves reconceptualizing the phenomenon itself. We shall therefore refer to probability statements within a theory as *intrinsic* probability and speak of the intrinsic use of probability. It is especially important to maintain a distinction between the intrinsic use of probability and the use of probability to deal with the conditions of observation, which we shall treat presently.

Sampling and Experimentation

Discussions about whether sociology can ever be a science and about whether it can be a science without experimentation often center on the competitive advantages of surveys and controlled experiments. In the former, probability considerations enter in the context of sampling from finite, existent populations, and in the latter they enter in the design and interpretation of randomized experiments. In the typology we are presenting here, these are auxiliary probability applications since they both refer primarily to the process of

4. These properties of scientific universes are not well recognized. For some discussion of them see Karl R. Popper (1959: esp. Chapter 1) and Braithwaite (1953: 123).

observation. The present concern is with the implications of each for theory and with the scope of induction they permit by virtue of being sampling procedures.

SAMPLING FROM FINITE, EXISTENT POPULATIONS

The mathematical theory of sampling from finite populations does not require any assumptions about the population sampled except that it remain fixed (or change predictably) during the course of our interest. We do not, for example, have to say anything about the way in which the population came into being. Given its existence and access to it, we can draw samples from the population by a method that has been previously verified to produce sampling distributions with predictable characteristics, and on the basis of this we can make some statements about the population by inspecting the sample. In doing so, we introduce the possibility of the two types of errors with which we are familiar. But it is conceivable that we can avoid sampling altogether by making a complete inspection of the population since it is finite and existent. To see the scientific relevance of a sample from such a population, then, we must understand how to use the information obtained by the complete inspection. It is clear that we must know the scientific relevance of the population itself.

In the context of theory construction we would regard any statement of fact about the population as a proposition deducible from a theory, i.e. as a hypothesis. In addition, since that statement is treated as a hypothesis, the elements to which it refers are designated by the coordinating definitions. Thus, the finite, existent population is interpreted as a collection of instances of the kind referred to by the coordinating definitions. By this interpretation the characteristics of the population become evidence for or against a hypothesis. Without some theoretical context, however, and without suitably generalized coordinating definitions, the statement of fact about the population has as yet no scientific relevance. It is merely a descriptive statement.

If some systematic theory is at issue and if the phenomenon is properly identified, we might resort to sampling as an efficiency measure. The scope of the induction provided by the sampling is, however, limited to the finite set of elements upon which the sampling procedure was employed. There is no procedure for *selecting* a probability sample from the infinite hypothetical universe specified by the coordinating definitions. To extend the scope of the sampling induction to the hypothetical universe requires us to *postulate* explicitly a specific connection between the existent elements and the hypothetical universe.

Since the connection between existing and future cases is a matter of postulation, whether a probability sampling scheme is required in choosing cases to observe depends upon the nature of the theory involved. If the theory is deterministic, the primary concern would be with choosing cases having the appropriate values of the independent variables specified by the theory. There is no need to know the proportionate representation of cases having those values; therefore, there is no need for probability sampling at all. For example, if the hypothesis asserts that all A is B, then any subset of A is also included in B, and so it is a matter of conveneince how we chose the subset of A's to observe. If it were our interest to know how many A's there are in the existing population, then, of course, we might choose to estimate this by a probability sampling procedure, but that number itself would have to be somehow consequential for a theory.

If the theory is probabilistic, we would be interested not only in making sure that the cases exhibit the appropriate values of the variables specified by the theory, but, in addition, we would be interested in the proportionate representation of the cases having these values, because it provides sample values of the probabilities at issue. We are obliged therefore to postulate a statistical relationship between the existing cases and the hypothetical pool. By virtue of this postulate a complete inspection of the existing cases would be interpreted as a probability sample of the hypothetical population. A probability sample of the existing cases would also be appropriate.[5]

It is virtually impossible in most social research to obtain such samples, however, and in practice the usual procedure is to get one's subjects as best one may, being as much concerned about the number of subjects as by the manner in which they are selected. These "convenience" samples are unsatisfactory from both standpoints: the representation of variables and the representation of populations. Unless these samples are huge they will usually not contain enough of the right combinations of variable values to make an adequate test of the theory. In addition, since they are not usually chosen by a randomizing procedure, they cannot be considered as probability samples representative of the existing population.

If we cannot afford both kinds of representation, the present writer feels that it is better to choose our cases so as to be sure to get the right combinations of variable values. Such sampling is termed judgment sampling and is usually eschewed on the advice of statisticians, but there seems to be no practicable way to improve on this in most instances. Such a procedure would greatly facilitate the direct replication of studies by keeping the sample size within reasonable limits and by an explicit statement of conditions, apart from possible selection biases, that would make the studies comparable.

RANDOMIZED EXPERIMENTATION

A fundamental criticism of Mill's methods of experimental inquiry is that to use them to discover the causes of a phenomenon or to prove that a particular factor is the cause of a phenomenon requires the *a priori* assumption that all

5. For a discussion of this type of sampling see W. Edwards Deming (1950: 252–259).

the possible causes are known and examinable. R. A. Fisher flatly denied the possibility of an exhaustive enumeration of possibly effective factors and proposed a formal argument, based upon the physical act of randomizing treatment, to account for the effects of all relevant factors, known or unknown, that might be present in the experimental situation (Fisher, 1935a, 5th ed.).

In a Fisherian experiment the experimenter exposes the experimental units or subjects to a pattern of treatment which has been randomly selected by the experimenter from all the possible patterns of treatment to which he might have exposed them. This universe of possible patterns of treatment is, of course, hypothetical in that it exists only symbolically. Once a particular pattern has been applied to a set of subjects its effects cannot (usually) be erased so that the same subjects may be given another selected pattern of treatment. If this could be done, there would be no need to randomize treatment if the objective is to satisfy the form of Mill's canons. The formal argument by which the effects of the randomized and controlled factors are accounted for includes a specification of the form of the functional relationships between the factors and the effects, for example the often referred to "linear hypothesis."

It is also relatively a matter of course to postulate a fundamental, irreducible variation in the experimental units. That is, an intrinsic statistical proposition is postulated, but the variation due to this is not usually separable from errors of observation or other probabilistic errors in conducting the experiment. As a matter of fact, a number of combinations of deterministic and statistical propositions could lead to the same deductions, and which is to be preferred among these is precisely the task involved in constructing any theory.

We see, then, that a Fisherian experiment involves a particular theoretical conception. This conception is by no means appropriate to all phenomena, but this fact has been obscured by the great attention paid to the act of randomization.

The scope of generalization or induction that is due to the act of randomization can only be subjects like those used in the experiment and under the conditions of the experiment, whatever they were. If the experimenter cannot provide us with the identifying characteristics of like subjects and of the same conditions, we cannot know where to apply the conclusions of the experiment.[6] The scope of the induction can be increased by choosing subjects at

6. In this connection Cochran and Cox wrote, ". . . Agricultural and field experiments are often repeated both at a number of places and for a number of years. . . . In experimental programs of this type, it is usually hoped that the places and years constitute a representative sample of the population of places and years to which the results will be applied. There are obvious practical difficulties in choosing places and years that can confidently be asserted to be such a representative sample, and sometimes little effort is made to ensure that this will be so. The hard fact is that any statistical inference made from an analysis of the data will apply only to the population (if one exists) of which the experiments are a random sample. If this population is vague and unreal, the analysis is likely to be a waste of time, at least from the strictly practical point of view" (1950: 411).

random from an existent population of subjects, and by choosing environmental conditions from an existent population of these, but then the induction from sampling would apply only to those finite, existent populations. Predictions to situations other than the sampled existent ones must be made on other grounds. This will require the *explicit* specification of conditions, or the adoption of some assumption to the effect that conditions will be presented (but not by a voluntary act of selection) to the experimenter in a random fashion. This all amounts to the assertion of a complex probabilistic theory which is to be verified by the experimentation. Without such a theory the act of randomization is useless.

The Verification of Hypotheses

THE SPECIFICATION OF CONDITIONS

In a sense, a theory says everything it has to say all at once, and to make it say something about a specific instance some parts of the theory have to be pinned down. This is accomplished by specifying the values of some of the variables in the theory and thereby deriving a specific consequence (Popper, 1959: §28). The examples given to illustrate this point are usually taken from physics (the law of freely falling bodies is asserted under the condition that the falling is done in a vacuum; the boiling point of water depends upon the pressure on the water, etc.), but it is possible to find examples from social research. In general, whenever a relationship between two variables is affected by a third variable, that variable becomes a condition for the relationship.

The significance of these remarks is that any test of a theory requires the specification of conditions under which certain results are expected. The researcher must ascertain that those conditions did in fact exist when the test was made, and to do this he must make observations on the conditioning variables. Thus observations enter into the testing situation at two points: the empirical determination of the specified conditions and the empirical determination of the hypothesized results.

ERRORS OF OBSERVATION

The fact that the act of making observations may introduce errors into the testing setting serves to complicate the task of verification. A hypothesis may be mistakenly rejected or accepted because of error introduced either in the empirical certification of the conditions or of the results or both. To deal with this we must learn something about the possible errors that may obscure the phenomena or make them appear different and thereby lead us astray in evaluating our hypotheses.

These errors, if they exist, are as much phenomena to be explained as the events to which they are connected. We proceed in the explanation of errors in the same way we do in the explanation of the phenomenon of primary interest, i.e. by the formulation and verification of propositions about the errors. This does not mean that the propositions formed to explain the errors have to be the same as those formed to explain the phenomenon studied. In fact, the explanation of errors may be formally independent of the substantive theory, so that a statistical theory of errors may be employed in the testing of deterministic hypotheses and so on, for all the combinations of types of errors and types of hypotheses. The theory of errors is auxiliary to the substantive theory of primary concern.

The intrusion of errors of observation makes for peculiar difficulties in the process of verification in that the same data are used to disentangle the workings out of two distinct theories. To cope with the complexities added by the intrusion of errors of observation, we must deal with the fundamental problem of verifying hypotheses in general. We shall, therefore, restrict ourselves to the brief comments we have already made about these errors and return to the main trend of our discussion.

THE EMPIRICAL IMPLICATIONS OF DETERMINISTIC AND PROBABILISTIC HYPOTHESES

The hypotheses tested empirically make assertions about the specific situation being observed. If the hypothesis is deterministic it asserts something of this sort: All A is B. From this we infer that if the instance before us is in fact A, then it must also be B. If, having determined that the instance is in fact A, we observe that it is not also B, we have a clear, logical contradiction of the hypothesis and would, therefore, regard it as false. If the instance is in fact both A and B, we would regard this as *consistent* with the hypothesis, but not as logical proof that the hypothesis is true, because the hypothesis refers to *all* instances of this kind and we have examined only one (or some) of them. From the standpoint of formal proof, nothing is changed if we should multiply the number of instances observed (i.e. enlarge the sample); the hypothesis is only not contradicted.

The testing of statistical hypotheses is even less definitive, for it turns out that they are not even formally refutable on material grounds. A statistical hypothesis asserts something about the proportion of an infinite collection of samples (or elements) that are of each possible sample type, but nothing at all about the order in which they will exist. All that a statistical hypothesis implies about *any particular sample* is that it be one of the logically possible ones, but not which one. For example, to assert that the probability of heads for a given coin is *p* implies no more about the empirical events than that any sample of tosses of the coin will consist of some proportion (*any* proportion, from zero to one) of heads. Thus, any result, any sample proportion of heads, is formally consistent with the hypothesis and none is formally inconsistent with it. If matters were left at this, we should have to conclude that experience is irrelevant to the testing of statistical hypotheses and by virtue of this they would not be acceptable as scientific hypotheses.

It is clear that if statistical hypotheses are to be admitted as scientific propositions, they can be made empirically responsible only by a logically arbitrary decision to regard certain empirical results as consistent and confirming and to regard other empirical results as contradictory and disconfirming, even though all these empirical results are *logically* consistent with the respective hypotheses. The problem becomes one of finding a reasonable plan for determining which results shall be considered contradictory and which consistent and of justifying that plan with respect to scientific objectives.

INDUCTIVE PROBABILITY

Because of the empirical uncertainty stemming from errors of observation and intrinsic statistical formulations, and because no scientific hypothesis can be established conclusively, the entire process of verification has become regarded as probabilistic. In a broad construction, inductive probability has been construed as appropriate to the verification of any hypothesis, deterministic as well as statistical. This conception leads to statements asserting that it is probably true that all A is B (in the deterministic case) or it is probably true that the probability of B given A is *p* (in the statistical case).

To the extent that the probability-of-hypotheses conception is primarily an expression of the attitude toward induction that the verification of any hypothesis is always provisional and depends upon future observation, there seems to be nothing seriously objectionable about it. More than this is intended, however, for there is some general connotation that hypotheses are more or less likely to be refuted in the future depending upon the *degree* of verification in the past. This in itself seems reasonable enough, except that the idea has been pursued further by various schemes to quantify the degree of verification in probability terms, and these have been far from consistent with each other, much less universally acceptable.

THE TEST OF SIGNIFICANCE

The modern theory of statistical inference, however, while still a probabilistic policy of induction, has avoided the idea of the probability of a hypothesis in its program of dealing with statistical hypotheses. It concerns itself instead with the *probability of a correct decision* regarding a hypothesis tested.

The rationale of a test of significance is, in effect, this: If we adopt a rule to reject a statistical hypothesis whenever certain of the logically possible results occur and not to reject it when any of the other logically possible results occur, then if we test the hypothesis over and over again *indefinitely* and evaluate it each time by this rule, we should find that we have made the correct decision in a predictable proportion of the tests, provided the hypothesis is in fact true. There will be, then, as many rejections as there are events of the first kinds and as many failures to reject as there are events of the remaining kinds, if the events in fact occur as the hypothesis specified.

There is a fundamental difficulty with the theory of tests of significance. The probability of error is a hypothetical construct referring to a hypothetical population of tests. The empirical interpretation usually given this conception suggests to us that we can attain a desired level of accuracy in deciding the truth of a hypothesis if we test the hypothesis over and over again indefinitely. But this amounts to a commitment to test the same hypothesis forever, regardless of the previous outcomes. Thus the decision to reject or not to reject a hypothesis seems to be no more than to hang a "tag" on the hypothesis and to have no consequence for our future behavior with regard to that hypothesis.

This is patently absurd and not in fact what scientists do. They do not test the same hypothesis over and over again. Rather, a great many hypotheses will eventually be tested various finite numbers of times by a great number of researchers. The argument for tests of significance is that it is the proportions of *these* tests to which the level of significance is referred. Unfortunately this construction does not solve the problem, for it merely exchanges a sample space of tests of a single hypothesis for a sample space of tests of a large number of hypotheses. The latter sample space is a composite of all the separate spaces of the respective hypotheses. The "batting average" associated with the test of significance is still a hypothetical construct referring to an infinite number of tests.

There is another problem associated with the test of significance. The particular level of significance chosen for an investigation is not a logical consequence of the theory of statistical inference. We are free to choose whatever level seems appropriate. To cope with this arbitrariness (from the formal point of view) we are informed to consider a second kind of error, that of failing to reject the hypothesis tested when it is in fact false. That is, we are informed to choose a level of significance that balances the probability of the first kind of error against the probability of the second kind of error by a consideration of the costs attendant upon them and the costs of more sensitive research designs.

In many practical applications, for example, the control of quality in manufacturing, the various costs can be expressed in terms of money or time or some such firm criterion. But in scientific research the calculation of costs is more complicated. To be sure we can compute the costs of taking various kinds and sizes of samples, but how are we to compute the costs of making an incorrect decision about a scientific hypothesis? Which should we fear most, rejecting a true hypothesis or failing to reject a false one and thereby rejecting a true alternative? There is as yet no firm criterion of costs and in view of this the levels of significance so often used in sociological research (the 5 percent and the 1 percent levels) can hardly be the outcome of objective calculations of costs.

We must be explicit in disclaiming a possible interpretation of our criticism of statistical inference as implying that we abandon probability statements altogether. The fact is that the success of other sciences in using intrinsic and auxiliary probability statements and the promising results of their application in learning theory and small groups research argue against any categorical exclusion of statistical hypotheses in social science.

We can see no alternative to considering the probability of occurrence of certain consequences assuming that a hypothesis tested is true as *relevant* to the decision to regard that hypothesis as true in fact. Still this does not lead inevitably to the imperative criteria expressed in the test of significance. The precision and empirical concreteness often associated with the test of significance are illusory and it would be a serious error to predicate our actions toward hypotheses on the test of significance as if it were a reliable arbiter of truth.

THE CRITERION OF SYSTEMATIC IMPORT[7]

Much of social research is informed by the strategy of searching for reliable empirical generalizations which, because of their empirical constancy, must be included as propositions within any theory that might later be formed to deal with the phenomena. In this strategy the test of significance is often viewed as a means of determining which generalizations have that compelling empirical property. Thus we encounter statements to the effect that the test of significance tells us whether in a particular study there is anything to explain, i.e. whether a relationship between variables "exists" and so whether there is anything to theorize about. In view of the interpretive and pragmatic ambiguities involved in the test of significance this seems to be a very risky policy.

If we take as our objective the construction and verification of systematic theory, then the inductive policy we adopt should be measured also against this criterion. We suggest the following considerations as an essential part of the inductive policy that sociologists might adopt in pursuit of this objective.

In the development of systematic theory, the purpose of research is not primarily to determine the empirical adequacy of a particular hypothesis. Its purpose is to test the coordinated formal system that produced the hypothesis as a theorem. The question of testing a hypothesis thus would not occur until that hypothesis had been set in an explicit deductive context. In this framework the process of testing a hypothesis is not an infinitely repeated act whose

7. This discussion is a highly selected and simplified version of a complex point of view. Braithwaite and Popper offer comprehensive statements of the position. For an earlier and perhaps most understandable statement, see Homer H. Dubs (1930).

outcome is the accumulation of "verdicts," as is implied in the literal inter-pretation of the test of significance. The empirical truth of the particular hypotheses the researcher checks upon is valuable chiefly for its instrumental use in determining what he should do with the deductive system by which he produced the hypothesis. It is this conceptual relevance that we refer to as the systematic import of an empirical result.

It is clear that in this formulation, the process of induction intimately in-volves the verification of hypotheses, but it would be too simple to regard it as solely this. The history of science clearly indicates that theories do not emerge full grown from the brow of Jove and present themselves in their entirety to be tested, but rather that the construction of verified theory is a crescive process. Often research is undertaken not to test a theory, in the sense of trying to reject it, but to extend it, to determine its scope of applica-bility, or to enlarge this scope by the introduction of modifications in the theory (Dubs, 1930: Chapter 7).

Thus to test a theory in the present framework means that there must be some conception of how the theory is to be treated as a result of the test. The alternative courses of action dependent upon the results of empirical investi-gation are seen as alternative steps in the process of theory construction and not simply the acceptance or rejection of a particular formulation. This implies that the contrasted alternatives in a test situation be constructive ones, for if induction were merely the elimination of alternatives we would establish any hypothesis by testing it against absurd or surely false formulations.

To the extent that we are comparing systems of hypotheses, the research situation can be made as finely structured as need be to decide between the systems. This is because many consequences can be deduced from each system and the comparison is made over all of these. We would never be in a position of deciding between alternatives on the basis of a single hypothesis.

These considerations would seem to be essential aspects of the process of induction, but they could only be a part of the process. What more is required has only partly been determined. For example, although a great deal has been written about the nature of theory and its role in induction, astonishingly little attention has been given the process of theory development itself.[8]

Summary and Conclusions

In attempting to clarify the role of probability in sociological research we have been led into a discussion of the nature of scientific theory and induction. We have tried to articulate the principle that since scientific induction is accom-plished through the construction and verification of deductive theories, the

8. For an important contribution to the understanding of the development of theory, see Berger *et al.* (1962).

primary concern of the social scientist ought to be the development of such theories.

We distinguish three major uses of probability in science: in intrinsic statistical hypotheses; in auxiliary statistical hypotheses, including the theory of errors of observation, sampling from finite populations, and randomized experimentation; and thirdly in inductive probability policies, particularly the theory of tests of significance.

We have tried to show that the hypothetical character of the risk probabilities associated with the level of significance and the pragmatic ambiguities of the rationale for choosing any particular level of significance seriously undermine its value in the evaluation of statistical hypotheses.

It is our belief that the great reliance placed by many sociologists on tests of significance is chiefly an attempt to provide scientific legitimacy to empirical research without adequate theoretical significance. Lest we be misunderstood we wish to emphasize the point that the construction of deductive theory is *developmental*, involving an *accumulation* of research experience. But it also involves formal reasoning, and the task of the scientist is to expose his thinking to a formal reconstruction, to formalize it as well as he can, so that it can contribute to the construction of a deductive system of some size and scope.

17. The Sacredness of .05:
A Note Concerning the Uses of Statistical Levels of Significance in Social Science

JAMES K. SKIPPER, JR.,
ANTHONY L. GUENTHER, and
GILBERT NASS

DECISIONS REGARDING the uses of statistical levels of significance have typically been rendered by social scientists with specialized training and considerable expertise. However, a strong case can be made for more involvement in such issues by researchers whose activities include tests of significance in an applied context. The issue of whether inference should be made at all has been rather thoroughly debated (Selvin, 1957 [Chapter 9]; Gold, 1958 [Chapter 10]; and Kish, 1959 [Chapter 15]),[1] but there are indications that the use of present "recommended" levels of significance is due for reassessment. This note attempts to state the problem, explore the issues involved, and suggest an alternative to current policy.

The choice of a statistical level of significance, that is, establishing the probability of rejecting the null hypothesis when in fact it is true, apparently demands little psychic energy on the part of researchers. Casual examination of the literature discloses that the common, arbitrary, and virtually sacred levels of .05, .01, and .001 are almost universally selected regardless of the nature and type of problem. Of these three, .05 is perhaps most sacred.

Reprinted from *The American Sociologist*, 2 (February, 1967), 16–18, by permission of the authors and the publisher. Copyright © 1967 by the American Sociological Association.

1. For a discussion of the more general question of the current usefulness of mathematics and statistics in social science, the recent comments by Parsons (1964), Sibley (1965), Selvin (1965), and Etzioni (1965) are of special interest.

Although statistically-inclined methodologists do not often explicitly recommend use of these arbitrary levels, their positions frequently suggest that these levels are *conventional*. Prominent textbooks bear this out:

> But the agony of making a fresh selection of the level of significance in each instance would be a painful business. Hence, the statistical worker finds welcome relief in the .05 *convention*, which prescribes that the null hypothesis is automatically to be rejected whenever the probability of being wrong in that decision is 5 percent or less (Mueller and Schuessler, 1961: 395 [Skipper *et al.* italics]).

> By locating the observed findings in the theoretical distribution of all possible findings, the investigator determines the probability of obtaining the finding by chance if the null hypothesis is actually correct. If this probability turns out to be small enough (less than a predetermined level such as .05 or .01), he then decides to reject the null hypothesis (Riley, 1963: 638).

> In the social sciences, it is more or less *conventional* to reject the null hypothesis when the statistical analysis indicates that the observed difference would not occur more than 5 times out of 100 by chance alone (Selltiz *et al.*, 1961: 418 [Skipper *et al.* italics]).

> A *convention* frequently followed is to state the result *significant* if the hypothesis is rejected with $\alpha = .05$ and *highly significant* if it is rejected with $\alpha = .01$ (Dixon and Massey, 1957: 91 [Skipper *et al.* italics]).

The current obsession with .05, it would seem, has the consequence of differentiating significant research findings and those best forgotten, published studies from unpublished ones, and renewal of grants from termination. It would not be difficult to document the joy experienced by a social scientist when his F ratio or t value yields significance at .05, nor his horror when the table reads "only" .10 or .06. One comes to internalize the difference between .05 and .06 as "right" *vs.* "wrong," "creditable" *vs.* "embarrassing," "success" *vs.* "failure." Tradition notwithstanding, there seems to be little justifiable reason for such a state of affairs.[2] We find it hard to believe that social scientists simply wish to avoid the inconvenience of selecting significance level as one of the parameters of the problem under investigation.

Most textbooks in research methods and statistics recommend that the levels of significance associated with tests on the data be set in advance of actual data collection. In part, this decision is one of minimizing the probability of committing Type I or Type II error. (Type I error occurs when one rejects the null hypothesis which in fact is true. Type II error is committed when one fails to reject the null hypothesis which in fact is false.) Since the two types of errors are inversely related to each other, it is impossible to minimize both of them at the same time without increasing the sample size.

2. The use of the arbitrary .05 level of significance seems to have originated with the statistician R. A. Fisher, who used this level with experimental situations in agriculture and biology. Subsequently it was adopted by social scientists. See R. A. Fisher (1935a), and the discussion by Quinn McNemar (1955: 64).

Therefore, it is the *nature of the problem under study which ought to dictate which type of error is to be minimized.*

Some social scientists feel that too much emphasis has been placed upon the level of significance of a test at the expense of *power* considerations. Representative of these is the managing editor of *Biometrica* who comments:

> The frequent use of the .05 and .01 levels of significance is a matter of convention having little scientific or logical basis. When the power of tests is likely to be low under these levels of significance, and when Type I and Type II errors are of approximately equal importance, the .30 and .20 levels of significance may be more appropriate than the .05 and .01 levels (Winer, 1962: 13).

It is appropriate to consider at this point an example illustrating the dependence of optimal alpha level upon the problem investigated. A social scientist may face the problem of whether to recommend a group of college seniors having high I.Q.'s but poor academic indexes, for graduate school. Controlling other relevant variables, he sets up H_0: There will be no difference in rate of achievement in graduate school between individuals with (a) high I.Q.'s but low grades, and (b) high I.Q.'s and outstanding grades. A decision must be made regarding the circumstances under which H_0 may be rejected. If a low level of alpha, say, .001 is established, the probability of unfairly discriminating in favor of the good students is minimized. On the other hand, $\alpha = .001$ implies a much greater probability of sending students to graduate school who could not do the work than would be the case with, e.g., $\alpha = .25$.

In our opinion, there is no "right" or "wrong" level here—the decision must be made in full consideration of parameters inherent in the problem itself. It is doubtful that setting *a priori* levels of .05, .01, or what have you settles the matter.

The issue becomes clearer when we compare the above problem with another. A decision may be needed, for example, whether to recommend that military personnel in combat zones are more effective when fighting alone or when accompanied by a "buddy." Again, controlling other relevant variables, H_0 states: There will be no difference in the combat effectiveness of military personnel fighting alone, as opposed to fighting with a "buddy." What alpha level is optimal for this hypothesis? In contrast with the first problem, a higher level (say, .20) would be justified, that is, we feel it is more desirable to risk rejection of the null hypothesis when it is true. In terms of the example, it is more desirable to place combat personnel with buddies when in fact they are no more effective fighters this way.

In the first example (college students with poor grades), it seems best to minimize Type I error, while in the second case (combat effectiveness) it is more important to minimize Type II error. Whether one agrees with the values we have employed in interpreting these examples or not is of little importance. The point is that blind adherence to the .05 level denies any consideration of alternative strategies, and it is a serious impediment to the interpretation of data.

There are those, of course, who may feel that the cases we have cited are essentially moot problems, since they believe social scientists *should* deal only in "pure" and not "applied" research. This view contends that one should design, execute, and report upon his research, but not serve to implement his findings. Only in his role as "moral man" should the scientist-as-layman take responsibility for practical decisions affecting the community.[3] Where one stands on the "pure or applied" issue is also of little consequence for this report. Either view implies a responsible consideration of the various factors underlying decisions of significance levels.

Even if one accepts the pure research orientation there are still important reasons for choosing optimal alpha levels. Blalock, for example, points out that usually the only real decision facing a social scientist is whether to publish or suppress his findings (Blalock, 1960: 135).[4] In this circumstance he suggests a rule of thumb: "The researcher should lean over backwards to prove himself wrong or to obtain results that he usually does not want to obtain" (Blalock, 1960: 125). He warns, however, that this too may involve the decision to minimize chances of making Type I or Type II error depending on the type of problem.[5] It is inadvisable, therefore, in any type of problem, to state *a priori* the "most appropriate" level of significance. The risk of so doing may be evinced by the paucity of criteria or guidelines for establishing levels of significance for different classes of problems.

Blalock's remarks are most appropriate for research which is designed to test hypotheses. There is, of course, much social science research which seeks to *develop* hypotheses rather than attempt verification. The rubrics pilot study and exploratory research are well known, and frequently provide for the collection and analysis of data not "explained" in advance by an explicit theoretical framework. Past experience has proven this to be an effective strategy and often such serendipitous findings outweigh the original purpose of the research (see Merton, 1957: 103–108). In cases where the goal is generation rather than testing hypotheses, it would seem advisable to tailor level of significance to the open-ended character of research design. Again, an arbitrary .05 level may not be optimal.

With the increasing involvement of social scientists in the very structures

3. A view most elaborately put forth by George Lundberg (1947). An antithetical position, of course, was set forth by Robert S. Lynd (1939).

4. Pragmatically, whether findings are acceptable may depend upon achieving significance at .05 or better.

5. McNemar suggests: "If the findings of a study are to be used as the basis either for theory and further hypotheses or for social action, it does not seem unreasonable to require a higher level of significance than the .05 level" (McNemar, 1955: 65).

and processes they study, e.g. complex organizations, ethnic relations, and political activity, it is expected that published research findings will be disseminated to audiences of laymen untrained in the interpretation of data. When the researcher places an asterisk (or maybe even *two* asterisks) after his obtained F, Chi-square, or t value, and thereby decrees *significance*, this may constitute the basis for social action by statistically unsophisticated decision makers. On the other hand, failure to achieve significance at the prearranged level may lead to immediate dismissal of the hypothesized relationship.[6]

As social scientists increasingly publish research findings in scholarly journals having a wider circulation than the scientific community of their own particular discipline, this problem becomes of increasing importance. Journals of practicing professions and occupations often reach an audience who is relatively unsophisticated in scientific methodology and techniques of statistical analysis. For example, in the area of medical sociology a study concerning interaction and communication in a hospital, say between nurses and patients, published in a sociological journal might reach anywhere from 2,000–12,000 subscribers, mostly sociologists and other social scientists. The same findings published in the *American Journal of Nursing* would come to the attention of more than 160,000 registered nurses, a portion of whom might try to make practical use of findings which are termed "significant." It is not at all improbable that this may be characteristic of many other areas of specialization.

Finally, we would like to suggest that a more suitable procedure for setting and reporting levels of significance might be: (1) For the social scientist to reflect upon the *arbitrary* nature of .05, .01, and .001 levels now in common use, and to recognize that selection of one of these is not a panacea for the interpretation of evidence. Different classes of research problems (or components thereof) may require different levels of alpha. (2) For the social scientist to recall that even well educated lay groups are inexpert in the interpretation of statistical significance. The demand by such groups upon professional researchers must be met by a response that is ethical and communicative. (3) For social scientists reporting research findings (especially in journals indigenous to their discipline) to do away with arbitrary levels of significance and the calling of one test result "significant" and another "not significant."[7] We recommend a procedure whereby the actual level of significance associated with each research finding be stated (Siegel, 1956: 9). In other words, if a

---

6. Yates points out that this may be equally true of the scientists themselves: ". . . Scientific workers have often regarded the execution of a test of significance on an experiment as the ultimate objective. Results are significant or not significant and this is the end of it" (Yates, 1951: 33).

7. Kish has recommended that: "*Significance* should stand for meaning and refer to substantive matter" (1957: 336 [Chapter 15: 139]). Kish's *Survey Sampling* (1965) treats the importance of significance under a variety of sampling designs.

finding yields significance at the .30 level, or the .09, or whatever, report it *at that level* together with other essential information (e.g. degrees of freedom) and let the reader determine whether for his purposes it has any practical significance. A statement of opinion regarding support or non-support of the data for the relevant hypothesis may be made by the researcher.

We believe that this type of research reporting would have several advantages over present procedure: (1) Non-social scientists using the professional literature would be spared the judgments of "significant" and "nonsignificant" as the basis for deciding whether findings have important implications in their applied situations. In a sense this would force laymen to acquire some expertise to offset an absence of "spoon-fed" decisions regarding data analysis. (2) Much the same function would be performed for individual social scientists whose time for critical thinking about verification of research propositions is frequently negligible. (3) It would tend to intensify the search for appropriate significance levels, given attributes of the research problem. If, in contrast with present policy, it were conventional that editorial readers for professional journals routinely asked: "What justification is there for this level of significance?" authors might be less likely to indiscriminately select an alpha level from the field of popular eligibles.

In summary, there is a need for social scientists to choose levels of significance with full awareness of the implications of Type I and Type II error for the problem under investigation. The current use of arbitrary levels of alpha, while appropriate for some designs, detracts from interpretive power in others. Moreover, the tendency to dichotomy resulting from judging some results "significant" and others "nonsignificant" can be misleading both to professional and lay audiences. It is suggested that a more rational approach might be to report the actual level of significance, placing the burden of interpretive skill upon the reader. Such a policy would also encourage social scientists to give higher priority to selecting appropriate levels of significance for a given problem.

18. Common
Misinterpretations of
Significance Levels
in Sociological Journals

THOMAS J. DUGGAN and
CHARLES W. DEAN

PERIODICALLY, THE USES and misuses of probability statistics in social and behavioral science research have been reviewed. For instance, Lewis and Burke (1949) pointed to several misuses of the Chi-square test and an article by Selvin (1957 [Chapter 9]) stimulated discussion on the general question of using statistics in social surveys. Most recently, Skipper, Guenther, and Nass (1967 [Chapter 17]) reviewed the discussion of substantive interpretation associated with significant levels. While such discussions have served to clarify some of the technical requirements and have corrected some of the misunderstandings often associated with the use of statistical tests, one crucial matter has received relatively little attention. This concerns the substantive interpretation of significance tests and the consequences of such interpretations.

The frequently used Chi-square test and the interpretations given to data analyzed by this statistic will serve to illustrate the problem. This statistic can be used to test goodness of fit or independence although it is the latter which is more frequently used in reporting research. Since this is a test of the independence of variables, significant values of Chi-square are often taken to indicate a dependence or relationship between variables. In interpreting such relationships, there are two serious problems which are often overlooked. The first concerns the strength of relationship and the second, the form of relationship.

Reprinted from *The American Sociologist*, 3 (February, 1968), 45–46, by permission of the authors and the publisher. Copyright © 1968 by the American Sociological Association.

Strength of Relationship

As to the first problem, if the Chi-square is significant at the chosen level, then the investigator routinely rejects the null hypothesis of independence and tentatively accepts the alternate hypothesis that the variables are dependent or are related. Regardless of how low the probability associated with the obtained value of Chi-square, nothing can be inferred about the strength or degree of that relationship. However, in practice, this point is often overlooked.

Consider Table 18.1, which was reported in a major sociological journal.[1] According to the authors' intepretation, the significance level of the Chi-square test was so high that if variables X and Y were not clearly separate measures, "We would suspect the relationship to be tautological." Since the authors failed to report the degree of association, Goodman and Kruskals'

Table 18.1.

	Variable X			
Variable Y	*Very high*	*High*	*Low*	*Very low*
High	5	24	17	9
Moderate	12	18	9	3
Low	19	22	16	2

$\chi^2 = 14.0$, $P < .05$, $G = -.30$.

gamma was computed. In this instance, gamma equalled $-.30$ which suggests a relationship which is far from tautological. The difference in the interpretation based on Chi-square and gamma should be noted and emphasized. In contrast to the above data, consider Table 18.2, which was presented in the same article.

Table 18.2.

	Variable X		
Variable Y	*High*	*Moderate*	*Low*
High	24	18	10
Moderate	19	12	7
Low	19	18	21

$\chi^2 = 6.2$, $P < .20$, $G = .22$.

1. Since the purpose of these tables is to illustrate peculiarities in the use of Chi-square rather than to criticize individual research, no tables will identify author, journal or original variables actually treated. However, all tables were reported in refereed sociological journals within a year prior to the writing of this piece.

In this case, the probability is such that the sociologist normally would accept the null hypothesis of independence. However, gamma was computed for these data and G equalled .22. Here, data nonsignificant according to Chi-square has a relationship only slightly lower than that of the preceding example, where it was concluded that the variables were highly related. To further demonstrate the need for sensitivity to the difference between significance level and strength of association, Table 18.3 was constructed.[2]

Table 18.3. A comparison of level of significance and strength of relationship (n = 45 articles).

Level of significance	Strength of relationship								
	.00 to .09	.10 to .19	.20 to .29	.30 to .39	.40 to .49	.50 to .59	.60 to .69	.70 to .79	.80 to .89
.001	3	2	2	8	..	1	..	2	3
.01	1	..	6	..	2	..	..	..	..
.05	4	3	..	..	..	..	..	..	..
.10	1	..	1	..	..	..	..	..	..
.20	1	1	1	..	..	..	..	..	..
.30+	..	3	..	..	..	..	..	..	..

Table 18.3 clearly demonstrates that while Chi-square, properly used, may be sensitive to the dependence of variables, after dependence is shown the usefulness of this statistic is exhausted. As the data of Table 18.3 show, significance at the .001 level could mean that the relationship between the variables could be less than .09 or more than .80. At the .001 level, the distribution of the strength of association appears to approach randomness. While these data do show that nonsignificance, or significance at or about the arbitrary .05 level, will usually result in a relationship which is consistently weaker, the relationship is as likely to be above .10 at the .30 level as at the .05 level. Still, all three tables reporting significance above the .30 level had gammas ranging between .10 and .19. In contrast, of the seven tables reporting significance at .05, only three had gammas in the .10 to .19 range, while four tables had gammas below .10. If more nonsignificant tables were reported in the journals, the distribution of measures of association would probably be even broader. Generally, the lower the significance level, the greater the probability of a low relationship, but this cannot be assumed. These data

2. These data were derived from major sociological journals published between 1955 and 1965 in a systematic search for three by three tables, both variables ordinal.

emphatically demonstrate that a measure of strength of association is necessary before statements about strength of relationship can be made.

These data illustrate the serious problem in interpreting significance levels of the Chi-square test of independence and indicate the need for a reminder that statistical significance is not equatable with practical significance. A significant Chi-square value, at best, permits one to say that *probably* there is some dependence between variables in the population, but the extent of dependence may be virtually zero *regardless of the significance level*. The consequences for understanding the phenomena under investigation and for the construction of theories require constant awareness of the limited interpretations which can be given to statistical significance.

Form of Relationship

The second problem refers to the form of relationship between variables. In using tables three by three or larger, users of the Chi-square are often prone to think and interpret results in terms of linear relationships, but the contingency table and the Chi-square statistic are not sensitive to and provide no basis for assuming the existence of this form of relationship.

The data of Table 18.4, also presented in a major sociological journal, illustrate the error in interpreting the direction of the relationship in linear terms. The author stated that the data of this table confirmed the hypothesis that the greater the degree of variable X, the greater the degree of variable Y.

Table 18.4.

| Variable Y | Variable X | | |
	Frequent	Occasional	Infrequent
High	3	9	6
Moderate	14	30	12
Low	17	12	6

$\chi^2 (4df) = 8.51; P < .05.$

An inspection of the table reveals that this is not the case. As Table 18.4 indicates, the largest number of subjects ranking in the "frequent" category of variable X rank in the low category of variable Y. However, the largest number of subjects in the "occasional" and "infrequent" categories of variable X rank in the "moderate" category of variable Y. Only those ranking "low" on variable Y are distributed in the expected pattern.

Another team of authors in an edition of another sociological journal presented data similar to those of Table 18.4 to test the hypothesis that the greater the degree of variable A, the higher the degree of variable B. They computed Chi-square values for their data table and stated, "The relationship shown is significant beyond the .001 level; therefore, the hypothesis is accepted." Throughout the article, the authors made similar statements from similar data about linear relationships.

While the above authors did not attempt to disguise their acceptance of linearity, frequently, other researchers state a linear hypothesis, present the data table, accept the hypothesis on the basis of the Chi-square probability, and then discuss only those proportions of the table which fit the linear model. This more subtle but equally erroneous procedure appears frequently in the sociological literature.

A linear relationship exists only if the pattern of concentration of subjects lies along a diagonal of the table. If this is not the case, the relationship cannot be interpreted as a linear one. If the phenomenon of possible nonlinearity is not taken into account or if the implication of linearity is made in interpreting Chi-square, serious consequences again arise in interpreting data and in developing explanatory theories. This problem can be averted by inspecting the data table, outlining the pattern of concentration, and describing the pattern.

Conclusion

To avoid these errors of confusing significance with strength of association and of misinterpreting form of relationship, two elementary safeguards can be exercised in reporting results. One is routinely to compute and report a measure of degree of association in addition to the statistical test whenever this is possible. The second safeguard is the introduction of care and caution in the verbal interpretation of data tables and the inferred association of variables.

In this day when computer technology is so drastically improving the analytical tools of the sociologist, it seems paradoxical that there is a need to remind researchers of such basic rules of interpretation.

19. Criteria for Selecting a Significance Level:

A Note on the Sacredness of .05

SANFORD LABOVITZ

Sociologists have now been formally warned that not only is .05 not sacred, but the selection of a significance level is a complex process. One major suggestion by Skipper *et al.* (1967 [Chapter 17]) is that the researcher should no longer choose a standard level, but report the obtained level, e.g. .40 or .003. On the one hand, this suggestion seems to involve less thinking than choosing the conventional .05 or .01. At least here there are two or three conventions on which to base a selection. On the other hand, reporting the obtained level and letting the reader figure out the significance is not entirely a new suggestion. Some authors already have been reporting the obtained level, and they do not appear to give any more consideration to the dynamics of a significance level than their more conventional colleagues. However, the article by Skipper and his associates may prove to be important, if it can sensitize sociologists to some of the serious problems involved with tests of significance.

The authors give three suggestions pertaining to significance levels and how to report them: (1) think and reflect on the arbitrary nature of conventional levels of significance, (2) report the actual level obtained, and (3) regardless of the level obtained, give an opinion on whether or not it supports the hypothesis. If these suggestions are followed extensively in research reporting, some

Reprinted from *The American Sociologist*, 3, 3 (August, 1968), 200–222, by permission of the author and the publisher. Copyright © 1968 by the American Sociological Association.

of the problems of interpreting significance tests should diminish. However, the authors do not adequately spell out the guide lines (criteria) leading to the selection of a significance level.[1] The following section specifies eleven criteria applicable to this problem.

Some Criteria to Consider in Choosing a Significance Level

The following is neither an exhaustive nor all inclusive classification scheme of criteria on which to select a significance level. However, it appears to represent the major dimensions that should be either explicitly or implicitly considered by researchers. There is no attempt to integrate the entire list, nor to rank order the criteria in terms of importance. To do either seems premature. Note that none of the criteria should be considered in isolation—each should constitute just one of several guide lines in selecting a significance level.

Eleven more or less independent criteria are delimited.

1. *Practical consequences.* The practicality of the problem refers to the gravity of available kinds of error on the basis of value orientations. Testing whether prefrontal lobotomy or sedation is the better method for curing patients is a grave choice if we value vitality and recognize the long lasting and extreme effects of lobotomy. In this example, a small error rate (level of significance) of perhaps .001 or less would be chosen so that it would be extremely difficult to reject the null hypothesis of no difference and accept lobotomy over sedation. On the other hand, if we were testing the difference between two types of sedation, perhaps a larger error rate would be chosen (.05), if there were few drastic or long range effects for either one.

2. *Plausibility of alternatives.* A test of hypothesis should not be considered in isolation. Unless the inquiry is in an area where virtually nothing is known, the available rationales and empirical evidence (from other studies) should be considered in interpreting a significance test. Suppose the results are directly opposed to existing theory and empirical evidence, or even "common sense." That is, the evidence against the conclusion is large, and there is no theoretical or empirical support for the finding. Under these conditions, it would probably be best to choose a small error rate (.01 or .001), because in all the

---

1. Besides the problem of criteria for significance levels, there are three general points that are not handled adequately by the authors. First, the authors do not place the arbitrariness of a significance level within the perspective of the general state of theory, knowledge, and evidence. Instead, they emphasize the single test. Actually, a single test, *whether or not* it reaches a predetermined significance level, leads to no major decision. Few, if any, researchers would accept or reject any statement on the basis of a single test. Second, Skipper *et al.* ignore cross-classification versus tests of significance arguments. While this is not their concern, their whole article is essentially meaningless if tests of significance are not applicable. Finally, the authors emphasize applied research and the lay, statistically unsophisticated audience. If statistically adroit colleagues are the prospective audience, perhaps their suggestions are less useful.

studies opposing the conclusion we are bound to find a few negative results on the basis of chance alone. We would hesitate to so easily reject the null hypothesis, when rejection is such a deviant result. On the other hand, if the evidence supports the conclusion, a larger significance level would be more appropriate, since now we are usually more willing to reject the null hypothesis of no difference.

3. *Power of the test—sample size.* The power of a test varies directly with sample size, that is, as N increases there is a greater probability of correctly rejecting the null hypothesis (in comparison to a specific alternative hypothesis). Moreover, the standard error varies inversely with sample size. Consequently, with a large N a small difference is likely to be statistically significant, while with a small N even large differences may not reach the predetermined level. Therefore, small error rates (.01 or .001) should usually accompany large N's and large error rates (.10 or .05) should be used for small N's.

4. *Power of the test—size of true difference.* The power of a test not only varies with sample size (and level of significance), but also with the size of the "true" difference, e.g. the magnitude of the difference between means. Therefore when the true difference is large, the probability of correctly rejecting the null hypothesis is also large, except if the sample size is small enough to offset this condition. A small error rate probably should be used when the difference is expected to be substantial. This conclusion is based on the rationale that if a large difference is expected and only a small difference is obtained, the null hypothesis of no difference should not be rejected.

5. *Type I vs. Type II error.* As pointed out by Skipper *et al.*, most textbooks emphasize the criterion of minimizing the probability of errors described as Type I (rejecting a true null) and Type II (failing to reject a false null). These errors, to some extent, vary inversely with one another. Consequently, minimizing one type of error tends to increase the other. To illustrate, a .05 significance level yields fewer Type II errors than the .01.

To digress on tests of hypotheses, a large significance level (.05) makes it easier to reject the null hypothesis and accept the original hypothesis set up by the researcher. The original hypothesis usually states a difference (and perhaps specifies the direction), while the null usually is stated in terms of no difference. Therefore, a large error rate increases the probability of accepting the researcher's hypothesis, but it also increases the probability of doing so incorrectly (Type I error). However, with a large error rate, there is a low probability that the original hypothesis is both correct and we failed to accept it. If we feel that the original hypothesis should not be accepted until a high level of certainty is reached, then many true original hypotheses are likely to be lying around that are not accepted (Type II error). Which error is best? Aside from our personal feelings on how a science should develop, at this point, the other alternatives listed should help solve the apparent dilemma.

6. *Convention.* Skipper *et al.* strongly argue against using conventional levels of significance such as .05 and .01. For the most part their conclusion seems justified, and the other criteria listed further indicate the limitations of using a conventional level. It is listed as a separate criterion primarily because (1) these conventions are used in sociology, and (2) they may be positively evaluated as yielding some consistency among research results. If most results are applied to a similar standard, readers have some idea of the comparability of results from one study to another. However, the disadvantages of a conventional level (such as not considering available evidence or the nature of the problem) well outweigh this factor. As a final remark, the selection of a conventional level may not rest on any sound rationale, but on such incidental factors as the particular field of social science, where an individual received his degree, or the journal under consideration.

7. *Degree of control in design.* It is well known that R. A. Fisher generally selected the .05 level in his agricultural experiments. These experiments were based on complex (e.g. latin square or factorial) designs that offered a high degree of control over the effects of extraneous factors. The effects of "other factors" were handled by randomizing plots of ground, rows and columns of products, etc. Under such highly controlled conditions Fisher seemed justified in using the larger error rate of .05 instead of .01 or lower. If other factors are controlled, the results of the experiment are likely to be due to the experimental variable or chance differences and not due to extraneous factors. Stated otherwise, a large amount of control in an experiment reduces alternative interpretations so that a larger level of significance can be tolerated. In designs of low control, perhaps a more stringent error rate should be selected (.01) since the alternative to chance differences could be due to extraneous factors as well as to the independent variable. Consequently, under low control conditions we should make it more difficult to reject the null hypotheses of no difference.

8. *Robustness of test.* Robustness is the ability of a statistical test to maintain its logically deduced conclusion when one or more assumptions have been violated. For example, Student's *t* and analysis of variance have been demonstrated to be robust under the conditions of nonnormality and heterogeneity. However, under these conditions the actual .02 level of significance may be met at the .01 level and the .10 at the .05. Consequently, depending on the statistical test in question, when the data do not meet all the assumptions, a small error rate should be chosen and interpreted as a larger one, e.g. .01 is interpreted as .02 or .05. On the other hand, if the data reasonably meet the assumptions, then a large error rate can be used with confidence.

9. *One-tail vs. two-tail tests.* As stated in most introductory statistics books, it is easier to reject the null hypothesis in a directional (one-tail test) as

opposed to a nondirectional (two-tail test) hypothesis. The z-score equivalents for a one-tail test are lower than those for two-tail (e.g. 1.65 as compared to 1.96 at the .05 level). It is reasoned that knowledge of the direction of the hypothesis should give the researcher the advantage of more easily rejecting the null and accepting the original hypothesis.

However, the notion of one-tail vs. two-tail is largely a myth, because it is based on the rationale that we either have absolutely no idea of the direction of the hypothesis or we have absolute knowledge of the direction. Either extreme alternative is an unlikely occurrence. It is most probable that we have some idea of the direction of the hypothesis, but there is a small to large amount of uncertainty in our reasoning. Consequently, we should neither accept the z-score equivalent of the one-tail or two-tail test, e.g. 1.96 or 1.65, but an intermediate score between the two values. At the .05 level if we are largely certain of the direction (that is, it is supported by previous research or sound rationale), then we should select a z-score closer to 1.65. If, on the other hand, there is a large degree of uncertainty, a z-score nearer to 1.96 would be more appropriate. This is the equivalent of saying that we should choose a larger or smaller error rate depending upon our degree of confidence in the direction of our hypothesis.

10. *Confidence interval.* A confidence interval not only provides a probability band containing some statistical measure or difference, but actually provides tests of hypotheses. Therefore, the difference between a test and an interval is not clearcut. The importance of considering the confidence interval as a criterion in selecting a level of significance depends on whether or not the problem requires a small or large interval. For a smaller interval a larger error rate is necessary (.05), while for larger intervals (in which there is more confidence that they contain the parameters) a smaller error rate is necessary (.01).

11. *Testing vs. developing hypotheses.* If testing a well reasoned and developed hypothesis that will distinguish between two theories, it seems logical to select a small level of significance. This is based on the notion that we want to be fairly sure if one theory is to be selected over another. On the other hand, if we are just exploring a set of interrelations for the purpose of developing hypotheses to be tested in another study, a larger error rate will tend to yield more hypotheses—any of which may be subsequently validated. Therefore, in this exploration stage perhaps the .10 or .20 level would be sufficient.

Caution should be used not to fall into the trap of thinking that the few "significant" relations out of many possible ones have truly reached the designated level. Out of twenty interrelations we are likely to find one significant at the .05 level on the basis of chance alone. However, we do not fall into this trap if the "significant" relations are subsequently tested.

Conclusion

In conclusion, Skipper *et al.* have performed a definite service to sociology if more of us probe deeper into the rationales behind significance levels and stop using an absolute standard as proof of a hypothesis. To buttress this position, eleven criteria are presented that hopefully will aid researchers in selecting an appropriate level. These criteria should not be viewed as definitive in any sense, and some undoubtedly are more important than others. I welcome any response on other possible criteria, and any thoughts on the evaluation of those presented above.

20. Statistical Tests and Substantive Significance

DAVID GOLD

THE CONTROVERSY over tests of significance seems to have reached a curious plateau since Selvin (1957 [Chapter 9]) first sparked the issue that had been building up for a number of years among social scientists. Currently, there are those who eschew the use of tests of significance; a footnote reference to Selvin has become standard,[1] despite the fact that his major argument for the inapplicability of tests, the existence of what he has referred to as "correlated biases" in nonexperimental research, has been shown by McGinnis (1958 [Chapter 14]) and Kish (1959 [Chapter 15]), among others, to be poorly explicated and essentially fallacious. There are also those, whose number is unquestionably greater than the former, who go on using tests of significance in more or less slavish manner, ignoring the warnings that have been posted by both sides to the controversy. It is noteworthy that footnote legitimation is rarely deemed necessary for the use of a test of significance, though the writer will often tell us where a discussion of the particular test he has used can be found.

Selvin argued that randomization of units of observation is an essential condition for the legitimate application of tests of significance, and McGinnis

Reprinted from *The American Sociologist*, 4, 1 (February, 1969), 42–46, by permission of the author and the publisher. Copyright © 1969 by the American Sociological Association.

1. Some also eschew measures of degree of association and incorrectly footnote Selvin as authority for this practice.

argued that ordinary probability sampling from a known population is acceptable. Presumably, neither would endorse the use of tests of significance in unknown probability sampling or nonsampling situations.[2] Yet casual inspection of almost any issue of the sociological journals currently will reveal numerous instances of use of tests of significance in cases in which there has been no random sampling at any conceivable level. The researcher may, in fact, often be more justified in referring to his set of observations as a *population* rather than a sample; at least, in these terms a population can be specified whereas no reasonable population can be specified if the observations are considered a sample.

In order to generalize meaningfully in statistical terms from a sample to a population, the analyst must be able to specify what represents the whole population from which the sample has been drawn.[3] A special complication in social research inheres in the distinction between a population of observations and a population of persons. In a sample survey, there can be a random selection of persons from a precisely specifiable population of persons, but this random selection does not necessarily represent a random sample from a population of observations that can be specified with similar precision. Only if there exists an invariant value, unrelated to the means of observation, attached to each person in the given population of persons is it reasonable to assume that a random sample of persons is adequate to provide a random sample of observations from a population of observations that can reasonably be specified. However, if the responses or behaviors of persons may vary *systematically* with time and place (within the time and place range to which conclusions apply) and means of observation, then a population to which the sample results can be generalized cannot be precisely defined. Even given that the sample of observations can be considered a random sample, which is certainly rarely clear, the population of relevance cannot be specified in the rigorous terms required by the statistical theory upon which generalizations from random samples are based. If, for example, the analyst is dealing with some phenomenon so complicated as a prejudice score and if we assume that to some extent the observations (scores) are a *systematic* function of the interaction between particular inverviewers and particular respondents, i.e. the score of given respondents might be systematically different with a different observer, then specification of the population of scores from which a sample, random or otherwise, has been obtained would not be practically possible nor, for that matter, particularly meaningful. Thus, in such circumstances, the use

2. This is perhaps unfair to both writers. Selvin has, in fact, suggested such use of tests of significance in the analysis of what he has referred to as "internal replications" in a paper delivered to the American Sociological Association meetings in Chicago in 1959. McGinnis, to the best of my knowledge, has simply not addressed himself to this matter.

3. Strictly speaking, of course, the population must be known in order to draw a probability sample in the first place.

of tests of significance does not permit generalization *statistically* to a known population.

This argument should not be confused with the issue of correspondence between statistical manipulations involving numbers and the properties and relations taken to be represented by such data and arithmetic operations.[4] The demand for a physical-mathematical isomorphism to warrant the use of statistical techniques seems to me unreasonable. The crucial question is whether the arithmetic operations performed upon numbers for which there does not exist a demonstrable isomorphism with the concepts of which the numbers are taken to be referents introduce an error of sufficient magnitude that false conclusions are apt to be drawn. There are two possible kinds of error—systematic and random. In the former case, a question of validity can be raised and treated like any other question of validity in empirical research. In the latter case, the effect of the error is known, for nonsystematic error can have only one effect—to reduce the size of any observed associations. Therefore, if, in spite of such effect, statistically significant associations of reasonably substantial size are observed, it cannot seriously be argued that a high association does not exist between the arithmetic operations, including the activity of deriving the numbers, and some variable being measured. Presumably, there is a strong case for that variable being what the investigator takes it to be; the data-producing activities have not been obtained out of thin air, they come from the theoretical notions and conceptualizations, either explicitly or implicitly, of the investigator. They are, of course, subject to the same questions of validity that may be raised with respect to any data-production process. It may well be that something is being measured other than that asserted by the investigator. However, he has produced an association whose existence, within probability bounds, has been established, though its meaning and interpretation can be questioned. It should be understood that the conclusion based on such observed and statistically significant association is always provisional and subject to destruction at the hands of some other researcher. One form of destruction can certainly be demonstration that the procedures followed from conceptualization to production and analysis of data are not logically consistent and do not adequately represent the theoretically asserted phenomena.

In short, I do not question the legitimacy of the use of statistical tests in terms of the use of number data and the required arithmetic operations. My question concerns sampling considerations in relation to generalization to a specified population. I have argued that even a strict random sample of persons cannot generally serve as an adequate basis for generalization of

4. This correspondence issue has been prominent among the arguments of ethnomethodologists to justify wholesale rejection of conventional sociological research (see Cicourel, 1964: Chapter 1).

observations. Nevertheless, random selection of sample units at some level of sampling is often a useful device to protect the investigator from his own possible biases; and it is generally necessary if the task of the investigator is to estimate the value of some characteristic in a specified population. But random sampling is by no means a necessary criterion for establishing the validity of a proposition statistically expressed. The validity we seek in social science research can come only from repeated observation under varying conditions of population.[5] A test of significance under the best of circumstances provides only an index of reliability, restricted by time, place, and people, so that we are in fact dealing with unique historical knowledge. Statistical analysis can be considered only a preliminary screening that any hypothesis must pass to merit further investigation. Presumably, if the hypothesis is false but has by chance or bias been accepted by one investigator, the natural operation of the scientific enterprise will more or less quickly lead to rejection or qualification by other investigators who attempt replication and find contrary results.[6]

Statistical Tests as Model

Given that tests of significance cannot commonly be legitimately used for purposes of generalization to a specified population, it is my contention, nonetheless, that a meaningful and useful interpretation can be given to a test of significance applied to any set of data, without regard to sampling considerations. A test of significance can be viewed as an attempt to fit observed data to a model. The model is that of a random process which, for a given set of data, can generate a sampling distribution of a statistic whose characteristics are known. A decision about the substantive significance or importance, *not statistical significance*, of an observed relationship can be made on the basis of the degree to which the model provides a good fit. In the absence of other explicit criteria, the degree of fit can be taken as an explicit minimum criterion which any relationship taken to be important must meet and which is superior to subjective variable judgments of importance.

Consider, for example, a cross-tabulation of two dichotomous variables for all units of observation in a given universe. Since this is not a sample, what meaning can a Chi-square test of independence have?

The model to which I refer supposes a sampling distribution generated entirely by the observed data on each variable. Imagine one variable represented

5. Indeed, it seems warranted to assert that there is always an inferential step beyond that of sample to population—that is, from the given population from which the investigator has drawn his data to the general hypothesis being tested.

6. Unfortunately, there is some question about the natural operation of the scientific enterprise in social research, but that is a matter not directly relevant to the logic of the argument I am developing here. As an especially relevant study of the natural operation, note Theodore D. Sterling (1959 [Chapter 29]).

by red and white beads and the other variable represented by blue and green beads. The number of each color of beads will correspond to the frequencies in each category of the two variables. For example, for one variable the number of red beads will correspond to the frequency in one category and the number of white beads will correspond to the frequency in the other category. Now assume that N (the total number of observations) random draws of pairs of beads, one from the red and white group and one from the blue and green, will be made until all the beads have been drawn. A cross-tabulation of red-and-white and blue-and-green can be made by tallying the N draws in Table 20.1 below. The marginal frequencies of this contingency table will correspond to those of the contingency table of the two dichotomous variables with which we are concerned. The cell frequencies will probably differ, though they could on a few occasions be the same.

Table 20.1.

	Red	White	Total
Blue			
Green			
Total			N

Now assume that this process of drawing N pairs of beads from the two groups of beads has been repeated for a very large number of times and that Chi-square has been computed for each resulting contingency table. The distribution of these Chi-square values would, in fact, be the sampling distribution of Chi-square with one degree of freedom. The particular Chi-square value that we have observed for our two dichotomous variables can now be assessed in terms of this model. We can determine the probability of observing the obtained Chi-square (or some larger value) by cross-tabulating the colored beads. If, for example, the probability is less than .05, this tells us that in the process of cross-tabulating the colored beads we would have obtained contingency tables resulting in Chi-square values as large as obtained (or larger) less than 5 percent of the time. Now we know as a matter of indisputable fact that there is no association between the red and white beads and the blue and green beads. Any association observed in any particular set of N drawings is the result of nothing other than our random pairing process. Therefore, we can say that only 5 per cent of the time could an association as large as that observed be obtained by randomly pairing the values of the two variables.

It becomes clear that lack of a statistically significant Chi-square is especially revealing, for this tells us that an association is no larger than would ordinarily be expected by randomly pairing the values of two variables (maintaining the proper frequencies for each category of each variable, i.e. keeping the marginals constant). With such results of statistical analysis, even though we know there are no random processes out there in the empirical world,[7] and even though we are not dealing with a sample but with all of a given universe, it is inconceivable that in nonexperimental social research any importance could be attributed to the association.

Unfortunately, the reverse is not true. Statistically significant results do not necessarily indicate that an association is substantively important. Statistical significance is only a necessary but not sufficient criterion of importance. An assessment of the magnitude of the association must still be made in some terms other than that of a statistical test of significance.

When lack of statistical significance by any test is found in a universe or a given set of data (keep in mind, not a sample), we can say that in the empirical world the association produced by nature is no greater than that produced by a chance (e.g. random pairing) process. And it would seem a fair rule of thumb that, given our present state of knowledge about associations among sociological variables, we cannot with any confidence attribute substantive importance to associations of such magnitude.

Confusion of Statistical and Substantive Significance

Now, it should be noted that in the controversy over the use of tests of significance, there is no disagreement that the confusion of statistical significance with substantive importance is a sin (Selvin, 1957 [Chapter 9]; Mc-Ginnis, 1958 [Chapter 14]; Kish, 1959 [Chapter 15]; Duggan and Dean, 1968 [Chapter 18]; and Bakan, 1967 [Chapter 25]). But, aside from exhortation to avoid sin in this respect, the analyst in social research has been offered few practical guides to avoid the primrose path. The major suggestion has been to attend to measures of degree of association as well as tests of significance. For when an investigator concludes that a statistically significant relationship exists between two variables, he generally means to suggest not only that he is confident that a relationship between the two variables exists in the universe from which the data have come, but that the *degree* of relationship is such that it is worth taking account of. Committed as we are in nonexperimental social research to notions of multiple causation, we cannot simply accept as substantively important any observed relationship that differs from zero, regardless of how confident we may be that such difference exists. Substantive importance involves statistical significance *plus* a sufficiently high degree of

7. There are, however, sources of error in any social research that are not systematically related to the "true" associations—sources such as response, interviewing, coding, punching, tabulating—the effect of which approximates that of a random process.

relationship. How high the degree of relationship must generally be is a matter of subjective judgment, but this judgment need not be completely arbitrary by any means. The nature of the problem should offer some basis for judgment, and common sense (general informal consensus) will indicate that the degree of relationship is too low in some cases and high enough in others. However, common sense in many instances will be greatly aided if an explicit measure of degree of association has been applied.

When dealing with two quantitative variables, the analyst is less apt to ignore the degree of relationship and thus ignore the matter of substantive significance. The use of a correlation coefficient calls direct attention to the degree of association. Even though he may be intent on determining the statistical significance of the correlation, he is almost forced to note the size of the r. And when, for example, a statistically significant r is .12, he is not generally going to attach a great deal of importance to the relationship.

When qualitative variables are involved in the analysis, the tendency to ignore the degree of the relationship increases greatly.[8] While it is generally necessary to compute a measure of degree of relationship when investigating the statistical significance of association between quantitative variables, this is not so when dealing with qualitative variables. In order to determine whether a statistically significant relationship exists in a contingency table, it is not necessary to compute a measure of degree of association. Typically, Chi-square is computed and nothing more.

When the analyst deals with one qualitative and one quantitative variable, the matter of degree of relationship seems to become lost entirely. The language of traditional statistical analysis is such that it apparently hardly ever occurs to the analyst to raise a question about degree of relationship in a case in which, say, the statistical significance of the difference between means has been determined. For example, an investigator demonstrates that there is a statistically significant difference between the mean scores of men and women on test A. He then concludes that sex is related to A. No attention has been given to degree of relationship. Yet, it is my contention, the investigator necessarily means that the relationship is sufficiently large, as well as that it exists in magnitude greater than zero.

Some Criteria for Substantive Importance

Some general criteria for evaluating the degree of relationship in order to attribute substantive significance will be offered here as illustrative of the approach that must be taken. In particular, these criteria apply to those analytic situations in which no explicit standard measures of degree of

8. For recent documentation see Duggan and Dean (1968 [Chapter 18]); for an earlier documentation see Gold (1957).

association are available, i.e. those cases in which the analyst is faced with a combination of quantitative and qualitative variables.

1. If a set of responses or items (weighted or unweighted) makes up the quantitative measure, e.g. scores on a test, the substantive significance of statistically significant differences between means can be evaluated by comparing item differences in the original data. The substantive significance will be evaluated differently in the case in which only one or two items seem to contribute to the statistical significance from one in which there appear to be differences with respect to many items.

If we assume that most of the items making up a score are more or less adequate indicators of whatever variable the score represents, then it follows that, even in the case in which there is no substantial difference between two groups on the given variable, just by the chances of sampling it might be expected that one or two of many indicators will show a somewhat extreme difference between the two groups. In effect, in such case the analyst may well be making the same sort of error as that involved in testing for the significance of only the largest observed difference between means among many possible comparisons. The probability of a Type I error has been increased well beyond that specified by the statistical test. Therefore, the analyst should not in general attribute substantive significance to differences between mean scores that can be accounted for by the differences on only one or two items.[9]

2. In some cases, particular ranges of mean scores have special significance. For example, given mean scores on an attitude test may indicate the neighborhood of the middle range of possible scores, and this middle has been explicitly labeled in the original items as the neutral or undecided position. Lower scores may indicate favoring, higher scores disfavoring. Mean scores that differ across the neutral neighborhood may in general be attributed greater significance substantively than mean scores that differ by the same amount on the same side of (or within) the neutral neighborhood. The rationale is simply that a difference between a mean score that represents a preponderance of favoring and a mean score that represents a preponderance of disfavoring is intuitively more meaningful, as well as probably more reliable, than a difference between mean scores, both of which represent a preponderance in the same direction.

This is not to deny the psychological fact that *change* of mean scores in the middle range can often be accomplished more easily than *change* of an extreme score in the direction of greater extremeness (Hovland *et al.*, 1949: 290–292). In such cases, the nature of the problem may well dictate that smaller differences in the extreme ranges will be considered substantively

---

9. In the case of items that form a Guttman scale, this reasoning may not be appropriate.

important, while larger differences in the less extreme ranges may not be considered important. The point to keep in mind is that the *meaning* of a given amount of unit measures can vary in terms of location on the scale of measure, and this can occur even with equal-interval units of measure. To borrow common physical examples, if you are interested in whether the water in your car radiator is going to freeze, a drop in temperature from 80° F. to 50° F. is quite unimportant, but a drop from 50° F. to 20° F. is quite important; or if the speed limit is 65 mph, an increase in speed from 65 to 75 mph has entirely different significance from an increase from 45 to 55 mph. Therefore, the analyst must always look to the nature of the problem at hand to interpret the substantive significance of given observed differences; and he cannot be bound by slavish notions that a unit difference is a unit difference is a unit difference.

3. Note the degree of relationship in similar cases,[10] particularly paying attention to the differences between those cases in which statistical significance exists and those in which statistical significance (by the same explicit test) does not exist. For example, if a series of independent variables is being investigated in relation to the same dependent variable, compare the degree of the several relationships (size of differences). Especially note the *pattern* of the relationship manifest in a case of statistical nonsignificance in comparison to the statistically significant relationship under consideration. If a series of statistical tests has been performed, it will be instructive to compare in each case the size of differences on the dependent variable between the extremes of the different independent variables. Thus it may be found in a given case that the extremes in a statistically significant outcome do not differ from the extremes in a statistically nonsignificant outcome (same tests, same degrees of freedom, etc., in both instances). This can well be taken as an indication of little substantive significance despite the existence of statistical significance.

In general, assuming reasonably comparable N's, the maximum observed difference among statistically nonsignificant differences should be less than that of any difference to which substantive significance is attributed.[11] If this rule is not followed, the analyst may well find himself in the position of accepting as important some associations whose magnitude is less than that of some associations that have been (by tests of significance) deemed unimportant. The point to keep in mind, again, is that demonstration that the size of an association probably differs from zero is not enough.[12] The importance of the amount of that difference from zero must also be evaluated.

10. For a somewhat similar suggestion in a different context, see Hyman (1955b: 126–131).
11. What is being suggested is that, in effect, differences be evaluated in terms of the maximum possible estimates of variation within subgroups.
12. Of course, as Kish (1959 [Chapter 15]) has pointed out, and apparently with little impact upon researchers reporting in sociological journals, confidence limits are generally more useful than the usual test of significance. This is particularly so for purposes of making the kind of judgment of importance I am here discussing.

4. In general, it is probably safe to assume that there is an inverse relationship between the crudity of observations or measurement and the size of association to which substantive importance should be attributed. The more crude or gross the measurements may be, the greater the probability that the magnitude of existing differences may be obscured. Therefore, in general, smaller differences based on relatively crude observations may justify attribution of substantive significance, while larger differences may be demanded from more refined observations in order to attribute substantive importance.[13]

It appears that there is a tendency in criticizing social research to discount associations in which one or both of the variables have been measured (categorized) crudely and with suspect reliability. Yet it seems to me, unless a question of *validity* can also be raised, the best guess would be that the given associations would be observed even more markedly if measured less crudely and with greater reliability; for the effect of crude measurement and unreliability would, in general, be expected to decrease any observed association.

Conclusion

It has been contended that a test of significance can be viewed as an indication of the probability that an observed association could be generated in a given set of data by a random process model, without respect to sampling considerations. Statistical significance, in these terms, provides an explicit criterion for attributing substantive importance to the observed association. However, statistical significance is only the minimal criterion, necessary but not sufficient. In addition, the analyst must attend to the size of the association and must also make this criterion for the acceptance of the importance of the association reasonably clear. Some rules of thumb along these lines, especially useful in assessing associations among mixed variables (qualitative and quantitative), have been suggested as illustrative of a general approach to be taken.

13. Refinement of measurements should not be confused with merely decreasing the size of the units of measure. The notion of refinement also involves the reliability with which each unit of observation is measured or categorized.

21. Significance Tests Reconsidered

DENTON E. MORRISON and RAMON E. HENKEL

SLIGHTLY OVER a decade ago Selvin (1957 [Chapter 9]) published an article critical of the use of significance tests in survey research.[1] His article soon brought two minor responses (Gold, 1958 [Chapter 10]; Beshers, 1958 [Chapter 12]) and a major one (McGinnis, 1958 [Chapter 14]) that took explicit issue with his position.[2] Delayed, more implicit reactions, both critical of Selvin's position (Kish, 1959 [Chapter 15]) and semi-supportive of at least the general tone of his conclusion (Camilleri, 1962 [Chapter 16]), followed. The published controversy over significance tests subsequently generally subsided in sociology for several years, but the recent appearance of several discussions indicates the issues are still viable (Gold, 1964; Skipper *et al.*, 1967 [Chapter 17]; Duggan and Dean, 1968 [Chapter 18]; Labovitz, 1968 [Chapter 19]; Galtung, 1967; Gold, 1969 [Chapter 20]; Reynolds, 1969). In psychology the tests have undergone a parallel but quite independent scrutiny,

Reprinted from *The American Sociologist*, 4 (May, 1969), 131–140, by permission of the authors and the publisher. Copyright © 1969 by the American Sociological Association.

1. Selvin's article was preceded by some briefer critiques by other sociologists (Zeisel, 1955 [Chapter 5]; Lipset *et al.*, 1956 [Chapter 6]; Kendall, 1957 [Chapter 7]; see also Davis, 1958 [Chapter 8]). Practitioners in other fields had much earlier questioned significance tests (for example, Berkson, 1942 [Chapter 28]).

2. Selvin (1958a [Chapter 11]; 1958b [Chapter 13]) wrote rejoinders to the minor responses to his article, and Hirschi and Selvin (1967) have recently expanded some of the points made by Selvin and his respondents.

which started, at least in systematic form, after the sociological controversy (Rozeboom, 1960 [Chapter 24]) and continues with vigor (Cohen, 1962; Bolles, 1962; Bakan, 1966, 1967 [Chapter 25]; Wilson *et al.*, 1967; Meehl, 1967 [Chapter 26]; Forge, 1967; Lykken, 1968 [Chapter 27]).

It is our belief that sociology will profit by still further reopening of this controversy. But, in contrast to (*not* as a specific response to) Gold (1969 [Chapter 20]), our view of the tests is basically negative.[3] Judging by research practice, there is every indication that many sociologists are not even aware of the issues surrounding use of significance tests, much less in agreement on their resolution. Researchers continue to use significance tests irrespective of the type of sample, type of research problem, or type of research design. In addition to important technical errors, fundamental errors in the philosophy of science are frequently involved in this indiscriminate use of the tests. In this paper we shall examine these errors. What we say is frankly polemical, though not original. We hope only to contribute to the debate started by Selvin and recently re-opened, with the aim of moving it toward closure.

It will be useful to comment on Selvin's article and some of the controversy it generated at particular points in our discussion, but we will not limit our remarks to either the older or newer aspects of the sociological debate. We think that Selvin was correct in being critical of significance tests but that the main technical reason for his skepticism was unfounded. Selvin's argument tended, in our view, to focus subsequent discussion away from the more pertinent problems in the use of significance tests. Selvin's critics properly rejected his main specific argument against use of the tests; in doing so they explicitly or implicitly endorsed their use, partly by missing or underemphasizing other important arguments against their use (many of which were mentioned by Selvin). Much of the initial controversy, then, focused on the wrong issues. It is thus not surprising that Selvin's article and the subsequent responses did little to enlighten or even make consistent the use of significance tests.

What Are Tests of Significance?

Because highly technical expositions of significance tests are readily available, we will proceed from an elementary and nontechnical statement that emphasizes the meaning and interpretation of the tests.

First and foremost, a test of significance is a formal procedure for making a decision between two hypotheses about some characteristic of a population (parameter) on the basis of knowledge obtained from a sample (sample

3. This paper is broader in focus than Gold's and, though revisions have been incorporated to respond directly to his position, this paper was originally written, and basically remains, independent of Gold's.

statistic) of that population. Typically, one of these hypotheses, the hypothesis tested, is termed the "null hypothesis," and the other is termed the "alternative." Substantively, null and alternative hypotheses may be any assertions about the population of interest, but typically in sociology the null hypothesis is an assertion that variables are unrelated or that groups do not differ with regard to some characteristic. Given this situation, two levels of substantive specificity for the alternative hypothesis are common in sociology. The less specific (and probably more common), the nondirectional alternative, simply asserts the presence of a difference or a relationship. A somewhat more specific alternative, the directional alternative, specifies whether a difference or a relationship will be positive or negative. More specific alternatives are possible but are rare in sociological research.

Either the null hypothesis or the alternative hypothesis may be derived from the substantive theory in question, though it is conventional that the theory being tested (that is, the researcher's hunch, or the possibility he thinks most interesting) is identified with the alternative hypothesis. This means that null hypotheses are usually set up for the express purpose of seeing if there is empirical warrant for their rejection or nullification so that the alternative hypothesis can be accepted—hence the term "null hypothesis." It is possible, however, to identify the theory being tested with the null hypothesis (see Binder, 1963; Wilson *et al.*, 1967). In this event, failure to reject the null hypothesis is interpreted as supporting the theory in question. Hereafter, when we speak of the null hypothesis we will refer only to the practice of identifying one's theory with the alternative hypothesis; we will use the term "typical null hypothesis" for the practice of offering no-difference or no-relationship null hypotheses.

Testing for significance involves a comparison of the difference between the sample statistic and the parameter specified by the null hypothesis with a theoretically determined sampling distribution. This comparison allows estimation of how often such a difference would occur if the difference were due to random errors in the sample selection process (sample error). The significance level that results from the comparison gives the relative frequency (probability) with which a sample statistic of the obtained size or more extreme size would be expected to occur over repeated trials (samples) utilizing the same probability sampling method on the same population if the hypothesized value for the population parameter (null hypothesis) were true. If, on the basis of the computed level of significance, the researcher is satisfied that it is improbable that over repeated samples he would obtain the statistic he did if the null hypothesis were true, he concludes he must reject the null hypothesis and accept the alternative. The researcher must decide what he will consider "improbable"—the *criterion* level of significance—on the basis of the percentage of the time he can afford wrong decisions to reject the null hypothesis on the basis of sampling error. The 5 percent level is conventionally considered the minimum level of significance acceptable. It is customary, of course, to refer to a *statistic* that would reject a null hypothesis at the 5 percent level as "significant."

At the risk of being too elementary, we must point out that knowledge of *only* the level of significance of a sample statistic tells *nothing* about the magnitude of the relationship or difference being studied, nor does it provide any clues as to its theoretical or other interpretation. The probability expressed in a significance level is what Camilleri has termed an "auxiliary" probability in contrast to an "intrinsic" probability (1962 [Chapter 16]). In auxiliary probability, the reference is to a *relation between a sample and a population* in terms of the relative frequency with which sampling error will lead to a wrong decision to reject the null hypothesis. Auxiliary probability is involved in tests of significance. Intrinsic probability refers to a *relationship between variables* that is expressed in probabilistic terms. Most of our substantive hypotheses in social science are intrinsically probabilistic. As Braithwaite says (1953: 115), we do not expect all things that are *A* also to be *B* (his universalistic hypothesis); rather, we expect only a certain proportion between 100 percent and 0 percent of the things that are *A* also to be *B* (his statistical or probabilistic hypothesis). Thus, when most social science hypotheses are subjected to significance tests, auxiliary *and* intrinsic probability notions are involved, but the notions are logically independent. The fact that, for certain ranges of sample size, level of significance is more or less closely correlated with level of association should not lead us to think that, in general, a high level of significance is closely reflective of a high level of association. The level of significance of a given sample statistic under the typical null hypothesis is a monotonic increasing function of the sample size. A correlation coefficient of .11 can be the basis for rejecting a typical null hypothesis at the .05 level of significance in one study, but in another study of the same population a correlation coefficient of .25 will not reject the null hypothesis at the same level of significance, if the sample sizes are respectively large enough (300) and small enough (50).

Summarizing the above, significance tests refer (1) to a procedure for deciding whether sample error shall be considered a probable or improbable source of difference between a hypothesized population parameter and a sample statistic when (2) the statistic is obtained by probability sampling from the population. The two major classes of errors in the use of significance tests can be discussed in terms of the two components of this summary statement. There are, then, (1) errors of the analytic nature of the inferences based on significance tests, and (2) errors of the method and scope of inferences based on significance tests. We start with consideration of the latter.

Errors of the Method and Scope of Inferences Based on Significance Tests

In the introduction to his article, Selvin dismissed concern with "technical errors" in the use of significance tests as well as concern with describing particular populations on the basis of a sample (1957: 520 [Chapter 9: 95]). In doing so, he failed to examine the two most flagrant and frequent misuses of the tests and, in effect, encouraged the continuance of these misuses. Although he subsequently put severe limitations on the conditions under which he thought significance tests can be legitimately used, he played down the importance of more basic technical errors and left wide open the possibility that significance tests are useful in assessing the probability of sampling error of a statistic that describes more than a particular population. Significance tests, however, are not legitimately used for any purpose other than that of assessing the sampling error of a statistic designed to describe a particular population on the basis of a probability sample.

Whatever the breadth of his interest, the researcher must always select his cases for study from an existent and finite population of cases. If he explicitly circumscribes some larger set of cases existing in time and place of which he considers his cases for study a sample, he has specified a population. When this is not done, the population is unspecified. The specified population, or the population that would seem to be implicitly specified by the characteristics of the cases when the population is unspecified, may or may not be coterminous with the population to which the researcher ultimately wishes or expects his findings to apply, i.e. his conceptual or target population. In contrast with conceptual populations, then, specified or unspecified populations are best termed "sampled" populations; they are the circumscribed or uncircumscribed, but nevertheless existent and finite, cases to which techniques of sampling are actually applied in a given research situation. The two classes of techniques for sampling are, of course, probability and nonprobability, leading to the following cross-classification:

Table 21.1.

Type of sampling technique	Type of population sampled	
	Specified	Unspecified
Probability	A	B
Nonprobability	C	D

In terms of this table, significance tests can in a technical, legitimate sense be used only on studies that fall in cell A. Additionally, for statistical

inference to be possible one must *first* specify the population and *then* probability sample from that population. The notions of sampling distribution and
sampling error have no meaning in statistical inference apart from the assumption of randomness in the sample selection procedure—randomness being a
central feature incorporated in all probability sampling designs.

Probably the most frequent misuses of the tests occur, then, in studies that
fall in cells C and D, that is, using the tests where the sampling methods are
inappropriate. Some notion of the extent of C and D errors in sociological
research is given in Figure 21.1, which is based on an analysis of all quantitative articles and research notes using significance tests which appeared in the

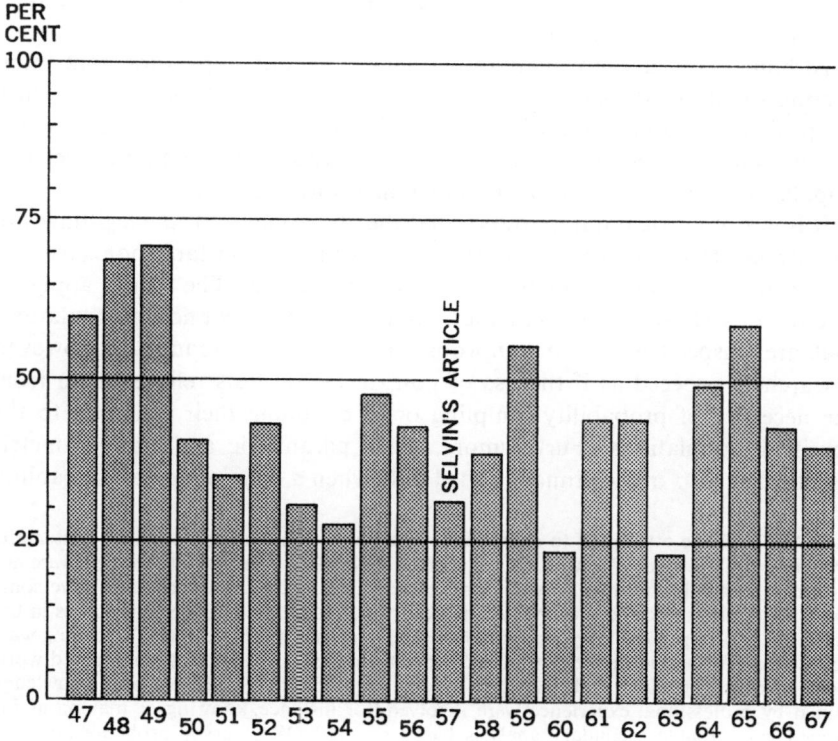

*Figure 21.1. Quantitative articles using tests of significance on nonprobability samples or in
nonrandomized experiments*, American Sociological Review, *1947–1967.*

American Sociological Review from 1947 through 1967. The data in Figure
21.1 (and Figure 21.2) should not be interpreted precisely, since the information given about sampling and use of significance tests in journal articles
make judgment equivocal in many cases. However, our crude and probably

conservative analysis indicates that an average of around 40 percent of the published articles in the *Review* that have used the tests have been characterized by this kind of misuse over the twenty-one-year period, with no sign of systematic improvement since Selvin's 1957 article.[4] Figure 21.2 shows, in addition, that after 1952 an average of half of the *Review's* quantitative articles and notes employed significance tests. Prior to 1953, the rate of usage was lower, but there is no indication of a systematic trend toward reduced usage since Selvin's 1957 article. We repeat: These data must be interpreted only as crude estimates. But we think it is highly improbable that additional analysis of the *Review* or other sociological journals would alter our *general* conclusion that both the use of significance tests and this (C and D) type of misuse are substantial and continuing.

Although we have no quantitative data, it is our impression that interpreting significance tests to apply to populations beyond those from which the probability sample was selected (cell B) is also a frequent misuse. Because it is technically impossible to select a probability sample from an unspecified population, cell B must in fact (i.e. demonstrably) be empty.

It is not of incidental importance that the above claim for the legitimacy of significance tests only for studies that fall in cell A of our table holds true for parametric *and* nonparametric tests of significance. The latter apply to populations whose variable distributions are unknown, but not to populations that are unspecified, or to nonprobability samples. Frequently, however, researchers proceed as if the use of nonparametric tests relieves them from the necessity of probability sampling or of confining their estimates to the specified populations. Furthermore, most parametric and nonparametric significance tests are legitimately used only when a certain *type* of probability

---

4. Although we attempted to construct a relatively exact coding procedure, systematic checking with "generous" in contrast to "exact" coding (i.e. whether data were or were not judged a probability sample or a randomized selection where the information gave some room for doubt) indicated that the two modes would result in consistent differences in the range of 10 to 15 percent per year in our description of misuse of the tests. The coding was, however, performed in the "exact" mode, since the coder, a graduate student, could work more reliably and comfortably in this mode: less subjective judgment and knowledge gained by professional experience were involved (for instance, knowing or making an informed guess that a secondary analysis based on an NORC survey probably involved probability sampling). But since we preferred our conclusions to be based on conservative estimates of misuse, we have sytematically subtracted 15 percent from each year's misuse as judged by the "exact" mode (i.e. reduced a given year's value from, for instance, 75 to 60 percent on Figure 21.1). All of the estimates are based on quantitative articles usable for this analysis, i.e. articles for which there was sufficient information to make *some* judgment of the probability or nonprobability nature of the data-gathering technique (whole populations are here classed as nonprobability samples). About 8 percent of the total of 917 quantitative articles were judged not usable. It should also be noted that the data reported in Figure 21.2 are conservative estimates. Use of the tests is sometimes not detected in a necessarily rapid visual scanning when such usage is mentioned only in the article's text or footnotes.

sampling is employed, namely simple random sampling. Considerable error in the computed level of significance can result from the use of formulas assuming simple random sampling when other probability sampling designs, particularly cluster (area probability) sampling, are used. Unfortunately, the proper formulas for significance tests when probability samples other than simple random samples are used are either unavailable or very difficult to obtain and use (Kish, 1957).

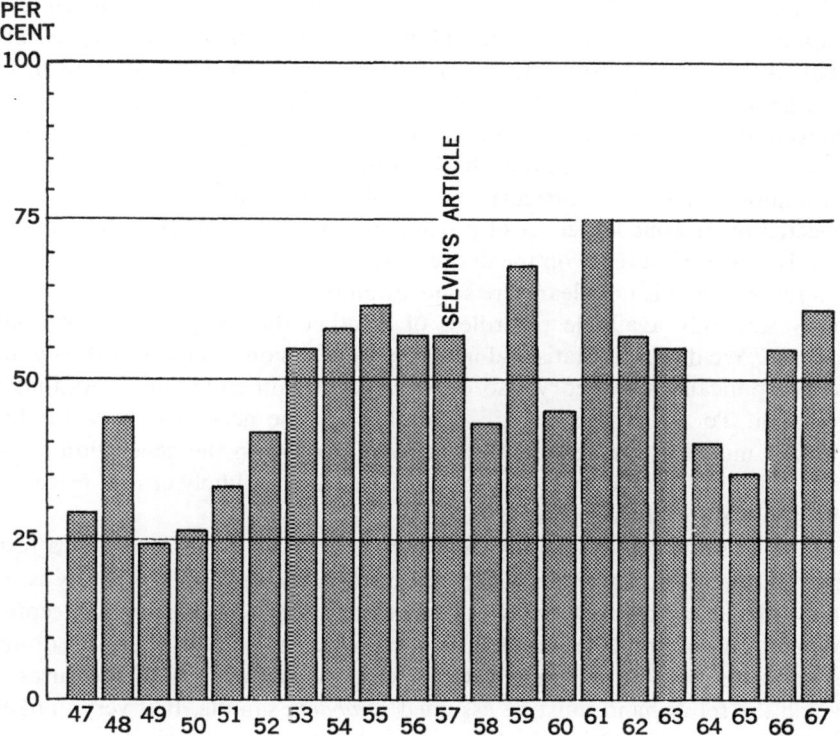

Figure 21.2. Quantitative articles using tests of significance, American Sociological Review, *1947–1967.*

Another common misuse of significance tests occurs when they are employed on a set of cases termed a "population" rather than a "sample." Were it literally true that the cases constitute a population, significance tests would be both inapplicable and unnecessary, since the probability relation of a sample and a population is by definition unity when they are the same. What seems more likely is that the researcher actually considers that his cases do not constitute his conceptual population, regardless of the fact that they exhaust some population that he has specified for study. Significance tests in such an instance are applicable only if the cases at hand have been selected by probability methods—a very unlikely possibility, given that the researcher terms his cases a "population." However, it has been argued that significance tests can nevertheless meaningfully be applied to such a set of cases or, more generally, to *any* set of cases, by treating the set (1) as a probability sample of a hypothetical universe of possibilities (Hagood and Price, 1952: 193–195, 287–294, 419–423; Blalock, 1960: 270), or (2) "as an indication of the probability that an observed association could be generated in the given set of data by a random process model" (Gold, 1969: 46 [Chapter 20: 181]; see also Blalock, 1960: 270), for instance, by repeated random redistribution of the marginal frequencies of a fourfold table into the cells. Although these two conceptions differ somewhat in that the first aims at generalization while the second disclaims this goal, their basic rationale is the same. Both require the assumption that the particular result observed was somehow randomly selected from some larger set of possible results. On what basis this assumption is warranted except on the desire of the researcher to apply the statistical inference model is not clear. Are some or all of the specific benefits of probability sampling available regardless of whether the sample is a probability sample? We doubt it. Statistical inference depends on a statistical theory, but to be applicable the theory also depends on certain empirical operations in research. To ask whether a given result could be generated by a random process model in the absence of a random process in the generation of the data is simply to raise an irrelevant question; an absolutely crucial feature of the application of the model is missing.

For the same reasons, to say that there is "measurement error" as a justification for using significance tests, even when the researcher literally is not interested in any cases other than those at hand (they constitute his conceptual population), is simply to say that in some sense the data on the cases do not exhaust his interest. There is interest in some population of measures of which the set at hand must be assumed a *random* sample. It is very unlikely that this can be convincingly demonstrated in most sociological research.

The most generous thing that can be said about the misuses of significance tests thus far discussed is that they are likely to be errors more of method than purpose. To say that a researcher misuses significance tests when he does not have a probability sample is not to say he necessarily behaves incorrectly in making or wanting to make generalizations of his results beyond the cases he studies. He must desire to do so if he is a scientist. However, he has no basis for doing so in terms of probability levels resulting from significance tests unless he has a probability sample. But, even in that case, his generalizations apply no further than his specified population. Such populations never exhaust the required logical scope of substantive scientific hypotheses. Scientific hypotheses are not bound by time and space but are posited as holding wherever instances of the phenomena in question occur under specified conditions. Finite populations, specified or unspecified, thus put technical limits on the cases available for any particular test of a hypothesis, but the scientist makes his generalization to an infinite conceptual universe (Camilleri, 1962: 171 [Chapter 16: 143–144]; Willer, 1967; Zetterberg, 1954: 54–58).

Similarly, to use significance tests to assess the substantive significance of a finding, the goal acclaimed by those who disclaim generalization as the reason for wholesale use of the tests (Gold, 1969 [Chapter 20]; Blalock, 1960: 270), is a mistake of method rather than purpose. The scientist must be concerned with assessing the substantive significance of his findings, but applying the tests to any set of data is neither technically appropriate nor, for reasons we elaborate in the next section of this paper, logically compelling.

The scientist is interested in particular events only insofar as they bear on his task of developing general principles. It is the social scientist's lack of theoretical development and of theoretical concern that make significance tests attractive. It is only to the extent that a scientific hypothesis states a specific expectation under clearly specified conditions that the scientist can know what cases to select to test it *or* how it is to be judged. When a typical null hypothesis is stated without specifying conditions (as is usual in social science), there is no basis for purposive selection of cases to test it, nor much basis for judging the results. In fact, *any* set of instances constitutes an adequate test of an unconditional hypothesis. Moreover, the data from virtually any sample will support a nondirectional alternative hypothesis (for instance, $A \neq B$) if the hypothesis is interpreted as being literally true. Consequently, we fall back on judging the evidence with significance tests and give ourselves a fine (though spurious) sense of accomplishment by applying the tests in their strongest form, by adopting high, arbitrary levels of significance. Sometimes our selection procedure makes the tests legitimate, sometimes not. Regardless, as our measures become more sensitive and our samples become larger, it becomes easier to reject our null hypotheses and easier to confirm our alternative hypotheses. As Meehl (1967 [Chapter 26]) has pointed out, this situation is just the opposite of that in the more developed sciences.

Perhaps a charitable view of at least the errors in cells B and D of our table would be that they often indicate the germ of concern with an infinite conceptual universe—the scientist's appropriate concern. Significance tests are, however, of no avail for this purpose, nor are the tests useful for the more modest purpose of generalizing to a specified or unspecified population when a nonprobability sampling method is employed. In substantial measure, then, misuses of the tests in these contexts represent errors of sample scope and of sample method, respectively. Unfortunately, such errors are often, as hinted above, compounded with misconceptions about the nature or meaning of the inferences based on the tests.

Errors of the Analytic Nature of Inferences Based on Significance Tests

THE SUBSTANTIVE SIGNIFICANCE ERROR

The error of confusing statistical significance with substantive significance, although widely recognized, remains commonplace (Duggan and Dean, 1968 [Chapter 18]). It is the error of confusing auxiliary probability for intrinsic probability. The practice of using statistical significance to judge substantive significance is often made as an outright and simple error of misinterpretation, wherein statistical significance is taken as *the* criterion for considering a finding substantively important. Although eschewing this, Gold (1969 [Chapter 20]) has recently re-issued a more systematic version of his previous rationale (1964) for using significance tests as a necessary, though not sufficient, condition for substantive significance (see also Davis, 1958 [Chapter 8]; Blalock, 1960: 270)—regardless of sampling considerations, since no intention of generalization is claimed. Thus, if a relationship or difference is so small as to be statistically insignificant under a typical null hypothesis, it is asserted that it cannot possibly be considered further as a candidate for interpretation as substantively significant.

There are problems with this viewpoint that argue strongly against its veracity. In the first place, the approach tends to foster an atheoretical view of sociology. It promotes an erroneous notion of the *sufficient* basis for substantive significance, since basic to the approach is the notion that substantive significance is in direct proportion to strength of a relationship or the size of a difference found when testing a typical null hypothesis. Thus, the approach would lead to an emphasis on prediction or "variance explained" in an actuarial sense rather than to concern with developing explanatory concepts and theories. The former emphasis puts variables in a strictly and prematurely quantitative competition, rather than in a competition based on consistency over replication, scope of coverage for the concepts, and deductive stature for the hypotheses, i.e. theory. It is the deductive feature in particular which eventually gives us a high degree of confidence in our hypotheses, that is, high *inductive* probability (Camilleri, 1962: 175–176 [Chapter 16: 150–151]), and relieves us of the infinite testing required by the statistical inference model. Theory in its most advanced form aims for logical consistency, economy, and accuracy in predicting the form and strength of empirical relationships *whatever* that strength happens to be, and not necessarily for accuracy in the sense of hypotheses wherein one variable "accounts for" a high proportion of the variance of another. Thus, the only source of necessary and sufficient criteria for the substantive significance of a particular finding is in the support it gives to specific expectations derived from a theory that provides an explanation. The broad range of evidence that will support our present vague alternative hypotheses is such that to label some arbitrary degree of that support "necessary" (through the significance level adopted) makes little sense. (Why one significance level is more appropriate than another is not discussed by Gold.)

Even if statistical significance were a necessary basis for judging substantive significance, use of this criterion would run into practical difficulty, since it disregards the notion of the *power* of the test of significance. As our tests become more powerful (through larger samples, improved measurement, and the like) there is the danger that statistical significance will imply nothing necessary or sufficient about the substantive import of a finding, even if this is defined in terms of size of association, given our typical null hypotheses and the alternatives presently common. Meehl, for instance, reports an example in which 91 percent of all the pairwise associations of 45 miscellaneous variables in a sample of 55,000 are significant (1967: 109 [Chapter 26: 259]).[5]

Thus, those who argue for statistical significance as a necessary or minimum criterion for substantive significance for *any* sets of cases (Gold, 1969 [Chapter 20]) show an atheoretical orientation in two ways: They disregard the technical requirements of statistical theory, and, in addition, provide a criterion of no logical or practical help in building social theory. They do not recognize what is necessary for social theory any better than they recognize what is necessary for statistical theory.

All the points above also, of course, argue against the practice of using significance as a necessary or sufficient criterion for substantive significance, even if the tests are technically warranted on sampling grounds, given typical null hypotheses, and the alternatives most common. Indeed, we usually know in advance of testing that the typical null hypothesis is false—which is all the significance test is able to tell us. It is, thus, not the confusion of statistical and substantive significance *per se* which causes the difficulty at the present stage of social science so much as the fact that the confusion occurs in the context of undeveloped theory. Because we have few substantively significant hypotheses (i.e. deduced, precise, and condition-specified), *findings or statistics* come to be referred to as "significant," which is particularly misleading. The only reasonable interpretation that can be made of such designations is that the statistic "signifies" a basis for rejecting the null hypothesis, given the probability level adopted. Also, the interpretation of a "significant" finding as a "precise" or "reliable" estimate of a parameter can be quite misleading. A correlation coefficient of .34 that is "significant" at the .05 level under the typical null hypothesis does not necessarily mean that the statistic itself is a

5. Meehl argues that directional hypotheses have an *a priori* probability of .5 of being supported as the power of the test approaches unity, *even though the hypotheses have no verisimilitude* (1967 [Chapter 26]; see also, Chandler, 1957 [Chapter 23]).

reliable or precise estimate of the parameter value (it *is* your best estimate, and may be a good estimate if your sample is of the right design and large enough). A subsequent correlation of .12 or correlations that average .15 on repeated samples may also fully vindicate the decision to reject the null hypothesis. What is "reliable" then, that is, what will be stable or consistent over repeated sampling, is your *decision* to reject the null hypothesis, not necessarily the statistic that led to that decision.

THE ERROR OF FIRM DECISIONS BASED ON HIGH, ARBITRARY LEVELS OF SIGNIFICANCE

Even if the general logic of statistical inference provided a sound basis for judging scientific hypotheses—this would be close to the case if the tests allowed generalization to more than a specified population *and* if our hypotheses were quite specific—it would not make sense to insist on arbitrary, consistently high levels of significance in scientific work. Significance exists on a probabilistic continuum, and to make discrete judgments at high levels of significance that dichotomize this continuum says more about particular decision needs than about the knowledge needs and knowledge characteristics of science. The only way a significance level can be decided upon rationally is in the situation where the cost of a wrong decision can be calculated. Where (1) there are no discrete, firm decisions necessary, or (2) the costs of making decision errors cannot be known, there is no reason for selecting any particular level of significance as a basis for a decision to reject or accept a hypothesis. These conditions are never the conditions of basic science. To insist on the .05 or .01 level is, then, to talk about the science of business, not the business of science. To say we want to be conservative, to guard against accepting more than 5 percent of our false alternative hypotheses as true (by rejecting less than 5 percent of our true null hypotheses) is nonsense in scientific research.

The reasons for this harsh claim exist quite apart from the fact that such "truth" or "falsity" can refer only to a particular (specified) population and, in current social science, to a hypothesis whose "truth" is often trivial. Why should we, even if auxiliary probability were an available and sufficient basis for judging a hypothesis on our infinite conceptual population, apply the language of "acceptance" and "rejection" when the empirical phenomena in question are continuous in nature? We have a whole range of more appropriate and less discrete language conventions available: "support," "lack of support," "weak support," "strong support," and so on. Why should we be arbitrarily, consistently, and rigidly conservative regardless of the state of knowledge on the problem or regardless of the use to be made of the knowledge (see Zeisel, 1955 [Chapter 5])? In basic science, adjustment of degree of belief based on the strength of the evidence rather than firm decision is the appropriate response.[6] As Rozeboom (1960: 423 [Chapter 25: 224]) states in his discussion of null hypothesis significance tests: ". . . A hypothesis is not something, like a piece of pie offered for dessert, which can be accepted or rejected by voluntary physical action. Acceptance or rejection of a hypothesis is a cognitive process, a *degree* of believing or disbelieving which, if rational, is not a matter of choice but determined solely by how likely it is, given the evidence, that the hypothesis is true."

The above does not mean that to make and report estimates of sampling error, or even to have one's conclusions guided by these, is irrelevant or inappropriate in basic scientific work. What it does mean is that consistently conservative, arbitrary rules for making discrete judgments about such error serve little purpose in scientific research. If, indeed, .05 (or any other level) is "sacred," as one recent article that is critical of this level suggests (Skipper *et al.*, 1967 [Chapter 17]), then what we do in sociology surely is much more akin to religion than science and we might as well forget empirical work and get on with the development of more rituals. We do not see how a conclusion about a typical null hypothesis that is based on *any* arbitrary level of significance can qualify as a scientific conclusion (see Sterling, 1959 [Chapter 29]; and Tullock, 1959 [Chapter 30]).

THE ERROR OF CAUSAL INFERENCE

Selvin (1957 [Chapter 9]) clearly was technically wrong in asserting that tests of significance are illegitimate in research designs unless control is maximum (as with randomization). McGinnis (1958 [Chapter 14]), Selvin's most systematic respondent, was correct in pointing out that, while randomization and random sampling have some differences in purpose and method, basic similarities also exist.

Randomization involves random administration of an independent variable and thus provides maximum assurance that "on the average" over repeated trials the independent variable is, in fact, independent of other independent variables, i.e. that they are "controlled." In a simple experimental design, equal size groups into which subjects are randomly assigned (randomized) from a larger pool (which may or may not be exhausted by the assignment procedure) will tend on the average over repeated assignments to be "equal" on any variable's value, and thus either group will provide equally good estimates of that variable's value in the larger pool. Thus, any difference in the groups on a particular variable in a given assignment will have some

---

6. Even in applied situations where the cost of a wrong decision is known to be practically nonexistent, *any* evidence against a null hypothesis and for an encouraging alternative may be the basis for a rational or at least defensible decision, such as the decision to use certain "unproved" drugs or methods on a patient whose disease is terminal and whose death is imminent.

calculable probability of being due to errors in the assignment procedure (which are by definition random errors, exactly equivalent to sampling errors). To the degree that this probability is low, we may have confidence that the difference is due to something about the groups other than the random errors of assignment, one alternative being a "treatment" that one group received (at random, often simultaneous with the assignment) and the other did not. Note, however, that the population to which the inference applies is only the specified population from which the subjects are randomly assigned. This population may be (and often is) a pool of convenient sophomores (sometimes matched carefully on age, sex, or other variables to increase assurance that the groups are equivalent at the beginning of the experiment), or any other specified population. Note, also, that rejecting the typical null hypothesis for the groups tells nothing, in itself, of the magnitude of the difference, or whether a causal interpretation is warranted.

Actually, inferring causality from significance is an error seldom made. Selvin does not appear particularly concerned with this problem. He is more basically concerned with the tendency of survey researchers to rely on the assessment of sampling error in evaluating their hypotheses and findings when so many other sources of nonrandom error are overlooked. In particular, Selvin thinks the independent variable and dependent variable relationships that interest us are often affected by other independent variables, i.e. the "other" variables are *conditions* for the relationships we study (Camilleri, 1962: 174 [Chapter 16: 148]). In Selvin's view, then, the priority problem in survey analysis is to assess whether and how such other independent variables constitute conditions that make the attribution of explanatory power to our independent variables of central interest an error (a "spurious" relationship). In fact, there is some room for doubt as to whether Selvin considered the use of significance tests in uncontrolled survey analysis a *technical* error. Certainly, he considered it an *analytic* error, and in so doing we believe he was at least heading in the right direction.

Selvin's article does, however, allow the interpretation that he viewed the use of significance tests in uncontrolled survey analysis as both an analytic error *and* a technical error. His more direct respondents (McGinnis, 1958 [Chapter 14]; Gold, 1958 [Chapter 10]; Beshers, 1958 [Chapter 12]) tended to address themselves mainly to the technical issue, and in doing so they may have had the effect, intentionally or unintentionally, of endorsing significance tests. Perhaps this interpretation of Selvin's article was encouraged by his additional and clearly technical error. He was properly criticized for his notion that when significance tests are used in fully controlled designs or analyses they allow estimates beyond some particular (specified) population (McGinnis, 1958: 411–412 [Chapter 14: 122]).

Despite his technical error, Selvin was, again, in a very general analytic sense correct in emphasizing that we want to generalize beyond particular populations, and that one aspect of doing so is to know the conditions under which our hypotheses hold. He was wrong in thinking that generalizations beyond particular populations are ever justified by significance tests, and, moreover, he was wrong in implying that (1) empirical analysis will be sufficient in revealing these conditions, (2) only relationships that hold when all relevant variables are controlled are of scientific interest, and (3) sampling distributions and sampling theory will provide the only probability considerations of relevance in assessing relationships when our controls are maximum. Selvin shows some reluctance to recognize that the intrinsic probabilistic character of most social science hypotheses will obtain regardless of their conditions. While empirical work with and without controls can give some clues to intrinsic probability values, empirical work will not be sufficient to constitute a theory involving such values. In a genuine theory, the intrinsic probabilities must be deducible from higher-level propositions.

Summary, Implications, Conclusions

In statistical analysis we are accustomed to thinking in terms of Type I and Type II errors that are compelled by the logic of significance tests. Our analysis suggests a different classification of errors (not to be confused with alpha and beta errors, which are synonymous with Type I and II errors):

Type A: Errors of sample method and scope of inference based on significance tests.
1. Sample method errors: using the tests on nonprobability samples.
2. Scope errors: estimating the parameters of populations beyond those probability-sampled.

Type B: Errors of the analytic nature of inferences based on significance tests.
1. Substantive significance errors: imputing substantive significance to significance tests.
2. Level and decision errors: arbitrarily or by convention adopting a high level of significance and subsequently making firm decisions about hypotheses on the basis of that level.
3. Causal inference errors: (a) confusing randomization with random sampling, (b) focusing on the assessment of sampling error before other sources of error are eliminated.

All these errors have important implications for the improved *use* of significance tests that flow directly from the analysis of this paper. It is our view that practice would be improved if a term like "auxiliary probability decision procedure," or "sample error decision procedure," were used instead of "significance test." In particular, we think much would be gained if the practice of labeling *findings* "significant" were eliminated in favor of phrases like, "this finding signifies a basis for rejection of the null hypothesis at the .05 level." Researchers have long recognized the unfortunate connotations and consequences of the term "significance," and we propose it is time for a change. The substitutes we offer (at least the latter one) sound cumbersome, but there is ample evidence that it is impossible to use the term "significant" in a statistical context and avoid the erroneous connotation of that term (for writers *and* readers). With regard to significance *level*, it can only be hoped that researchers will *report* levels for the probability of sampling error for their null hypotheses, rather than using high, arbitrary levels for rejecting or accepting hypotheses where the research context does not call for discrete decisions.

Apart from implications for improved *use*, however, our analysis, like Selvin's, more basically questions the general *utility* of the tests in basic (*not* applied) scientific research. The tests provide neither the necessary nor the sufficient scope or type of knowledge that basic scientific social research requires. In fact, there is every evidence that significance tests have been a genuine block to achieving such knowledge. As Meehl has said of the researcher who relies on (typical) null hypothesis significance tests, "His true position is that of a potent-but-sterile intellectual rake, who leaves in his merry path a long train of ravished maidens, but no viable scientific offspring" (Meehl, 1967: 114 [Chapter 26: 265]).

Alas, statistical inference is not scientific inference. To have the latter we will have to have much more than the façade that claims of significance provide. But how *is* scientific inference possible if significance tests are of little help? This question leads us beyond the scope of this paper, but we have offered some hints: replication over diverse samples as well as internally, the use of abstract concepts, and the incorporation of such concepts in deductive theories with the conditions of their validity specified. There are, of course, no computational formulas for scientific inference: the questions are much more difficult and the answers much less definite than those of statistical inference. In the absence of such computations, as Hogben points out, "Sociologists will have to use their brains" (1957: 30 [Chapter 1: 21]). We agree with Hogben that "science will not suffer" (1957: 30 [Chapter 1: 21]).

22. Proof? No.
Evidence? Yes.
The Significance of Tests
of Significance

ROBERT F. WINCH and
DONALD T. CAMPBELL

TO DO OR NOT TO DO a test of significance—that is a question that divides men
of good will and sound competence. We believe that although unreasonable
claims are sometimes made for the test of significance and that although
many have sinned in implicitly treating statistical significance as proof of a
favored explanation, still the social scientist is better off for using the significance test than for ignoring it. More precisely, it is our judgment that although
the test of significance is irrelevant to the interpretation of the cause of a
difference, still it does provide a relevant and useful way of assessing the relative likelihood that a real difference exists and is worthy of interpretive
attention, as opposed to the hypothesis that the set of data could be a haphazard arrangement.

In this paper we differ from Selvin (1957 [Chapter 9]) in advocating the use
of tests of significance in nonexperimental settings. Even though we believe
the significance test is of critical importance in weighing the plausibility that a
relationship exists, we advocate its use in a perspective that demotes it to a
relatively minor role in the valid interpretation of sociological comparisons.
In particular, our view is critical of much past practice with respect to the
following three basic errors in the use of tests of significance:

Reprinted from *The American Sociologist*, 4, 2 (May, 1969), 140–143, by permission
of the authors and the publisher. Copyright © 1969 by the American Sociological
Association.

1. The interpretation of the significant outcome of a test as proof of a given interpretation of a relationship.
2. Equating statistical significance with substantive significance, a point recently emphasized by Gold (1969 [Chapter 20]).
3. The use of the wrong error term in "dredging" operations and other multiple comparisons, an issue that has been treated by Selvin (1968) and Ryan (1959). Such procedures often involve hundreds of comparisons and employ error terms that are far too lenient with respect to Type I error.

What is the nature of evidential inference in sociology? The place to begin, we believe, is with the distinction between true experimentation and quasi-experimental analyses. Only if the sociologist can apply experimental treatments independently of prior states of his units of observation (actors, social systems, etc.) is he able to conduct a true experiment. Because of democratic ideology and a host of other considerations, true experimentation is unusual in sociology. Therefore, our attention turns to the problems of inference in other empirical studies, which we shall call quasi-experimental analyses.

The inability to use true experimentation results in a lack of control over a wide variety of largely unknown and unenumerable potential influences. These uncontrolled variables throw doubt on any finding that a difference observed between two subsamples results from the variable used by the researcher in classifying his sample of observations into one or the other of the two subsamples. Accordingly, the empirical sociologist is confronted with the necessity to identify and to find justification for rejecting as many of these potential influences as he can.

As we consider the logical problem of scrutinizing the validity of the relationship between the researcher's variable of classification and his quantitative results, it is useful to distinguish between internal and external validity. Internal validity addresses the question as to whether or not the experimental variable (here called the variable of classification) made a difference in the specific instance under consideration. External validity asks about the range of generalizability. To what populations, settings, treatment variables, and measurement variables can this effect be generalized? Nine threats to internal validity and six to external validity have been distinguished in the outline below. Out of this formidable list of fifteen threats to the relationship, it is seen that only A.1., instability, is illuminated by the test of significance. On the other hand, the fact that instability is only one of fifteen threats to the validity of a relationship certainly does not lead to the conclusion that such tests are pointless. Rather it is our purpose to show that such tests are desirable but should also be seen in their proper context and relative importance.

FACTORS THREATENING THE INTERPRETABILITY OF AN EMPIRICAL
RELATIONSHIP[1]

A. Threats to internal validity shared by true experiments and quasi-experiments.

 1. *Instability:* unreliability of measures, fluctuations in sampling persons or components, autonomous instability of repeated or "equivalent" measures. (This is the only threat to which statistical tests of significance are relevant.)

B. Threats to internal validity that are unique to quasi-experiments.

 1. *History:* events, other than the experimental treatment, occurring between pre-test and post-test and thus providing alternate explanations of effects.

 2. *Maturation:* processes within the respondents or observed social units producing changes as a function of the passage of time *per se*, such as growth, fatigue, secular trends, etc.

 3. *Testing:* the effect of taking a test upon the scores of a second testing. The effect of publication of a social indicator upon subsequent readings of that indicator.

 4. *Instrumentation:* changes in the calibration of a measuring instrument or changes in the observers or scorers used producing changes in the obtained measurements.

 5. *Statistical regression artifacts:* pseudo-shifts occurring when persons or treatment units have been selected upon the basis of their extreme scores.

 6. *Selection:* biases resulting from differential recruitment of comparison groups, producing different mean levels on the measure of effects.

 7. *Experimental mortality:* the differential loss of respondents from comparison groups.

 8. *Selection-maturation interaction:* selection biases resulting in differential rates of "maturation" or autonomous change.

C. Threats to external validity shared by true and quasi-experiments.

 1. *Interaction effects of testing:* the effects of a pre-test in increasing or decreasing the respondent's sensitivity or responsiveness to the experimental variable, thus making the results obtained for a pre-tested population unrepresentative of the effects of the experimental variable for the unpre-tested universe from which the experimental respondents were selected.

 2. *Interaction of selection and experimental treatment:* unrepresentative responsiveness of the treated population.

1. Adapted from Donald T. Campbell (1969).

3. *Reactive effects of experimental arrangements:* "artificiality"; conditions making the experimental setting atypical of conditions of regular application of the treatment; "Hawthorne effects."
4. *Multiple-treatment interference:* where multiple treatments are jointly applied, effects atypical of the separate application of the treatments.
5. *Irrelevant responsiveness of measures:* apparent effects produced by inclusion of irrelevant components in complex measures.
6. *Irrelevant replicability of treatments:* failure of replications of complex treatments to include those components actually responsible for the effects.

Perhaps we can communicate more clearly our view concerning the place of tests of significance by a couple of examples. Let us begin by imagining a researcher who has just completed a true experiment. He has now to decide whether or not the experimental treatment has made any difference. Is the difference observed of such magnitude that it could easily have happened by chance? We realize that this rival interpretation can never be completely ruled out, but its plausibility can be nonsubjectively examined and quantitatively assessed.

To consider the hypothesis of chance with a minimum of parametric assumptions, let us invoke a method of randomization rather than employing a traditional test of significance. Let us assume our researcher's experimental and control groups each had an N of 20. Assigning to each subject the score he received in the "after" condition, the experimenter could have his computer programmed to partition the 40 scores randomly into two sets of 20 and to compute the mean difference resulting from this random partitioning. This process would be done again and again, generating a literal random sample of mean differences. Let us say that 10,000 such differences are thus computed. If the observed difference lies well within this generated set of chance differences, it is indeed of a size that could have been generated by chance. A nonpresumptive descriptive P value comparing the experimental difference with this random experience is immediately available. We can assert that differences as large as that observed were obtained as often, say, as 50 times out of the 10,000 trials.

A second method of randomization might be to have the computer programmed so as to produce every possible pair of subsamples and to construct a sampling distribution of the differences between the means of all pairs of subsamples.[2] (On the method of randomization, see: Bradley, 1968: 68–86;

2. For $N_1 = N_2 = 20$ and $N_1 + N_2 = N = 40$, the number of pairs of subsamples in this sampling distribution would be $\binom{N}{N_1} = \binom{40}{20} = \frac{(8.15915)(10^{47})}{[(2.43290)(10^{18})]^2}$, and of course this large number would become much larger with the increase in size of sample.

Chung and Fraser, 1958; Fisher, 1935a, 3rd ed.: 43–47; Wilks, 1962: 462–468.) Each random sample of 10,000 or any other number of trials, as proposed in the preceding paragraph, is an estimation of this procedure and converges upon it as the number of trials increases. A statistical cost of this procedure is that since the variance would be calculated about the mean of the 40 observations, the error term would be larger, and the test less sensitive, than in the t-test for a pair of sample means, wherein the error term is based upon "within" variation.

From this perspective and for this use the conventional statistical procedures can be interpreted as low-cost approximations of the literal randomization achieved at the cost of assuming such parameters as normality, independence, sampling from an infinite universe, etc.

At the end of a true experiment *we know* that the research subjects are not randomly equivalent samples from the same universe of attributes, for, in fact, the experimentals have systematically different life histories from the controls. What we ask is whether, on a given measure, they differ from the controls more than two randomly assigned groups would differ from each other.

Now let us depart from the consideration of the true experiment and consider a survey of the students in a classroom. In this quasi-experimental analysis, let us assume that we are interested in a difference between the mean of men and the mean of women (or between Catholics and Protestants, or between freshmen and seniors) and that we are tempted to discuss this difference or to learn something from it. Here we are beset with many additional plausible rival interpretations beyond those in the case of a true experiment listed in our outline. But one of these rivals is again that it is an unsystematic difference, a trivial difference of the kind that *could have occurred by chance*.

In this quasi-experimental setting of the classroom, it is unlikely that the two sets we wish to compare will be of equal size. But of course there is nothing in the procedure proposed that requires $N_1 = N_2$. Let us suppose that there are 25 women and 15 men in the class. Now the computer is directed to produce a sampling distribution of differences between pairs of subsample means with $N_1 = 25$ and $N_2 = 15$. As before, the computer might be programmed to make 10,000 such trials or to run all possible pairs of subsamples.[3] Again, having generated our distribution of haphazard differences, we can assess the probability that our observed male-female difference is of a magnitude that would occur frequently by chance. If so, we rightly find other interpretations of the difference gratuitous.

---

3. In the latter case, the number of subsamples would be a little smaller than in our case of equal N's. It would be

$$\binom{40}{25} = \frac{(8.15915)(10^{47})}{(1.55112)(10^{25})(1.30767)(10^{12})}.$$

The reader may be hanging back unpersuaded that our reasoning is legitimate. We have specified neither a universe nor a sampling plan. We have not alleged—legitimately or otherwise—that our class constitutes a random sample. What can our test of significance demonstrate? Let us be clear about its logic. We have said the following:

1. Let us assume that our two subsamples constitute a single homogeneous set with respect to the variable under consideration. Our plan is to examine the plausibility of this assumption. Or, if we wish to label this a null hypothesis, our task is to assess the probability of this null hypothesis.

2. Let us divide that set into two subsets of the specified size,[4] and let us do so in such a way that the number of pairs of subsets in the resulting distribution is large, whether it be done by a specified number of randomizations or by exhausting the set of possible subsets.

3. By considering each pair of subsets to be equally probable we construct a sampling distribution of the differences of pairs of subsample means.

Then we are permitted to ask our question: Is our variable of classification one that orders our data into subsamples giving a mean difference that is well within the chance distribution, or did it produce a difference that is—with respect to our sampling distribution—quite unusual? If the former, it is quite possible that the variable did not produce a real difference; if the latter, explaining the difference away by chance is quite implausible, and other "causal" explanations are needed. In the quasi-experimental setting, many of these other causal explanations are the *other* threats to validity, which must be ruled out on other grounds. But the critical point is that the test of significance registers the degree to which there is any point in going about the task of excluding the other threats to validity.

Now we may extend our example of data of the classroom to the kind of problem in political science where the researcher exhausts the known universe. Is it legitimate to apply a test of significance where the researcher's data exhaust the specified universe? If the problem is stated in the conventional way, the answer is clearly "No." According to conventional reasoning, one is inquiring as to the probability that the observed difference between subsample means may have occurred in a universe where the true difference—or population value—is zero. The reason for a negative answer in this case is of course that the observed difference *is* the population value.

But we elect to phrase the question differently: If we assume the set to be homogeneous, what is the probability that dividing the set into two subsets on the basis of a variable of classification that makes no real difference would give a difference between subsample means as great as that observed? With this reasoning, there is every justification to run a test of significance. Thus, if

4. Whether or not the subsets are of equal size is determined, of course, by the distribution of the classifying variable in the sample.

a political scientist compares the stability of all countries with press censorship with the stability of all those without, it is a plausible rival explanation of a difference that it is of the magnitude that would appear frequently by chance (Gold, 1964).

In these two examples and in general in studies that are not true experiments, the establishing of a statistically significant difference goes but one step toward establishing an interpretation of that difference. That step is to exclude the hypothesis of chance. Moreover, the decision as to the plausibility of chance is made by formal, objective, communicable, and reproducible procedures rather than by intuition, with the consequent inconsistency that Davis (1958 [Chapter 8]) notes in the absence of tests of significance. Once chance has been adjudged improbable, the resulting difference may not be due to sex *per se* but to a bias in recruiting from each sex in an 8 A.M. class on this particular subject, and so on down through a long list of rival hypotheses. Or the censorship may be a symptom rather than a cause of the instability of nations, or each may be co-symptoms of the expectations trap of newly decolonialized nations, etc., etc.[5]

Finally, no study, whether a true experiment or not, ever proves a theory; it merely probes it. True experiments, as well as other designs, are vulnerable to the threats of external validity. Generalizability involves considerably more than the relation of the sample of research subjects to some population. Generalization involves research subjects, measurement variables, conceptual variables, and test settings. If a second study should happen to replicate a first with respect to all four categories of features just noted and if it should yield similar results, the second study would provide confirmatory evidence of the findings of the first. If a third study should deviate from the first in some respects—e.g. using college seniors where the first had used sophomores, and/or using a new scale for the same conceptual variable—and if it produced results similar to those in the first study, this third study would help not merely to confirm the original relationship but to extend its generalizability with respect to categories of research subjects and measurement variables. By such accretion we acquire confidence in the range of generalizability.

Summary and Conclusions

Since it is central to the scientific endeavor to seek for relationships among variables, the paradigm of the scientific problem is that in which observations are classified by one variable and are measured or classified with respect to another. Irrespective of the design of the study and whether or not it is a true experiment, the researcher is then confronted with the question as to whether

5. For categories of these hypotheses, see our outline. See also Campbell and Stanley (1963).

the covariation in his data is so slight as to be haphazard or so great as to be systematic. This problem, labeled "instability" in our outline of fifteen factors threatening the interpretability of an empirical relationship, is the only one that is illuminated by the test of significance.

The method of randomization dispenses with the nagging problems of normality, probability sampling, etc. Tests of significance are seen as low-cost approximations of this model, which provide greater precision at the cost of more assumptions.

Some critics of tests of significance seem to be saying that since these tests do not dispose of all rival hypotheses, they are useless and misleading and should be abandoned. We reason that it is very important to have a formal and nonsubjective way of deciding whether a given set of data shows haphazard or systematic variation. If it is haphazard, there is no reason to engage in further analysis and to be concerned about other threats to validity; if it is systematic, our outline shows that the analysis is not concluded with the test of significance but is just getting under way. And we believe it is important not to leave the determination of what is a systematic or haphazard arrangement of data to the intuition of the investigator.

PART THREE

Criticism by Psychologists

INTRODUCTION

D ISCUSSIONS OF ISSUES concerning the proper technique for use of significance tests have been common in psychology for many years (see, for example, Lewis and Burke, 1949). In general, we have not included in this volume essays that do not go considerably beyond technical issues, but Chandler's brief note on "The Statistical Concepts of Confidence and Significance" [Chapter 23] borders on an exception to this rule. His succinct clarification of the frequent confusion between significance tests and confidence intervals as well as his description of the meaning and role of power in significance testing provide the grounds for understanding better other papers in this section and elsewhere in the book. Chandler points out the magnitude of the confusion on these points in a sober tone, but his paper is characteristic of the implicit optimism of the many earlier and strictly technical attampts to teach better *use* of the tests. In contrast, an important segment of the psychological literature of the last decade is marked by strong pessimism about the *use and purpose* of the tests, as is shown in the papers that follow Chandler's in this section.

Psychological discussions that seriously question the research contribution of the tests started after Selvin's [Chapter 9] major scrutiny of the tests in sociology with the publication of Rozeboom's "The Fallacy of the Null Hypothesis Significance Test" [Chapter 24]. Rozeboom's essay and the three essays that follow all deal with statistical issues and relate them to the full range of philosophy of science issues, with basically negative conclusions regarding the role of the tests in scientific inference. In contrast with the extended and direct debate in sociology, no essays in psychology that explicitly and systematically defend the tests against these attacks have been published, although clearly part of the motivation for the later essays is that many psychologists have paid no heed to Rozeboom's points and continue to give implicit support to the tests in their teaching and in everyday research practice. The extent to which the later writers of the psychology essays are aware and appreciative of earlier criticisms of the tests in their own discipline is somewhat greater than in sociology, but, like most of the sociologists, the psychologists exhibit little if any awareness of discussions in their sister discipline.

Rozeboom's essay deals with the issues of meaning, level, and power (implicitly) as these issues relate to the broader issues of the role of the tests in the form and particularly the process and purpose of scientific inference. Like Hogben, Rozeboom endorses essentially the Neyman-Pearson-Wald notion that a significance test is actually a null hypothesis *decision* test. Rozeboom thinks that such a formal process for making decisions means nothing in theory or in practice in science. As an audience, scientists by and large adjust their beliefs about a hypothesis in informal ways on the basis of evidence, regardless of the formal decisions to reject or accept hypotheses made by individual researchers. Rozeboom, like Hogben, thinks that the firm decisions about hypotheses derived from conventional significance tests do not provide clues as to the probable validity of hypotheses, but, unlike Hogben, he thinks (with certain important reservations) that scientists should work toward expressing their degree of belief about hypotheses in probability terms. Moreover, Rozeboom sees no alternative other than to derive probability statements from observed outcome probabilities under given hypotheses. He clearly allows the possibility that some of the basic theoretical notions associated with statistical inference, such as Bayes' Theorem, confidence intervals,[1] or simply reporting the exact probability level from a significance test, may be valuable in this.

Bakan's paper, "The Test of Significance in Psychological Research" [Chapter 25], is a general, an eclectic, and, like Rozeboom's, a powerful indictment of the tests. It explores nearly the full range of statistical issues and relates them to each philosophy of science issue. He concludes that the tests have little, if anything, to contribute to scientific inference. Bakan implicitly endorses Rozeboom's view that the tests are appropriate for making null hypothesis decisions but not for the more general process by which scientific knowledge is generated. In addition to describing the general

---

1. Rozeboom's discussion of the relationship of significance testing and confidence intervals complements Chandler's as an important clarification of this topic.

inappropriateness of significance testing for scientific inference, Bakan shows that the "automatic inference" feature of decision making with the tests is largely illusory because decisions about level, sample size, and one- or two-tailed tests make results far from independent of the choices of the particular investigator. Further, Bakan describes how conventional significance levels influence researchers' decisions to submit findings for publication and editors' decisions to publish findings, and he demonstrates how these decisions are dysfunctional for the development of scientific knowledge (see also Sterling [Chapter 29]; Tullock [Chapter 30]). His review of the problems and contributions of Fisher's approach and of the controversy between Fisher and the Neyman-Pearson-Wald position is intrinsically valuable and also supplements understanding of the more detailed explorations of this topic in Hogben's essays. One of Bakan's central points is that, contrary to the Fisherian view of induction commonly practiced in psychology, the nature and form of scientific hypotheses require that the investigator make an inference from his population to his hypothesis ("induction to the general)" in addition to whatever statistical inference he makes from the sample to the population ("induction to the aggregate"). This point is different from Camilleri's [Chapter 16] notion of the importance of knowing the relevance of the *population* studied, but Camilleri's views on this topic and on the relevance of statistical inference in deterministic and probabilistic theory nicely complement Bakan's notions.

In his "Theory Testing in Psychology and Physics: A Methodological Paradox" [Chapter 26] Meehl deals with the issue of power and relates it to the broader issues of form and process. He points out, as several other authors have, that virtually all point null (no-difference, no-relationship) hypotheses will be rejected, given a powerful enough significance test.[2] He goes considerably beyond this assertion, however, to demonstrate that as the power of our test increases toward the maximum by improving design or technique or by increasing the sample size, we also know in advance that the probability of rejecting a directional null hypothesis (for example, $A \geqslant B$) approaches .50 even if our directional alternative hypothesis $(A < B)$ is totally without empirical merit. This, in Meehl's view, is a paradox, since in physics increasing the power of a test to the maximum decreases the *a priori* probability of a successful experimental outcome because the null hypotheses of physics specify exact values of parameters or precise functional relationships. The result is that when a null hypothesis in physics is rejected by a more powerful significance test there is a greater increment in the corroboration of the substantive theory (because the theory has cleared a more difficult hurdle), while in psychology the rejection of a directional null hypothesis by

2. The power of a test is its ability to let the researcher avoid accepting false null hypotheses, the familiar Type II error.

a more powerful test means a weaker corroboration of the substantive theory (because the theory has cleared an easier hurdle). Meehl also points out how the structure and practice of research in psychology (as in the other behavioral sciences) encourage and reward and thus perpetuate such nondevelopment of theory by significance testing. His argument, however, is not a blanket condemnation of significance testing as a mode of inference in the behavioral sciences but an argument against the tests in the context of the current *form* of conventional null hypotheses. His is also an argument *for* the development of theory. While Meehl realizes that theory such as that in physics must be constructed independently of significance tests, he does not, like Camilleri, urge the development of deductive theory for deriving exact hypotheses for testing. Rather, he simply presents the paradox as a bothersome and damaging problem that psychologists and philosophers of science should address.

In "Statistical Significance in Psychological Research" [Chapter 27], Lykken uses an extended example of a published study to explore the issue of the meaning of the tests in relation to the philosophy of science issues. He seeks to demonstrate that the probabilities expressed in significance tests can contribute little to the central task of scientic activity: establishing the general confidence warranted in a hypothesis.[3] An important aspect of his argument is his distinction between literal replication, operational replication, and constructive replication. Scientific inference, Lykken argues, must depend ultimately on constructive replication (use of the same concepts and hypotheses, but different methods), while significance tests give probability information only for literal replications (identical methods, more of the same subjects)—a point that bears critically on some of the arguments made by sociological defenders of the tests. Lykken also follows Baken, Sterling [Chapter 29], and Tullock [Chapter 30], in concluding that significance tests are seldom a valid guide for publication decisions.

Rozeboom, Bakan, Meehl, and Lykken share some particular points of criticism, but each attacks the tests from his special perspective to arrive at the common conclusion that the tests contribute little to and in many ways detract from the development of scientific theory in psychology. Together their arguments constitute a challenge to the tests that has yet to be answered by the faithful.

3. Lykken uses the term "confidence" in a general way to speak of the probability of a hypothesis. He possibly violates Chandler's plea for "pristine terminology" but he does not exhibit the technical confusion of confidence and significance described by Chandler. Rather, Lykken recognizes that whatever significance is called in statistical jargon it does *not* establish the general confidence in hypotheses he considers the hallmark of scientific hypotheses.

23. The Statistical Concepts of Confidence and Significance

ROBERT E. CHANDLER

RECENTLY THERE HAVE BEEN at least three different book reviewers who have commented on the confusion that currently exists in the psychological literature regarding the statistical concepts of confidence and significance (Chandler, 1957a; Milton, 1956; Walker, 1956). Although this confusion can be partially explained as a semantic problem, it behooves the psychologist to examine these two concepts rather closely and to adopt pristine terminology for the benefit of beginning students and individuals of other disciplines that draw rather heavily upon the psychological literature.

Confidence and Confidence Coefficients

Confidence, a concept customarily reserved for discussions of interval estimation, is the faith which one is willing to place in a statement that an interval established by a sampling process actually contains or bounds a parameter of interest. One generally expresses this faith statistically by affixing to each interval a confidence coefficient, or confidence probability, which can be written as $1 - \epsilon$, where $\epsilon = p/100$ for $0 \leqslant p \leqslant 100$, and p is usually taken to be a very small number (Anderson and Bancroft, 1952; Cramér,

Reprinted from *Psychological Bulletin*, 54, 5 (September, 1957), 429–430, by permission of the author and the publisher. Copyright © 1957 by the American Psychological Association, Inc.

1946; Hoel, 1954; Mood, 1950). For example, if $p = 5$ the confidence coefficient would be .95, and one would refer to the interval with which this coefficient is associated as the 95 percent confidence interval.

The confidence coefficient is frequently interpreted in the following manner: If one were to draw samples of size K from a population of N elements (K naturally being $< N$) and from each sample establish a 95 percent confidence interval on some specified parameter of the population, then in the long run about 95 percent of the totality of these intervals would actually contain the parameter of interest, and approximately 100 ϵ percent, or 5 percent, of them would not (Cramér, 1946). This interpretation is correct, but of course assumes $\binom{N}{K}$ to be a rather large number.[1]

Significance and Significance Levels

Significance, as contrasted to confidence, is given to the testing of hypotheses. Here one makes a statement, i.e. states a hypothesis, which will hereafter in this discussion be represented as H, that may be either true or false and then takes action on this H by accepting or rejecting it. Clearly, any one of the following actions is a likely outcome as a result of testing an H: (*a*) rejection of a false H; (*b*) acceptance of a true H; (*c*) rejection of a true H; or (*d*) acceptance of a false H. It is quite evident that actions (*a*) and (*b*) are desirable, while (*c*) and (*d*) carry the connotation of committing an error—*c* being the familiar Type I error or an error of the first kind, while *d* is called a Type II error or an error of the second kind (Dixon and Massey, 1957; Mood, 1950).

When one tests an H, the probability that he will take action *c* is defined as the significance level, which we will represent as α (Mood, 1950). Although α is generally of the same order of magnitude as ϵ, α and ϵ differ in the amount of information which they convey, for while ϵ completely tells all there is to know about "being wrong" in interval estimation, α only gives information about a very particular type of error, i.e. the action described by *c*. To emphasize the contrast made here between α and ϵ, one merely needs to examine the other type of error that can be made in the test of an H.

For this purpose, let β represent the probability that action *d* is taken, i.e. a Type II error is committed; then, by definition, $1 - \beta$ is known as the power of the statistical test or the probability that action *a* will occur (Dixon and Massey, 1957; Mood, 1950). Although texts in psychological statistics do not seem to place a great deal of emphasis upon the power of a test, power is the basic concept responsible for one's employing statistical tests as a basis for taking action on an H. If this were not so, to test an H at the 5 percent level of significance, one could simply draw from a box of 100 beads—95

---

1. The notation $\binom{N}{K}$ is used here as a combinational symbol to indicate the number of ways that K objects can be selected from N.

white and 5 red—a bead at random and adopt the convention that he would reject the *H* whenever a red bead appeared. With such a test, one can readily see that not only α but also $1 - \beta$ always equals .05 or $\beta = .95$. It is this large value of β that precludes one's employing the bead-box test. For an excellent discussion of β and its relation to the alternative *H* against which one might be testing, the reader is referred to Dixon and Massey (1957: 244–261).

Summary and Discussion

The admixing of the concepts of confidence and significance has become so prevalent in the psychological literature that one typically reads statements, in the reports of psychological research, indicating that certain experimental results were significant at, say, the 5 percent "level of confidence."

It may be that this confusion arises from the fact that one can utilize a confidence interval as a significance test (see e.g. Edwards, 1954: 241) and in doing so may hastily, but incorrectly, conclude that there is no difference between the two concepts.

Inasmuch as explicit terminology is needed to convey the probabilities of committing statistical errors in the respective areas of interval estimation and testing of hypotheses, the concept of confidence should never be associated with the statistical test of an *H* regardless of the nature of the test being employed.

24. The Fallacy of the Null Hypothesis Significance Test

WILLIAM W. ROZEBOOM

THE THEORY OF PROBABILITY and statistical inference is various things to various people. To the mathematician, it is an intricate formal calculus, to be explored and developed with little professional concern for any empirical significance that might attach to the terms and propositions involved. To the philosopher, it is an embarrassing mystery whose justification and conceptual clarification have remained stubbornly refractory to philosophical insight. (A famous philosophical epigram has it that induction [a special case of statistical inference] is the glory of science and the scandal of philosophy.) To the experimental scientist, however, statistical inference is a research instrument, a processing device by which unwieldy masses of raw data may be refined into a product more suitable for assimilation into the corpus of science, and in this lies both strength and weakness. It is strength in that, as an ultimate *consumer* of statistical methods, the experimentalist is in a position to demand that the techniques made available to him conform to his actual needs. But it is also weakness in that, in his need for the tools constructed by a highly technical formal discipline, the experimentalist, who has specialized along other lines, seldom feels competent to extend criticisms or even comments; he is much more likely to make an unquestioning application

Reprinted from *Psychological Bulletin*, 57 (September, 1960), 416–428, by permission of the author and the publisher. Copyright © 1960 by the American Psychological Association, Inc.

of procedures learned more or less by rote from persons assumed to be more knowledgeable of statistics than he. There is, of course, nothing surprising or reprehensible about this—one need not understand the principles of a complicated tool in order to make effective use of it, and the research scientist can no more be expected to have sophistication in the theory of statistical inference than he can be held responsible for the principles of the computers, signal generators, timers, and other complex modern instruments to which he may have recourse during an experiment. Nonetheless, this leaves him particularly vulnerable to misinterpretation of his aims by those who build his instruments, not to mention the ever present dangers of selecting an inappropriate or outmoded tool for the job at hand, misusing the proper tool, or improvising a tool of unknown adequacy to meet a problem not conforming to the simple theoretical situations in terms of which existent instruments have been analyzed. Further, since behaviors once exercised tend to crystallize into habits and eventually traditions, it should come as no surprise to find that the tribal rituals for data-processing passed along in graduate courses in experimental method should contain elements justified more by custom than by reason.

In this paper, I wish to examine a dogma of inferential procedure which, for psychologists at least, has attained the status of a religious conviction. The dogma to be scrutinized is the "null hypothesis significance test" orthodoxy that passing statistical judgment on a scientific hypothesis by means of experimental observation is a decision procedure wherein one rejects or accepts a null hypothesis according to whether or not the value of a sample statistic yielded by an experiment falls within a certain predetermined "rejection region" of its possible values. The thesis to be advanced is that despite the awesome pre-eminence this method has attained in our experimental journals and textbooks of applied statistics, it is based upon a fundamental misunderstanding of the nature of rational inference, and is seldom if ever appropriate to the aims of scientific research. This is not a particularly original view—traditional null hypothesis procedure has already been superceded in modern statistical theory by a variety of more satisfactory inferential techniques. But the perceptual defenses of psychologists are particularly efficient when dealing with matters of methodology, and so the statistical folkways of a more primitive past continue to dominate the local scene.

To examine the method in question in greater detail and expose some of the discomfitures to which it gives rise, let us begin with a hypothetical case study.

A Case Study in Null Hypothesis Procedure; or, A Quorum of Embarrassments

Suppose that, according to the theory of behavior, T_0, held by most right-minded, respectable behaviorists, the extent to which a certain behavioral manipulation M facilitates learning in a certain complex learning situation C should be null. That is, if "ϕ" designates the degree to which manipulation M facilitates the acquisition of habit H under circumstances C, it follows from the orthodox theory T_0 that $\phi = 0$. Also suppose, however, that a few radicals have persistently advocated an alternative theory T_1 which entails, among other things, that the facilitation of H by M in circumstances C should be appreciably greater than zero, the precise extent being dependent upon the values of certain parameters in C. Finally, suppose that Igor Hopewell, graduate student in psychology, has staked his dissertation hopes on an experimental test of T_0 against T_1 on the basis of their differential predictions about the value of ϕ.

Now, if Hopewell is to carry out his assessment of the comparative merits of T_0 and T_1 in this way, there is nothing for him to do but submit a number of Ss to manipulation M under circumstances C and compare their efficiency at acquiring habit H with that of comparable Ss who, under circumstances C, have *not* been exposed to manipulation M. The difference, d, between experimental and control Ss in average learning efficiency may then be taken as an operational measure of the degree, ϕ, to which M influences acquisition of H in circumstances C. Unfortunately, however, as any experienced researcher knows to his sorrow, the interpretation of such an observed statistic is not quite so simple as that. For the observed dependent variable d, which is actually a performance measure, is a function not only of the extent to which M influences acquisition of H, but of many additional major and minor factors as well. Some of these, such as deprivations, species, age, laboratory conditions, etc., can be removed from consideration by holding them essentially constant. Others, however, are not so easily controlled, especially those customarily subsumed under the headings of "individual differences" and "errors of measurement." To curtail a long mathematical story, it turns out that with suitable (possibly justified) assumptions about the distributions of values for these uncontrolled variables, the manner in which they influence the dependent variable, and the way in which experimental and control Ss were selected and manipulated, the observed sample statistic d may be regarded as the value of a normally distributed random variate whose average value is ϕ and whose variance, which is independent of ϕ, is unbiasedly estimated by the square of another sample statistic, s, computed from the data of the experiment.[1]

The import of these statistical considerations for Hopewell's dissertation, of course, is that he will not be permitted to reason in any simple way from the observed d to a conclusion about the comparative merits of T_0 and T_1. To conclude that T_0, rather than T_1, is correct, he must argue that $\phi = 0$, rather

1. s is here the estimate of the standard error of the difference in means, not the estimate of the individual SD.

than $\phi > 0$. But the observed d, whatever its value, is logically compatible both with the hypothesis that $\phi = 0$ and the hypothesis that $\phi > 0$. How, then, can Hopewell use his data to make a comparison of T_0 and T_1? As a well-trained student, what he *does*, of course, is to divide d by s to obtain what, under H_0, is a t statistic, consult a table of the t distributions under the appropriate degrees-of-freedom, and announce his experiment as disconfirming or supporting T_0, respectively, according to whether or not the discrepancy between d and the zero value expected under T_0 is "statistically significant"—i.e. whether or not the observed value of d/s falls outside of the interval between two extreme percentiles (usually the 2.5th and 97.5th) of the t distribution with that df. If asked by his dissertation committee to justify this behavior, Hopewell would rationalize something like the following (the more honest reply, that this is what he has been taught to do, not being considered appropriate to such occasions):

> In deciding whether or not T_0 is correct, I can make two types of mistakes: I can reject T_0 when it is in fact correct [Type I error], or I can accept T_0 when in fact it is false [Type II error]. As a scientist, I have a professional obligation to be cautious, but a 5 percent chance of error is not unduly risky. Now if all my statistical background assumptions are correct, then, if it is really true that $\phi = 0$ as T_0 says, there is only one chance in 20 that my observed statistic d/s will be smaller than $t_{.025}$ or larger than $t_{.975}$, where by the latter I mean, respectively, the 2.5th and 97.5th percentiles of the t distribution with the same degrees-of-freedom as in my experiment. Therefore, if I reject T_0 when d/s is smaller than $t_{.025}$ or larger than $t_{.975}$, and accept T_0 otherwise, there is only a 5 percent chance that I will reject T_0 incorrectly.

If asked about his Type II error, and why he did not choose some other rejection region, say between $t_{.475}$, and $t_{.525}$, which would yield the same probability of Type I error, Hopewell should reply that, although he has no way to compute his probability of Type II error under the assumptions traditionally authorized by null hypothesis procedure, it is presumably minimized by taking the rejection region at the extremes of the t distribution.

Let us suppose that, for Hopewell's data, $d = 8.50$, $s = 5.00$, and $df = 20$. Then $t_{.975} = 2.09$ and the acceptance region for the null hypothesis $\phi = 0$ is $-2.09 < d/s < 2.09$, or $-10.45 < d < 10.45$. Since d does fall within this region, standard null hypothesis decision procedure, which I shall henceforth abbreviate "NHD," dictates that the experiment is to be reported as supporting theory T_0. (Although many persons would like to conceive NHD testing to authorize only rejection of the hypothesis, not, in addition, its acceptance when the test statistic fails to fall in the rejection region, if failure to reject were not taken as grounds for acceptance, then NHD procedure would involve no Type II error, and no justification would be given for taking the rejection region at the extremes of the distribution rather than in its middle.)

But even as Hopewell reaffirms T_0 in his dissertation, he begins to feel uneasy. In fact, several disquieting thoughts occur to him:

1. Although his test statistic falls within the orthodox acceptance region, a value this divergent from the expected zero should nonetheless be encountered less than once in 10. To argue in favor of a hypothesis on the basis of data ascribed a p value no greater than .10 (i.e. 10 percent) by that hypothesis certainly does not seem to be one of the more impressive displays of scientific caution.

2. After some belated reflection on the details of theory T_1, Hopewell observes that T_1 not only predicts that $\phi > 0$, but with a few simplifying assumptions no more questionable than is par for this sort of course, the value that ϕ should have can actually be computed. Suppose the value derived from T_1 in this way is $\phi = 10.0$. Then, rather than taking $\phi = 0$ as the null hypothesis, one might just as well take $\phi = 10.0$; for under the latter, $(d - 10.0)/s$ is a 20 df t statistic, giving a two-tailed, 95 percent significance, acceptance region for $(d - 10.0)/s$ between $-.209$ and 2.09. That is, if one lets T_1 provide the null hypothesis, it is accepted or rejected according to whether or not $-.45 < d < 20.45$, and by this latter test, therefore, Hopewell's data must be taken to support T_1—in fact, the likelihood under T_1 of obtaining a test statistic this divergent from the expected 10.0 is a most satisfactory three chances in four. Thus it occurs to Hopewell that had he chosen to cast his professional lot with the T_1-ists by selecting $\phi = 10.0$ as his null hypothesis, he could have made a strong argument in favor of T_1 by precisely the same line of statistical reasoning he has used to support T_0 under $\phi = 0$ as the null hypothesis. That is, he could have made an argument that persons partial to T_1 would regard as strong. For behaviorists who are already convinced that T_0 is correct would howl that since T_0 is the dominant theory, only $\phi = 0$ is a legitimate null hypothesis. (And is it not strange that what constitutes a valid statistical argument should be dependent upon the majority opinion about behavior theory?)

3. According to the NHD test of a hypothesis, only two possible final outcomes of the experiment are recognized—either the hypothesis is rejected or it is accepted. In Hopewell's experiment, all possible values of d/s between -2.09 and 2.09 have the same interpretive significance, namely, indicating that $\phi = 0$, while, conversely, all possible values of d/s greater than 2.09 are equally taken to signify that $\phi \neq 0$. But Hopewell finds this disturbing, for, of the various possible values that d/s might have had, the significance of $d/s = 1.70$ for the comparative merits of T_0 and T_1 should surely be more similar to that of, say, $d/s = 2.10$ than to that of, say, $d/s = -1.70$.

4. In somewhat similar vein, it also occurs to Hopewell that had he opted for a somewhat riskier confidence level, say a Type I error of 10 percent rather than 5 percent, d/s would have fallen outside the region of acceptance

and T_0 would have been rejected. Now surely the degree to which a datum corroborates or impugns a proposition should be independent of the datum-assessor's personal temerity. Yet, according to orthodox significance-test procedure, whether or not a given experimental outcome supports or disconfirms the hypothesis in question depends crucially upon the assessor's tolerance for Type I risk.

Despite his inexperience, Igor Hopewell is a sound experimentalist at heart, and the more he reflects on these statistics, the more dissatisfied with his conclusions he becomes. So while the exigencies of graduate circumstances and publication requirements urge that his dissertation be written as a confirmation of T_0, he nonetheless resolves to keep an open mind on the issue, even carrying out further research if opportunity permits. And reading his experimental report, so of course would we—has any responsible scientist ever made up his mind about such a matter on the basis of a single experiment? Yet in this obvious way we reveal how little our actual inferential behavior corresponds to the statistical procedure to which we pay lip-service. For if we did, in fact, accept or reject the null hypothesis according to whether the sample statistic falls in the acceptance or in the rejection region, then there would be no replications of experimental designs, no multiplicity of experimental approaches to an important hypothesis—a single experiment would, by definition of the method, make up our mind about the hypothesis in question. And the fact that, in actual practice, a single finding seldom even tempts us to such closure of judgment reveals how little the conventional model of hypothesis testing fits our actual evaluative behavior.

Decisions vs. Degrees of Belief

By now, it should be obvious that something is radically amiss with the traditional NHD assessment of an experiment's theoretical import. Actually, one does not have to look far in order to find the trouble—it is simply a basic misconception about the purpose of a scientific experiment. The null hypothesis significance test treats acceptance or rejection of a hypothesis as though these were *decisions* one makes on the basis of the experimental data—i.e. that we elect to adopt one belief, rather than another, as a result of an experimental outcome. *But the primary aim of a scientific experiment is not to precipitate decisions, but to make an appropriate adjustment in the degree to which one accepts, or believes, the hypothesis or hypotheses being tested.* And even if the purpose of the experiment *were* to reach a decision, it could not be a decision to accept or reject the hypothesis, for decisions are voluntary commitments to action—i.e. are *motor* sets—whereas acceptance or rejection of a hypothesis is a *cognitive* state, which may provide the basis for rational decisions but is not itself arrived at by such a decision (except perhaps indirectly in that a decision may initiate further experiences which influence the belief).

The situation, in other words, is as follows: As scientists, it is our professional obligation to reason from available data to explanations and generalities—i.e. beliefs—which are supported by these data. But belief in (i.e. acceptance of) a proposition is not an all-or-none affair; rather, it is a matter of degree, and the extent to which a person believes or accepts a proposition translates pragmatically into the extent to which he is willing to commit himself to the behavioral adjustments prescribed for him by the meaning of that proposition. For example, if that inveterate gambler, Unfortunate Q. Smith, has complete confidence that War Biscuit will win the fifth race at Belmont, he will be willing to accept any odds to place a bet on War Biscuit to win; for if he is absolutely *certain* that War Biscuit will win, then odds are irrelevant—it is simply a matter of arranging to collect some winnings after the race. On the other hand, the more that Smith has doubts about War Biscuit's prospects, the higher the odds he will demand before betting. That is, the *extent* to which Smith accepts or rejects the hypothesis that War Biscuit will win the fifth at Belmont is an important determinant of his betting decisions for that race.

Now, although a scientist's data supply *evidence* for the conclusions he draws from them, only in the unlikely case where the conclusions are logically deducible from or logically incompatible with the data do the data warrant that the conclusions be entirely accepted or rejected. Thus, e.g. the fact that War Biscuit has won all 16 of his previous starts is strong evidence in favor of his winning the fifth at Belmont, but by no means warrants the unreserved acceptance of this hypothesis. More generally, the data available confer upon the conclusions a certain *appropriate degree of belief*, and it is the inferential task of the scientist to pass from the data of his experiment to whatever *extent* of belief these and other available information justify in the hypothesis under investigation. In particular, the proper inferential procedure is *not* (except in the deductive case) a matter of deciding to accept (without qualification) or reject (without qualification) the hypothesis: Even if adoption of a belief were a matter of voluntary action—which it is not—neither such extremes of belief or disbelief are appropriate to the data at hand. As an example of the disastrous consequences of an inferential procedure which yields only two judgment values, acceptance and rejection, consider how sad the plight of Smith would be if, whenever weighing the prospects for a given race, he always worked himself into either supreme confidence or utter disbelief that a certain horse will win. Smith would rapidly impoverish himself by accepting excessively low odds on horses he is certain will win, and failing to accept highly favorable odds on horses he is sure will lose. In fact, Smith's two judgment values need not be *extreme* acceptance and rejection in order for his inferential procedure to be maladaptive. All that is required is that the degree of belief arrived at be in general inappropriate to the likelihood conferred on the hypothesis by the data.

Now, the notion of "degree of belief appropriate to the data at hand" has an unpleasantly vague, subjective feel about it which makes it unpalatable for inclusion in a formalized theory of inference. Fortunately, a little reflection about this phrase reveals it to be intimately connected with another concept relating conclusion to evidence which, though likewise in serious need of conceptual clarification, has the virtues both of intellectual respectability and statistical familiarity. I refer, of course, to the *likelihood*, or *probability*, conferred upon a hypothesis by available evidence. Why should not Smith *feel* certain, in view of the data available, that War Biscuit will win the fifth at Belmont? Because it *is* not certain that War Biscuit will win. More generally, what determines how strongly we should accept or reject a proposition is the probability given to this hypothesis by the information at hand. For, while our voluntary actions (i.e. decisions) are determined by our intensities of belief in the relevant propositions, not by their actual probabilities, expected utility is maximized when the cognitive weights given to potential but not yet known-for-certain pay-off events are represented in the decision procedure by the probabilities of these events. We may thus relinquish the concept of "appropriate degree of belief" in favor of "probability of the hypothesis," and our earlier contention about the nature of data-processing may be rephrased to say that the proper inferential task of the experimental scientist is not a simple acceptance or rejection of the tested hypothesis, but determination of the probability conferred upon it by the experimental outcome. This likelihood of the hypothesis relative to whatever data are available at the moment will be an important determinant for decisions which must currently be made, but is not itself such a decision and is entirely subject to revision in the light of additional information.

In brief, what is being argued is that the scientist, whose task is not to prescribe actions but to establish rational beliefs upon which to base them, is fundamentally and inescapably committed to an explicit concern with the problem of inverse probability. What he wants to know is how plausible are his hypotheses, and he is interested in the probability ascribed by a hypothesis to an observed experimental outcome only to the extent he is able to reason backwards to the likelihood of the hypothesis, given this outcome. Put crudely, no matter how improbable an observation may be under the hypothesis (and when there are an infinite number of possible outcomes, the probability of any particular one of these is, usually, infinitely small—the familiar p value for an observed statistic under a hypothesis H is not actually the probability of that outcome under H, but a partial integral of the probability-density function of possible outcomes under H), it is still confirmatory (or at least nondisconfirmatory, if one argues from the data to rejection of the background assumptions) so long as the likelihood of the observation is even smaller under the alternative hypotheses. To be sure, the theory of hypothesis-likelihood and inverse probability is as yet far from the level of development at which it can furnish the research scientist with inferential tools he can apply mechanically to obtain a definite likelihood estimate. But to the extent a statistical method does not at least move in the *direction* of computing the probability of the hypothesis, given the observation, that method is not truly a method of *inference*, and is unsuited for the scientist's cognitive ends.

The Methodological Status of the Null Hypothesis Significance Test

The preceding arguments have, in one form or another, raised several doubts about the appropriateness of conventional significance-test decision procedure for the aims it is supposed to achieve. It is now time to bring these charges together in an explicit bill of indictment.

1. The null hypothesis significance test treats "acceptance" or "rejection" of a hypothesis as though these were decisions one makes. But a hypothesis is not something, like a piece of pie offered for dessert, which can be accepted or rejected by a voluntary physical action. Acceptance or rejection of a hypothesis is a cognitive process, a *degree* of believing or disbelieving which, if rational, is not a matter of choice but determined solely by how likely it is, given the evidence, that the hypothesis is true.

2. It might be argued that the NHD test may nonetheless be regarded as a legitimate decision procedure if we translate "acceptance (rejection) of the hypothesis" as meaning "acting as though the hypothesis were true (false)." And to be sure, there are many occasions on which one must base a course of action on the credibility of a scientific hypothesis. (Should these data be published? Should I devote my research resources to and become identified professionally with this theory? Can we test this new Z bomb without exterminating all life on earth?) But such a move to salvage the traditional procedure only raises two further objections. (*a*) While the scientist—i.e. the person—must indeed make decisions, his *science* is a systematized body of (probable) *knowledge*, not an accumulation of decisions. The end product of a scientific investigation is a degree of confidence in some set of propositions, which then constitutes a *basis* for decisions. (*b*) Decision theory shows the NHD test to be woefully inadequate as a decision procedure. In order to decide most effectively when or when not to act as though a hypothesis is correct, one must know both the probability of the hypothesis under the data available and the utilities of the various decision outcomes (i.e. the values of accepting the hypothesis when it is true, of accepting it when it is false, of

rejecting it when it is true, and of rejecting it when it is false). But traditional NHD procedure pays no attention to utilities at all and considers the probability of the hypothesis, given the data—i.e. the inverse probability—only in the most rudimentary way (by taking the rejection region at the extremes of the distribution rather than in its middle). Failure of the traditional significance test to deal with inverse probabilities invalidates it not only as a method of rational inference, but *also* as a useful decision procedure.

3. The traditional NHD test unrealistically limits the significance of an experimental outcome to a mere two alternatives, confirmation or disconfirmation of the null hypothesis. Moreover, the transition from confirmation to disconfirmation as a function of the data is discontinuous—an arbitrarily small difference in the value of the test statistic can change its significance from confirmatory to disconfirmatory. Finally, the point at which this transition occurs is entirely gratuitous. There is absolutely no reason (at least provided by the method) why the point of statistical "significance" should be set at the 95 percent level, rather than, say, the 94 percent or 96 percent level. Nor does the fact that we sometimes select a 99 percent level of significance, rather than the usual 95 percent level, mitigate this objection—one is as arbitrary as the other.

4. The null hypothesis significance test introduces a strong bias in favor of one out of what may be a large number of reasonable alternatives. When sampling a distribution of unknown mean μ, different assumptions about the value of μ furnish an infinite number of alternative null hypotheses by which we might assess the sample mean, and whichever hypothesis is selected is thereby given an enormous, in some cases almost insurmountable, advantage over its competitors. That is, NHD procedure involves an inferential double standard—the favored hypothesis is held innocent unless proved guilty, while any alternative is held guilty until no choice remains but to judge it innocent. What is objectionable here is not that some hypotheses are held more resistant to experimental extinction than others, but that the differential weighing is an all-or-none side effect of a personal choice, and especially, that the method *necessitates* one hypothesis being favored over all the others. In the classical theory of inverse probability, on the other hand, all hypotheses are treated on a par, each receiving a weight (i.e. its "*a priori*" probability) which reflects the credibility of that hypothesis on grounds other than the data being assessed.

5. Finally, if anything can reveal the practical irrelevance of the conventional significance test, it should be its failure to see genuine application to the inferential behavior of the research scientist. Who has ever given up a hypothesis just because one experiment yielded a test statistic in the rejection region? And what scientist in his right mind would ever feel there to be an appreciable difference between the interpretive significance of data, say, for

which one-tailed $p = .04$ and that of data for which $p = .06$, even though the point of "significance" has been set at $p = .05$? In fact, the reader may well feel undisturbed by the charges raised here against traditional NHD procedure precisely because, without perhaps realizing it, he has never taken the method seriously anyway. Paradoxically, it is often the most firmly institutionalized tenet of faith that is most susceptible to untroubled disregard—in our culture, one must early learn to live with sacrosanct verbal formulas whose import for practical behavior is seldom heeded. I suspect that the primary reasons why null hypothesis significance testing has attained its current ritualistic status are (*a*) the surcease of methodological insecurity afforded by having an inferential algorithm on the books, and (*b*) the fact that a by-product of the algorithm is so useful, and its end product so obviously inappropriate, that the latter can be ignored without even noticing that this has, in fact, been done. What has given the traditional method its spurious feel of usefulness is that the *first*, and by far most laborious, step in the procedure, namely, estimating the probability of the experimental outcome under the assumption that a certain hypothesis is correct, is also a crucial first step toward what one is genuinely concerned with, namely, an idea of the likelihood of that hypothesis, given this experimental outcome. Having obtained this most valuable statistical information under pretext of carrying through a conventional significance test, it is then tempting, though of course quite inappropriate, to heap honor and gratitude upon the method while overlooking that its actual *result*, namely, a decision to accept or reject, is not used at all.

Toward a More Realistic Appraisal of Experimental Data

So far, my arguments have tended to be aggressively critical—one can hardly avoid polemics when butchering sacred cows. But my purpose is not just to be contentious, but to help clear the way for more realistic techniques of data assessment, and the time has now arrived for some constructive suggestions. Little of what follows pretends to any originality; I merely urge that ongoing developments along these lines should receive maximal encouragement.

For the statistical theoretician, the following problems would seem to be eminently worthy of research:

1. Of supreme importance for the theory of probability is analysis of what we mean by a proposition's "probability," relative to the evidence provided. Most serious students of the philosophical foundations of probability and statistics agree (see Braithwaite, 1953: 119f) that the probability of a proposition (e.g. the probability that the General Theory of Relativity is correct) does not, *prima facie*, seem to be the same sort of thing as the probability of an event-class (e.g. the probability of getting a head when this coin is tossed).

Do the statistical concepts and formulas which have been developed for probabilities of the latter kind also apply to hypothesis likelihoods? In particular, are the probabilities of hypotheses quantifiable at all, and for the theory of inverse probability, do Bayes' theorem and its probability-density refinements apply to hypothesis probabilities? These and similar questions are urgently in need of clarification.

2. If we are willing to assume that Bayes' theorem, or something like it, holds for hypothesis probabilities, there is much that can be done to develop the classical theory of inverse probability. While computation of inverse probabilities turns essentially upon the parametric *a priori* probability function, which states the probability of each alternative hypothesis in the set under consideration prior to the outcome of the experiment, it should be possible to develop theorems which are invariant over important sub-classes of *a priori* probability functions. In particular, the difference between the *a priori* probability function and the "*a posteriori*" probability function (i.e. the probabilities of the alternative hypotheses after the experiment), perhaps analyzed as a difference in "information," should be a potentially fruitful source of concepts with which to explore such matters as the "power" or "efficiency" of various statistics, the acquisition of inductive knowledge through repeated experimentation, etc. Another problem which seems to me to have considerable import, though not one about which I am sanguine, is whether inverse-probability theory can significantly be extended to hypothesis-probabilities, given knowledge which is only probabilistic. That is, can a theory of sentences of form "The probability of hypothesis *H*, given that *E* is the case, is *p*," be generalized to a theory of sentences of form "The probability of hypothesis *H*, given that the probability of *E* is *q*, is *p*"? Such a theory would seem to be necessary, e.g. if we are to cope adequately with the uncertainty attached to the background assumptions which always accompany a statistical analysis.

My suggestions for applied statistical analysis turn on the fact that, while what is desired is the *a posteriori* probabilities of the various alternative hypotheses under consideration, computation of these by classical theory necessitates the corresponding *a priori* probability distribution, and in the more immediate future, at least, information about this will exist only as a subjective feel, differing from one person to the next, about the credibilities of the various hypotheses.

3. Whenever possible, the basic statistical report should be in the form of a *confidence interval*. Briefly, a confidence interval is a subset of the alternative hypotheses computed from the experimental data in such a way that for a selected confidence level α, the probability that the true hypothesis is included in a set so obtained is α. Typically, an α-level confidence interval consists of those hypotheses under which the *p* value for the experimental

outcome is larger than $1 - \alpha$ (a feature of confidence intervals which is some-times confused with their definition), in which case the confidence-interval report is similar to a simultaneous null hypothesis significance test of each hypothesis in the total set of alternatives. Confidence intervals are the closest we can at present come to quantitative assessment of hypothesis-probabilities (see *technical note*, below), and are currently our most effective way to elimi-nate hypotheses from practical consideration—if we choose to act as though none of the hypotheses not included in a 95 percent confidence interval are correct, we stand only a 5 percent chance of error. (Note, moreover, that this probability of error pertains to the incorrect simultaneous "rejection" of a major part of the total set of alternative hypotheses, not just to the incorrect rejection of one as in the NHD method, and is a *total* likelihood of error, not just of Type I error.) The confidence interval is also a simple and effective way to convey that all-important statistical datum, the conditional proba-bility (or probability density) function—i.e. the probability (probability density) of the observed outcome under each alternative hypothesis—since for a given kind of observed statistic and method of confidence-interval determination, there will be a fixed relation between the parameters of the confidence interval and those of the conditional probability (probability density) function, with the end-points of the confidence interval typically marking the points at which the conditional probability (probability density) function sinks below a certain small value related to the parameter α. The confidence-interval report is not biased toward some favored hypothesis, as is the null hypothesis significance test, but makes an impartial simultaneous evaluation of all the alternatives under consideration. Nor does the confi-dence interval involve an arbitrary decision as does the NHD test. Although one person may prefer to report, say, 95 percent confidence intervals while another favors 99 percent confidence intervals, there is no conflict here, for these are simply two ways to convey the same information. An experimental report can, with complete consistency and some benefit, simultaneously present several confidence intervals for the parameter being estimated. On the other hand, different choices of significance level in the NHD method is a clash of incompatible decisions, as attested by the fact than an NHD analysis which simultaneously presented two different significance levels would yield a logi-cally inconsistent conclusion when the observed statistic has a value in the ac-ceptance region of one significance level and in the rejection region of the other.

Technical note: One of the more important problems now confronting theoreti-cal statistics is exploration and clarification of the relationships among inverse probabilities derived from confidence-interval theory, fiducial-probability theory (a special case of the former in which the estimator is a sufficient statistic), and classical (i.e. Bayes') inverse-probability theory. While the interpretation of con-fidence intervals is tricky, it would be a mistake to conclude, as the cautionary remarks usually accompanying discussions of confidence intervals sometimes seem to imply, that the confidence-level α of a given confidence interval I should not really be construed as a probability that the true hypothesis, H, belongs to the set I. Nonetheless, if I is an α-level confidence interval, the probability that H belongs to I as computed by Bayes' theorem given an *a priori* probability distribution will, in general, *not* be equal to α, nor is the difference necessarily a small one—it is easy to construct examples where the *a posteriori* probability that H belongs to I is either 0 or 1. Obviously, when different techniques for computing the probability that H belongs to I yield such different answers, a reconciliation is demanded. In this instance, however, the apparent disagreement is largely if not entirely spurious, resulting from differences in the evidence relative to which the probability that H belongs to I is computed. And if this is, in fact, the correct explanation, then fiducial probability furnishes a partial solution to an outstanding difficulty in the Bayes approach. A major weakness of the latter has always been the problem of what to assume for the *a priori* distribution when no pre-experimental information is available other than that supporting the background assumptions which delimit the set of hypotheses under consideration. The traditional assumption (made hesitantly by Bayes, less hesitantly by his successors) has been the "principle of insufficient reason," namely, that given no knowledge at all, all alternatives are equally likely. But not only is it difficult to give a convincing argument for this assumption, it does not even yield a unique *a priori* probability distribution over a continuum of alternative hypotheses, since there are many ways to express such a continuous set, and what is an equi-likelihood *a priori* distribution under one of these does not necessarily transform into the same under another. Now, a fiducial probability distribution determined over a set of alternative hypotheses by an experimental observation is a measure of the likelihoods of these hypotheses relative to all the information contained in the experimental data, but based on no pre-experimental information beyond the background assumptions restricting the possibilities to this particular set of hypotheses. Therefore, it seems reasonable to postulate that the no-knowledge *a priori* distribution in classical inverse probability theory should be that distribution which, when experimental data capable of yielding a fiducial argument are now given, results in an *a posteriori* distribution identical with the corresponding fiducial distribution.

4. While a confidence-interval analysis treats all the alternative hypotheses with glacial impartiality, it nonetheless frequently occurs that our interest is focused on a certain selection from the set of possibilities. In such case, the statistical analysis should also report, when computable, the precise p value of the experimental outcome, or better, though less familiarly, the probability density at that outcome, under each of the major hypotheses; for these figures will permit an immediate judgment as to which of the hypotheses is most favored by the data. In fact, an even more interesting assessment of the post-experimental credibilities of the hypotheses is then possible through use of "likelihood ratios" if one is willing to put his pre-experimental feelings about

their relative likelihoods into a quantitative estimate. For, let $Pr(H,d)$, $Pr(d,H)$, and $Pr(H)$ be, respectively, the probability of a hypothesis H in light of the experimental data d (added to the information already available), the probability of data d under hypothesis H, and the pre-experimental (i.e. *a priori*) probability of H. Then, for two alternative hypotheses H_0 and H_1, it follows by classical theory that

$$\frac{Pr(H_0, d)}{Pr(H_1, d)} = \frac{Pr(H_0)}{Pr(H_1)} \times \frac{Pr(d, H_0)}{Pr(d, H_1)} \qquad [1]^2$$

Therefore, if the experimental report includes the probability (or probability density) of the data under H_0 and H_1, respectively, and its reader can quantify his feelings about the relative pre-experimental merits of H_0 and H_1 (i.e. $Pr(H_0)/Pr(H_1)$), he can then determine the judgment he should make about the relative merits of H_0 and H_1 in light of these new data.

5. Finally, experimental journals should allow the researcher much more latitude in publishing his statistics in whichever form seems most insightful, especially those forms developed by the modern theory of estimates. In particular, the stranglehold that conventional null hypothesis significance testing has clamped on publication standards must be broken. Currently justifiable inferential algorithm carries us only through computation of conditional probabilities; from there, it is for everyman's clinical judgment and methodological conscience to see him through to a final appraisal. Insistence that published data must have the biases of the NHD method built into the report, thus seducing the unwary reader into a perhaps highly inappropriate interpretation of the data, is a professional disservice of the first magnitude.

Summary

The traditional null hypothesis significance-test method, more appropriately called "null hypothesis decision [NHD] procedure," of statistical analysis is here vigorously excoriated for its inappropriateness as a method of *inference*. While several serious objections to the method are raised, its most basic error lies in mistaking the aim of a scientific investigation to be a *decision*, rather than a *cognitive* evaluation of propositions. It is further argued that the proper application of statistics to scientific inference is irrevocably committed to extensive consideration of inverse probabilities, and to further this end, certain suggestions are offered, both for the development of statistical theory and for more illuminating application of statistical analysis to empirical data.

2. When the numbers of alternative hypotheses and possible experimental outcomes are transfinite, $Pr(d, H) = Pr(H, d) = Pr(H) = 0$ in most cases. If so, the probability ratios in Formula 1 are replaced with the corresponding probability-density ratios. It should be mentioned that this formula rather idealistically presupposes there to be no doubt about the correctness of the background statistical assumptions.

25. The Test of Significance in Psychological Research

DAVID BAKAN

THE VAST MAJORITY of investigations which pass for research in the field of psychology today entail the use of statistical tests of significance. Most characteristically, when a psychologist finds a problem he wishes to investigate he converts his intuitions and hypotheses into procedures which will yield a test of significance and will characteristically allow the result of the test of significance to bear the essential responsibility for the conclusions he will draw.

I will attempt to show that the test of significance does not provide the information concerning psychological phenomena characteristically attributed to it; and that, futhermore, a great deal of mischief has been associated with its use. If the test of significance does not yield the expected information concerning the psychological phenomena under investigation, we may well speak of a crisis (Hogben, 1957 [see Chapters 1, 2, 3]); for then a good deal of the research of the last several decades must be questioned. What will be said in this paper is hardly original. It is, in a certain sense, what "everybody knows." To say it "out loud" is, as it were, to assume the role of the child who pointed out that the emperor really had no clothes on. Little of what is

Reprinted from Chapter 1 of David Bakan, *On Method*, San Francisco: Jossey-Bass, 1967, 1–29, by permission of the author and the publisher. Copyright © 1967 by Jossey-Bass, Inc. An earlier version of this essay appeared in *Psychological Bulletin*, 66 (December, 1966), 423–437. Copyright © 1966 by the American Psychological Association, Inc.

contained in this paper is not already available in the literature, and the literature will be cited.

Lest what is being said here be misunderstood, some clarification needs to be made at the outset. It is not a blanket criticism of statistics, of mathematics, or, for that matter, even of the test of significance when it can be appropriately used, as in certain decision situations. The argument is rather that the test of significance has been carrying too much of the burden of scientific inference. It may well be the case that wise and ingenious investigators can find their way to reasonable conclusions from data because and in spite of their procedures. Too often, however, even wise and ingenious investigators, for varieties of reasons not the least of which are the editorial policies of our major psychological journals, which we will discuss below, tend to credit the test of significance with properties it does not have.

The test of significance has as its aim obtaining information concerning a characteristic of a *population* which is itself not directly observable, whether for practical or more intrinsic reasons. What is observable is the *sample*. The work assigned to the test of significance is that of aiding in making inferences from the observed sample to the unobserved population.

The critical assumption involved in testing significance is that, if the experiment is conducted properly, the characteristics of the population have a designably determinative influence on samples drawn from it; that, for example, the mean of a population has a determinative influence on the mean of a sample drawn from it. Thus if P, the population characteristic, has a determination influence on S, the sample characteristic, then there is some license for making inferences from S to P.

If the determinative influence of P on S could be put in the form of simple logical implication, that P implies S, the problem would be quite simple. For, then we would have the simple situation: If P implies S, and if S is false, P is false. There are some limited instances in which this logic applies directly in sampling. For example, if the range of values in the population is between 3 and 9 (P), then the range of values in any sample must be between 3 and 9 (S). Should we find a value in a sample of, say, 10, it would mean that S is false; and we could assert that P is false.

It is clear from this, however, that, strictly speaking, one can only go from the denial of S to the denial of P; and not from the assertion of S to the assertion of P. It is within this context of simple logical implication that the Fisher school of statisticians has made important contributions—and it is extremely important to recognize this as the context.

In contrast, approaches based on the theorem of Bayes' (Edwards *et al.*, 1963; Keynes, 1921; Savage, 1954; Schlaifer, 1959) would allow inferences to P from S even when S is not denied, as S adding something to the credibility of P when S is found to be the case. One of the most viable alternatives to the use of the test of significance involves the theorem of Bayes'; and the paper by Edwards *et al.* (1963) is particularly directed to the attention of psychologists for use in psychological research.

The notion of the null hypothesis[1] promoted by Fisher (1935a, 4th ed.) constituted an advance within this context of simple logical implication. It allowed experimenters to set up a null hypothesis complementary to the hypothesis that the investigator was interested in, and provided him with a way of positively confirming his hypothesis. Thus, for example, the investigator might have the hypothesis that, say, normals differ from schizophrenics. He would then set up the null hypothesis that the means in the population of all normals and all schizophrenics were equal. Thus, the rejection of the null hypothesis constituted a way of asserting that the means of the populations of normals and schizophrenics were different, a seemingly reasonable device whereby to affirm a logical antecedent.

The model of simple logical implication for making inferences from S to P has another difficulty which the Fisher approach sought to overcome. This is that it is rarely meaningful to set up any simple "P implies S" model for parameters that we are interested in. In the case of the mean, for example, it is rather that P has a determinative influence on the frequency of any specific S. But one experiment does not provide many values of S to allow the study of their frequencies. It gives us only one value of S. The sampling distribution is conceived which specifies the relative frequencies of all possible values of S. Then, with the help of an adopted level of significance, we could, in effect, say that S was false; that is, any S which fell in a region whose relative theoretical frequency under the null hypothesis was, say, 5 percent would be considered false. If such an S actually occurred, we would be in a position to declare P to be false, still within the model of simple logical implication.

It is important to recognize that one of the essential features of the Fisher approach is what may be called the "once-ness" of the experiment; the inference model takes as critical that the experiment has been conducted once. If an S which has a low probability under the null hypothesis actually occurs, it is taken that the null hypothesis is false. As Fisher (1935a, 4th ed.) put it, why should the theoretically rare event under the null hypothesis actually occur to "us"? If it does occur, we take it that the null hypothesis is false.

1. There is some confusion in the literature concerning the meaning of the term "null hypothesis." Fisher used the term to designate any exact hypothesis that we might be interested in disproving, and "null" was used in the sense of that which is to be nullified (see, for example, Berkson, 1942 [Chapter 28]). It has, however, also been used to indicate a parameter of zero (see, for example, Lindquist, 1940): the difference between the population means is zero, or the correlation coefficient in the population is zero, the difference in proportions in the population is zero, etc. Since both meanings are usually intended in psychological research, it causes little difficulty.

Basic is the idea that "the theoretically unusual does not happen to me."[2] It should be noted that the referent for all probability considerations is neither in the population itself nor the subjective confidence of the investigator. It is rather in a hypothetical population of experiments all conducted in the same manner, but only one of which is actually conducted. Thus, of course, the probability of falsely rejecting the null hypothesis if it were true is exactly that value which has been taken as the level of significance. Replication of the experiment vitiates the validity of the inference model, unless the replication itself is taken into account in the model and the probabilities of the model modified accordingly (as is done in various designs which entail replication, where, however, the total experiment, including the replications, is again considered as *one* experiment). According to Fisher (1935a, 4th ed.), "It is an essential characteristic of experimentation that it is carried out with limited resources." In the Fisher approach, the "limited resources" is not only a making of the best out of a limited situation, but is rather an integral feature of the inference model itself. Lest he be done a complete injustice, it should be pointed out that he did say, "In relation to the test of significance, we may say that a phenomenon is experimentally demonstrable when we know how to conduct an experiment which will rarely fail to give us statistically significant results." However, although Fisher himself believes this, it is not built into the inference model.[3]

As already indicated, research workers in the field of psychology place a heavy burden on the test of significance. Let us consider some of the difficulties associated with the null hypothesis.

1. *The a priori reasons for believing that the null hypothesis is generally false anyway.* One of the common experiences of research workers is the very high frequency with which significant results are obtained with large samples. Some years ago, the author had occasion to run a number of tests of significance on a battery of tests collected on about 60,000 subjects from all over the United States. Every test came out significant. Dividing the cards by such

2. I playfully once conducted the following "experiment": Suppose, I said, that every coin has associated with it a "spirit"; and suppose, furthermore, that if the spirit is implored properly, the coin will veer head or tail as one requests of the spirit. I thus invoked the spirit to make the coin fall head. I threw it once; it came up head. I did it again; it came up head again. I did this six times, and got six heads. Under the null hypothesis the probability of occurrence of six heads is $(\frac{1}{2})^6 = .016$, significant at the 2 percent level of significance. I have never repeated the experiment. But, then, the logic of the inference model does not really demand that I do! It may be objected that the coin, or my tossing, or even my observation was biased. But I submit that such things were in all likelihood not as involved in the result as corresponding things in most psychological research.

3. Possibly not even this criterion is sound. It may be that a number of statistically significant results which are borderline "speak for the null hypothesis rather than against it" (Edwards *et al.*, 1963). If the null hypothesis were really false, then, with an increase in the number of instances in which it can be rejected, there should be some substantial proportion of more dramatic rejections rather than borderline rejections.

arbitrary criteria as east versus west of the Mississippi River, Maine versus the rest of the country, North versus South, etc., all produced significant differences in means. In some instances, the differences in the sample means were quite small, but nonetheless, the *p* values were all very low. Nunnally (1960) has reported a similar experience involving correlation coefficients on 700 subjects. Joseph Berkson made the observation almost 30 years ago in connection with Chi-square:

> I believe that an observant statistician who has had any considerable experience with applying the Chi-square test repeatedly will agree with my statement that, as a matter of observation, when the numbers in the data are quite large, the *P*'s tend to come out small. Having observed this, and on reflection, I make the following dogmatic statement, referring for illustration to the normal curve: "If the normal curve is fitted to a body of data representing any real observations whatever of quantities in the physical world, then if the number of observations is extremely large—for instance, on an order of 200,000—the Chi-square *P* will be small beyond any usual limit of significance."
>
> This dogmatic statement is made on the basis of an extrapolation of the observation referred to and can also be defended as a prediction from *a priori* considerations. For we may assume that it is practically certain that any series of real observations does not actually follow a normal curve *with absolute exactitude* in all respects, and no matter how small the discrepancy between the normal curve and the true curve of observations, the Chi-square *P* will be small if the sample has a sufficiently large number of observations in it.
>
> If this be so, then we have something here that is apt to trouble the conscience of a reflective statistician using the Chi-square test. For I suppose it would be agreed by statisticians that a large sample is always better than a small sample. If, then, we know in advance the *P* that will result from an application of a Chi-square test to a large sample, there would seem to be no use in doing it on a smaller one. But since the result of the former test is known, it is no test at all (Berkson, 1938: 526–527).

As one group of authors has put it, "In typical applications . . . the null hypothesis . . . is known by all concerned to be false from the outset" (Edwards *et al.*, 1963). The fact of the matter is that there is really no good reason to expect the null hypothesis to be true in any population. Why should the mean, say, of all scores east of the Mississippi be identical to all scores west of the Mississippi? Why should any correlation coefficient be exactly .00 in the population? Why should we expect the ratio of males to females to be exactly 50:50 in any population? Or why should different drugs have exactly the same effect on any population parameter (Smith, 1960)? A glance at any set of statistics on total populations will quickly confirm the rarity of the null hypothesis in nature.

The reason the null hypothesis is characteristically rejected with large samples was made patent by the theoretical work of Neyman and Pearson

(1933). The probability of rejecting the null hypothesis is a function of five factors: whether the test is one- or two-tailed, the level of significance, the standard deviation, the amount of deviation from the null hypothesis, and the number of observations. The choice of a one- or two-tailed test is the investigator's; the level of significance is also based on the choice of the investigator; the standard deviation is a given of the situation and is characteristically reasonably well estimated; the deviation from the null hypothesis is what is unknown; and the choice of the number of cases in psychological work is characteristically arbitrary or expediential. Should there be any deviation from the null hypothesis in the population, no matter how small—and we have little doubt but that such a deviation usually exists—a sufficiently large number of observations will lead to the rejection of the null hypothesis. As Nunnally put it,

> If the null hypothesis is not rejected, it is usually because the N is too small. If enough data are gathered, the hypothesis will generally be rejected. If rejection of the null hypothesis were the real intention in psychological experiments, there usually would be no need to gather data (1960: 643).

2. *Type I error and publication practices.* The Type I error is the error of rejecting the null hypothesis when it is indeed true, and its probability is the level of significance. Later in this paper we will discuss the distinction between sharp and loose null hypotheses. The sharp null hypothesis, which we have been discussing, is an exact value for the null hypothesis as, for example, the difference between population means being precisely zero. A loose null hypothesis is one in which it is conceived of as being "around" null. Sharp null hypotheses, as we have indicated, rarely exist in nature. Assuming that loose null hypotheses are not rare and that their testing may make sense under some circumstances, let us consider the role of the publication practices of our journals in their connection.

It is the practice of editors of our psychological journals, receiving many more papers than they can possibly publish, to use the magnitude of the p values reported as one criterion for acceptance or rejection of a study. For example, consider the following statement made by Arthur W. Melton on completing twelve years as editor of the *Journal of Experimental Psychology*, certainly one of the most prestigious and scientifically meticulous psychological journals. In listing the criteria by which articles were evaluated, he said:

> The next step in the assessment of an article involved a judgment with respect to the confidence to be placed in the findings—confidence that the results of the experiment would be repeatable under the conditions described. In editing the *Journal* there has been a strong reluctance to accept and publish results related to the principal concern of the research when those results were significant at the .05 level, whether by one- or two-tailed test. This has not implied a slavish

worship of the .01 level, as some critics may have implied. Rather, it reflects a belief that it is the responsibility of the investigator in a science to reveal his effect in such a way that no reasonable man would be in a position to discredit the results by saying that they were the product of the way the ball bounces (1962: 553–554).

His clearly expressed opinion that nonsignificant results should not take up the space of the journals is shared by most editors of psychological journals. It is important to point out that I am not advocating a change in policy in this connection. In the total research enterprise where so much of the load for making inferences concerning the nature of phenomena is carried by the test of significance, the editors can do little else. The point is rather that the situation in regard to publication makes manifest the difficulties in connection with the overemphasis on the test of significance as a principal basis for making inferences.

McNemar (1960) has rightly pointed out that not only do journal editors reject papers in which the results are not significant, but that papers in which significance has not been obtained are not submitted, that investigators select out their significant findings for inclusion in their reports, and that theory-oriented research workers tend to discard data which do not work to confirm their theories. The result of all of this is that "published results are more likely to involve false rejection of null hypotheses than indicated by the stated levels of significance," that is, published results which are significant may well have Type I errors in them far in excess of, say, the 5 percent which we may allow ourselves.

The suspicion that the Type I error may well be plaguing our literature is given confirmation in an analysis of articles published in the *Journal of Abnormal and Social Psychology* for one complete year (Cohen, 1962). Analyzing seventy studies in which significant results were obtained with respect to the power of the statistical tests used, Cohen found that power, the probability of rejecting the null hypothesis when the null hypothesis was false, was characteristically meager. Theoretically, with such tests, one should not often expect significant results even when the null hypothesis was false. Yet, there they were! Even if deviations from null existed in the relevant populations, the investigations were characteristically not powerful enough to have detected them. This strongly suggests that there is something additional associated with these rejections of the null hypotheses in question. It strongly points to the possibility that the manner in which studies get published is associated with the findings; that the very publication practices themselves are part and parcel of the probabilistic processes on which we base our conclusions concerning the nature of psychological phenomena. Our total research enterprise is, at least in part, a kind of scientific roulette, in which the "lucky," or constant player, "wins," that is, get his paper or papers published. And certainly, going from 5 percent to 1 percent does not eliminate the possibility that it is "the way the ball bounces," to use Melton's phrase. It changes the odds in this roulette, but it does not make it less a game of roulette.

The damage to the scientific enterprise is compounded by the fact that the publication of "significant" results tends to stop further investigation. If the publication of papers containing Type I errors tended to foster further investigation so that the psychological phenomena with which we are concerned would be further probed by others, it would not be too bad. But it does not. Quite the contrary. As Lindquist (1940) has correctly pointed out, the danger to science of the Type I error is much more serious than the Type II error—for when a Type I error is committed, it has the effect of stopping investigation. A highly significant result appears definitive, as Melton's comments indicate. In the twelve years that he edited the *Journal of Experimental Psychology*, he sought to select papers which were worthy of being placed in the "archives," as he put it. Even the strict repetition of an experiment and not getting significance in the same way does not speak against the result already reported in the literature. For failing to get significance, speaking strictly within the inference model, only means that that experiment is inconclusive; whereas the study already reported in the literature, with a low p value, is regarded as conclusive. Thus we tend to place in the archives studies with a relatively high number of Type I errors, or, at any rate, studies which reflect small deviations from null in the respective populations; and we act in such a fashion as to reduce the likelihood of their correction. From time to time the suggestion has arisen that journals should open their pages for "negative results," so called. What is characteristically meant is that the null hypothesis has not been rejected at a conventional level of significance. This is hardly a solution to the problem simply because a failure to reject the null hypothesis is not a "negative result." It is only an instance in which the experiment is inconclusive.

To make this point clearer let us consider the odd case in which the null hypothesis may actually be true; say, the difference between means of a given measure of two identifiable groups in the population is precisely zero. Let us imagine that over the world there are one hundred experimenters who have independently embarked on testing this particular null hypothesis. By the theory under which the whole test of significance is conceived, approximately ninety-five of these experimenters would wind up by not being able to reject the null hypothesis, that is, their results would not be significant. It is not likely that they would write up their experiments and submit them to any journals. However, approximately five of these experimenters would find that their observed difference in means is significant at the 5 percent level of significance. It is likely that they would write up their experiments and submit them for publication. Indeed, one might imagine interesting quarrels arising among them concerning priority of discovery, if the differences came out in the same direction, and controversy, if the differences came out in different directions. In the former instance, the psychological community might even take it as evidence of "replicability" of the phenomenon, in the latter instance as evidence that the scientific method is "self-corrective." The other ninety-five experimenters would wonder what they did wrong. And this is in the odd instance in which the true difference between means in the population is precisely zero!

The psychological literature is filled with misinterpretations of the nature of the test of significance. One may be tempted to attribute this to such things as lack of proper education, the simple fact that humans may err, and the prevailing tendency to take a cookbook approach in which the mathematical and philosophical framework out of which the tests of significance emerge are ignored; that, in other words, these misinterpretations are somehow the result of simple intellectual inadequacy on the part of psychologists. However, such an explanation is hardly tenable. Graduate schools are adamant with respect to statistical education. Any number of psychologists have taken out substantial amounts of time to equip themselves mathematically and philosophically. Psychologists as a group do a great deal of mutual criticism. Editorial reviews prior to publication are carried out with eminent conscientiousness. There is even a substantial literature devoted to various kinds of "misuse" of statistical procedures, to which not a little attention has been paid.

It is rather that the test of significance is profoundly interwoven with other strands of the psychological research enterprise in such a way that it constitutes a critical part of the total cultural-scientific tapestry. To pull out the strand of the test of significance would seem to make the whole tapestry fall apart. In the face of the intrinsic difficulties that the test of significance provides, we rather attempt to make an "adjustment" by attributing to the test of significance characteristics which it does not have and overlook characteristics that it does have. The difficulty is that the test of significance can, especially when not considered too carefully, do *some* work; for, after all, the results of the test of significance *are* related to the phenomena in which we are interested. One may well ask whether we do not have here, perhaps, an instance of the phenomenon that learning under partial reinforcement is very highly resistant to extinction. Some of these misinterpretations are as follows:

1. *Taking the* p *value as a "measure" of significance.* A common misinterpretation of the test of significance is to regard it as a "measure" of significance. It is interpreted as the answer to the question "How significant is it?" A *p* value of .05 is thought of as less significant than a *p* value of .01, and so on. The characteristic practice on the part of psychologists is to compute, say, a *t,* and then "look up" the significance in the table, taking the *p* value as a

function of t, and thereby a "measure" of significance. Indeed, since the p value is inversely related to the magnitude of, say, the difference between means in the sample, it can function as a kind of "standard score" measure for a variety of different experiments. Mathematically, the t is actually very similar to a "standard score," entailing a deviation in the numerator, and a function of the variation in the denominator; and the p value is a "function" of t. If this use were explicit, it would perhaps not be too bad. But it must be remembered that this is using the p value as a statistic descriptive of the sample alone and does not automatically give an inference to the population. There is even the practice of using tests of significance in studies of total populations, in which the observations cannot by any stretch of the imagination be thought of as having been randomly selected from any designable population.[4] Using the p value in this way, in which the statistical inference model is even hinted at, is completely indefensible; for the single function of the statistical inference model is making inferences to populations from samples.

The practice of "looking up" the p value for the t, which has even been advocated in some of our statistical handbooks (e.g. Lacey, 1953; Underwood et al., 1954), rather than looking up the t for a given p value, violates the inference model. The inference model is based on the presumption that one initially adopts a level of significance as the specification of that probability which is too low to occur to "us," as Fisher has put it, in this one instance, and under the null hypothesis. A purist might speak of the "delicate problem . . . of fudging with *a posteriori* alpha values (levels of significance)" (Kaiser, 1960), as though the levels of significance were initially decided upon, but rarely do psychological research workers or editors take the level of significance as other than a "measure."

But taken as a "measure," it is only a measure of the sample. Psychologists often erroneously believe that the p value is "the probability that the results are due to chance," as Wilson (1961) has pointed out; that a p value of .05 means that the chances are .95 that the scientific hypothesis is correct, as Bolles (1962) has pointed out; that it is a measure of the power to "predict" the behavior of a population (Underwood et al., 1954); and that it is a measure of the "confidence that the results of the experiment would be repeatable under the conditions described," as Melton (1962) put it. Unfortunately, none of these intepretations is within the inference model of the test of significance. Some of our statistical handbooks have "allowed" misinterpretation. For example, in discussing the erroneous rhetoric associated with talking of the "probability" of a population parameter (in the inference model there is no probability associated with something which is either true or false), Lindquist (1940) said, "For most practical purposes, the end result

---

4. It was decided not to cite any specific studies to exemplify points such as this one. The reader will undoubtedly be able to supply them for himself.

is the same as if the 'level of confidence' type of interpretation is employed."
Ferguson (1959) wrote, "The .05 and .01 probability levels are descriptive of
our degree of confidence." There is little question but that sizable differences,
correlations, etc., in samples, especially samples of reasonable size, speak
more strongly of sizable differences, correlations, etc., in the population; and
there is little question but that if there is a real and strong effect in the population, it will continue to manifest itself in further sampling. However, these
are inferences which *we* may make. They are outside the inference model
associated with the test of significance. The p value within the inference model
is only the value which we take to be as how improbable an event could be
under the null hypothesis, which we judge will not take place to "us," in this
one experiment. It is not a "measure" of the goodness of the other inferences
which we might make. It is an *a priori* condition that we set up whereby we
decide whether or not we will reject the null hypothesis, not a measure of
significance.

There is a study in the literature (Rosenthal and Gaito, 1963) which points
up sharply the lack of understanding on the part of psychologists of the
meaning of the test of significance. The subjects were nine members of the
psychology department faculty, all holding doctoral degrees, and ten graduate
students, at the University of North Dakota; and there is little reason to
believe that this group of psychologists was more or less sophisticated than
any other. They were asked to rate their degree of belief or confidence in
results of hypothetical studies for a variety of p values, and for N's of 10 and
100. That there should be a relationship between the average rated confidence
or belief and p value, as they found, is to be expected. What is shocking is that
these psychologists indicated substantially greater confidence or belief in
results associated with the larger sample size for the same p values. According
to the theory, especially as this has been amplified by Neyman and Pearson
(1933), the probability of rejecting the null hypothesis for any given deviation
from null and p value increases as a function of the number of observations.
The rejection of the null hypothesis when the number of cases is small speaks
for a more dramatic effect in the population; and if the p value is the same,
the probability of committing a Type I error remains the same. Thus one can
be more confident with a small N than a large N. The question is, how could
a group of psychologists be so wrong? I believe that this wrongness is based
on the commonly held belief that the p value is a "measure" of degree of
confidence. Thus, the reasoning behind such a wrong set of answers by these
psychologists may well have been something like this: The p value is a
measure of confidence; but a larger number of cases also increases confidence;
therefore, for any given p value, the degree of confidence should be higher for
the larger N. The wrong conclusion arises from the erroneous character of the
first premise, and from the failure to recognize that the p value is a function of

sample size for any given deviation from null in the population. The author knows of instances in which editors of very reputable psychological journals have rejected papers in which the *p* values and *N*'s were small on the grounds that there were not enough observations, clearly demonstrating that the same mode of thought is operating in them. Indeed, rejecting the null hypothesis with a small *N* is indicative of a strong deviation from null in the population, the mathematics of the test of significance having already taken into account the smallness of the sample. Increasing the *N* increases the probability of rejecting the null hypothesis; and in these studies rejected for small sample size, that task has already been accomplished. These editors are, of course, in some sense the ultimate "teachers" of the profession; and they have been teaching something which is patently wrong.

2. *Automaticity of inference.* What may be considered to be a dream, fantasy, or ideal in the culture of psychology is that of achieving complete automaticity of inference. The making of inductive generalizations is always somewhat risky. In Fisher's *The Design of Experiments* (1935a, 4th ed.), he made the claim that the methods of induction could be made rigorous, exemplified by the procedures which he was setting forth. This is indeed quite correct in the sense indicated earlier. In a later paper, he made explicit what was strongly hinted at in his earlier writing: that the methods which he proposed constituted a relatively complete specification of the process of induction:

> That such a process induction existed and was possible to normal minds, has been understood for centuries; it is only with the recent development of statistical science that an analytic account can now be given, about as satisfying and complete, at least, as that given traditionally of the deductive processes (Fisher, 1955: 74).

Psychologists certainly took the procedures associated with the *t* test, *F* test, and so on, in this manner. Instead of having to engage in inference themselves, they had but to "run the tests" for the purpose of making inferences, since, as it appeared, the statistical tests were analytic analogues of inductive inference. The "operationist" orientation among psychologists, which recognized the contingency of knowledge on the knowledge-getting operations and advocated their specification, could, it would seem, "operationalize" the inferential processes simply by reporting the details of the statistical analysis. It thus removed the burden of responsibility, the chance of being wrong, the necessity for making inductive inferences, from the shoulders of the investigator and placed them on the tests of significance. The contingency of the conclusion upon the experimenter's decision of the level of significance was managed in two ways: the first, by resting on a kind of social agreement that 5 percent was good, and 1 percent better; the second in the manner which has already been discussed, by not making a decision of the level of significance, but only reporting the *p* value as a "result" and a presumably objective "measure" of degree of confidence. But that the probability of getting significance is also contingent upon the number of observations has been handled largely by ignoring it.

A crisis was experienced among psychologists when the matter of the one-versus the two-tailed test came into prominence; for here the contingency of the result of a test of significance on a decision of the investigator was simply too conspicuous to be ignored. An investigator, say, was interested in the difference between two groups on some measure. He collected his data, found that Mean A was greater than Mean B in the sample, and ran the ordinary two-tailed *t* test; and, let us say, it was not significant. Then he bethought himself. The two-tailed test tested against *two* alternatives, that the population Mean A was greater than population Mean B and vice versa. But then, he really wanted to know whether Mean A was greater than Mean B. Thus, he could run a one-tailed test. He did this and found, since the one-tailed test is more powerful, that his difference was now significant.

Now here there was a difficulty. The test of significance is not nearly so automatic an inference process as had been thought. It is manifestly contingent on the decision of the investigator as to whether to run a one- or a two-tailed test. And, somehow, making the decision *after* the data were collected and the means computed seemed like "cheating." How should this be handled? Should there be some central registry in which one registers one's decision to run a one- or two-tailed test before collecting the data? Should one, as one eminent psychologist once suggested to me, send oneself a letter so that the postmark would prove that one had predecided to run a one-tailed test? The literature on ways of handling this difficulty has grown quite a bit in the strain to overcome somehow this particular clear contingency of the results of a test of significance on the decision of the investigator. The author will not attempt here to review this literature, except to cite one very competent paper which points up the intrinsic difficulty associated with this problem, the *reductio ad absurdum* to which one comes. Kaiser, early in his paper, distinguished between the logic associated with the test of significance and other forms of inference, a distinction which, incidentally, Fisher would hardly have allowed: "The arguments developed in this paper are based on logical considerations in statistical inference. (We do not, of course, suggest that statistical inference is the only basis for scientific inference.)" But then, having taken the position that he is going to follow the logic of statistical inference relentlessly, he said: "*we cannot logically make a directional statistical decision or statement when the null hypothesis is rejected on the basis of the direction of the difference in the observed sample means*" (Kaiser, 1960: 161). One really needs to strike oneself in the head! If Sample Mean A is greater than Sample Mean B, and there is reason to reject the null hypothesis, in what other direction can it reasonably be? What kind of logic is it that leads one to believe that it could be otherwise than that Population Mean A is greater than Population Mean B? We do not know whether Kaiser intended his paper as a *reductio ad absurdum*, but it certainly turned out that way.

The issue of the one- versus the two-tailed test genuinely challenges the presumptive "objectivity" characteristically attributed to the test of significance. On the one hand, it makes patent what was the case under any circumstances (at the least in the choice of level of significance, and the choice of the number of cases in the sample), that the conclusion is contingent upon the decision of the investigator. An astute investigator, who foresaw the results, and who therefore predecided to use a one-tailed test, will get one *p* value. The less astute but honorable investigator, who did not foresee the results, would feel obliged to use a two-tailed test, and would get another *p* value. On the other hand, if one decides to be relentlessly logical within the logic of statistical inference, one winds up with the kind of absurdity which we have cited above.

3. *The confusion of induction to the aggregate with induction to the general.* Consider a not atypical investigation of the following sort: A group of, say, twenty normals and a group of, say, twenty schizophrenics are given a test. The tests are scored, and a *t* test is run, and it is found that the means differ significantly at some level of significance, say 1 percent. What inference can be drawn? As we have already indicated, the investigator could have insured this result by choosing a sufficiently large number of cases. Suppose we overlook this objection, which we can to some extent, by saying that the difference between the means in the population must have been large enough to have manifested itself with only forty cases. But still, what do we know from this? The only inference which this allows is that the mean of all normals is different from the mean of all schizophrenics in the populations from which the samples have presumably been drawn at random. (Rarely is the criterion of randomness satisfied. But let us overlook this objection too.)

The common rhetoric in which such results are discussed is in the form, "Schizophrenics differ from normals in such and such ways." The sense that both the reader and the writer have of this rhetoric is that it has been justified by the finding of significance. Yet clearly it does not mean *all* schizophrenics and *all* normals. All that the test of significance justifies is that measures of central tendency of the aggregates differ in the populations. The test of significance has *not* addressed itself to anything about the schizophrenia or normality which characterizes *each* member of the respective populations. Now it is certainly possible for an investigator to develop a hypothesis about the nature of schizophrenia from which he may infer that there should be differences between the means in the populations; and his finding of a significant difference in the means of his sample would add to the credibility of the former. However, that 1 percent which he obtained in his study bears only on the means of the populations and is not a "measure" of the confidence that he may have in his hypothesis concerning the nature of schizophrenia. There are two inferences that he must make. One is that of the sample to the population, for which the test of significance is of some use. The other is from his inference concerning the population to his hypothesis concerning the nature of schizophrenia. The p value does not bear on this second inference. The psychological literature is filled with assertions which confound these two inferential processes.

Or consider another hardly atypical style of research. Say an experimenter divides forty subjects at random into two groups of twenty subjects each. One group is assigned to one condition and the other to another condition, perhaps, say, massing and distribution of trials. The subjects are given a learning task, one group under massed conditions, the other under distributed conditions. The experimenter runs a t test on the learning measure and again, say, finds that the difference is significant at the 1 percent level of significance. He may then say in his report, being more careful than the psychologist who was studying the difference between normals and schizophrenics (being more "scientific" than his clinically interested colleague), that "the mean in the population of learning under massed conditions is lower than the mean in the population of learning under distributed conditions," feeling that he can say this with a good deal of certainty because of his test of significance. But here too (like his clinical colleague) he has made two inferences, and not one, and the 1 percent bears on the one but not the other. The statistical inference model certainly allows him to make his statement for the population, but only for *that* learning task, and the p value is appropriate only to that. But the generalization to "massed conditions" and "distributed conditions" beyond that particular learning task is a second inference with respect to which the p value is not relevant. The psychological literature is plagued with any number of instances in which the rhetoric indicates that the p value does bear on this second inference.

Part of the blame for this confusion can be ascribed to Fisher who, in *The Design of Experiments* (1935a, 4th ed.) suggested that the mathematical methods which he proposed were exhaustive of scientific induction, and that the principles he was advancing were "common to all experimentation." What he failed to see and to say was that after an inference was made concerning a population parameter, one still needed to engage in induction to obtain meaningful scientific propositions.

To regard the methods of statistical inference as exhaustive of the inductive inferences called for in experimentation is completely confounding. When the test of significance has been run, the necessity for induction has hardly been completely satisfied. However, the research worker knows this, in some sense, and proceeds, as he should, to make further inductive inferences. He is, however, still ensnarled in his test of significance and the presumption that it is the whole of his inductive activity, and thus mistakenly takes a low p value for the measure of the validity of his other inductions.

The seriousness of this confusion may be seen by again referring to the Rosenthal and Gaito study and the remark by Berkson which indicate that research workers believe that a large sample is better than a small sample. We need to refine the rhetoric somewhat. Induction consists in making inferences from the particular to the general. It is certainly the case that, as confirming particulars are added, the credibility of the general is increased. However, the addition of observations to a sample is, in the context of statistical inference, not the addition of particulars but the modification of what is one particular in the inference model, the sample aggregate. In the context of statistical inference, it is not necessarily true that "a large sample is better than a small sample." For, as has been already indicated, obtaining a significant result with a small sample suggests a larger deviation from null in the population and may be considered more meaningful. Thus more particulars are better than fewer particulars in the making of an inductive inference; but not necessarily a larger sample.

In the marriage of psychological research and statistical inference, psychology brought its own reasons for accepting this confusion, reasons which inhere in the history of psychology. Measurement pyschology arises out of two radically different traditions, as has been pointed out by Guilford (1936) and Cronbach (1957), and the matter of putting them together raised certain difficulties. The one tradition seeks to find propositions concerning the nature of man in general—propositions of a general nature, with each individual a particular in which the general is manifest. This is the kind of psychology associated with the traditional experimental psychology of Fechner, Ebbinghaus, Wundt, and Titchener. It seeks to find the laws which characterize the "generalized, normal, human, adult mind" (Boring, 1950). The research strategy associated with this kind of psychology is straightforwardly inductive. It seeks inductive generalizations which will apply to every member of a designated class. A single particular in which a generalization fails forces a rejection of the generalization, calling for either a redefinition of the class to which it applies or a modification of the generalization. The other tradition is the psychology of individual differences, which has its roots more in England and the United States than on the Continent. We may recall that when the young American, James McKeen Cattell, who invented the term "mental test," came to Wundt with his own problem of individual differences, it was regarded by Wundt as *ganz Amerikanisch* (Boring, 1950).

The basic datum for an individual-differences approach is not anything that characterizes each of two subjects, but the difference between them. For this latter tradition, it is the aggregate which is of interest, and not the general. One of the most unfortunate characteristics of many studies in psychology, especially in experimental psychology, is that the data are treated as aggregates while the experimenter is trying to infer general propositions. There is hardly an issue of most of the major psychological journals reporting experimentation in which this confusion does not appear several times, and in which the test of significance, which has some value in connection with the study of aggregates, is not interpreted as a measure of the credibility of the general proposition in which the investigator is interested. Roberts and Wist (n.d.) examined sixty articles from psychological literature from the point of view of the aggregate-general distinction. In twenty-five of the articles it was unambiguous that the authors had drawn general-type conclusions from aggregate-type data.

Thus, what took place historically in psychology is that instead of attempting to synthesize the two traditional approaches to psychological phenomena, which is both possible and desirable, a syncretic combination took place of the methods appropriate to the study of aggregates with the aims of a psychology which sought for general propositions. One of the most overworked terms, which added not a little to the essential confusion, was "error," which was a kind of umbrella term for (at the least) variation among scores from different individuals, variation among measurements for the same individual, and variation among samples.

Let us add another historical note. In 1936, Guilford published his well-known *Psychometric Methods*. In this book, which became a kind of "bible" for many psychologists, he made a noble effort at a "Rapprochement of Psychophysical and Test Methods." He observed, quite properly, that mathematical developments in each of the two fields might be of value in the other, that "Both psychophysics and mental testing have rested upon the same fundamental statistical devices." There is no question of the truth of this. However, what he failed to emphasize sufficiently was that mathematics is so abstract that the same mathematics is applicable to rather different fields of investigation without there being any necessary further identity between them. (One would not, for example, argue that business and genetics are essentially the same because the same arithmetic is applicable to market research and in the investigation of the facts of heredity.) A critical point of contact between the two traditions was in connection with scaling, in which Cattell's principle that "equally often noticed differences are equal unless always or never noticed" (Guilford, 1936) was adopted as a fundamental assumption. The "equally often noticed differences" is, of course, based on aggregates. By means of this assumption, one could collapse the distinction between the two areas of investigation. Indeed, this is not really too bad if one is alert to the fact that it is an assumption, one which even has considerable pragmatic value. As a set of techniques whereby data could be analyzed, that is, as a set of techniques whereby one could describe one's findings and then make inductions about the nature of the psychological phenomena, what Guilford put together in his book was eminently valuable. However, around this time the work of Fisher and his school was coming to the attention of psychologists. It was attractive for several reasons. It offered advice for handling "small samples." It offered a number of eminently ingenious new ways of organizing and extracting information from data. It offered ways by which several variables could be analyzed simultaneously, away from the old notion that one had to keep everything constant and vary only one variable at a time. It showed how the effect of the "interaction" of variables could be assessed. But it also claimed to have mathematized induction! The Fisher approach was thus "bought," and psychologists got a theory of induction in the bargain, a theory which seemed to exhaust the inductive processes. Whereas the question of the "reliability" of statistics had been a matter of concern for some time before (although frequently very garbled), it had not carried the burden of induction to the degree that it did with the Fisher approach. With the acceptance of the Fisher approach the psychological research worker also accepted, and then overused, the test of significance, employing it as the measure of the significance, in the largest sense of the word, of his research efforts.

Earlier, a distinction was made between sharp and loose null hypotheses. One of the major difficulties associated with the Fisher approach is the problem presented by sharp null hypotheses; for, as we have already seen, there is reason to believe that the existence of sharp null hypotheses is characteristically unlikely. There have been some efforts to correct for this difficulty by proposing the use of loose null hypotheses: in place of a single point, a region being considered null. Hodges and Lehmann (1954) have proposed a distinction between "statistical significance," which entails the sharp hypothesis, and "material significance," in which one tests the hypothesis of a deviation of a stated amount from the null point instead of the null point itself. Edwards (1950) has suggested the notion of "practical significance" in which one takes into account the meaning, in some practical sense, of the magnitude of the deviation from null together with the number of observations which have been involved in getting statistical significance. Binder (1963) has equally argued that a subset of parameters be equated with the null hypothesis. Essentially what has been suggested is that the investigator make some kind of a decision concerning "How much, say, of a difference makes a difference?" The difficulty with this solution, which is certainly a sound one technically, is that in psychological research we do not often have very good grounds for answering this question. This is partly due to the inadequacies of psychological measurement, but mostly due to the fact that the answer to the question of "How much of a difference makes a difference?" is not forth-coming outside of some particular practical context. The question calls forth another question, "How much of a difference makes a difference *for what?*"

This brings us to one of the major issues within the field of statistics itself. The problems of the research psychologist do not generally lie within prac-tical contexts. He is rather interested in making assertions concerning psycho-logical functions which have a reasonable amount of credibility associated with them. He is more concerned with "What is the case?" than with "What is wise to do?" (see Rozeboom, 1960 [Chapter 24]).

It is here that the decision-theory approach of Neyman, Pearson, and Wald (Neyman, 1937, 1957; Neyman & Pearson, 1933; Wald, 1939, 1950, 1955) becomes relevant. The decision-theory school, still basing itself on some basic notions of the Fisher approach, deviated from it in several respects:

1. In Fisher's inference model, the two alternatives between which one chose on the basis of an experiment were "reject" and "inconclusive." As he said in *The Design of Experiments* (1935a, 4th ed.), "The null hypothesis is never proved or established, but is possibly disproved, in the course of experi-mentation." In the decision-theory approach, the two alternatives are rather "reject" and "accept."

2. Whereas in the Fisher approach the interpretation of the test of signifi-cance critically depends on having one sample from a hypothetical population of experiments, the decision-theory approach conceives of, is applicable to, and is sensible with respect to numerous repetitions of the experiment.

3. The decision-theory approach added the notions of the Type II error (which can be made only if the null hypothesis is accepted) and power as significant features of their model.

4. The decision-theory model gave a significant place to the matter of what is concretely lost if an error is made in the practical context, on the presump-tion that "accept" entailed one concrete action, and "reject" another. It is in these actions and their consequences that there is a basis for deciding on a level of confidence. The Fisher approach has little to say about the consequences.

As it has turned out, the field of application par excellence for the decision-theory approach has been the sampling inspection of mass-produced items. In sampling inspection, the acceptable deviation from null can be specified; both "accept" and "reject" are appropriate categories; the alternative courses of action can be clearly specified; there is a definite measure of loss for each possible action; and the choice can be regarded as one of a series of such choices, so that one can minimize the overall loss (see Barnard, 1954). Where the aim is only the acquisition of knowledge without regard to a specific practical context, these conditions do not often prevail. Many psychologists who learned about analysis of variance from books such as those by Snedecor

(1937, 4th ed.) found the examples involving hog weights, etc., somewhat annoying. The decision-theory school makes it clear that such practical contexts are not only "examples" given for pedagogical purposes but actually are essential features of the methods themselves.

The contributions of the decision-theory school essentially revealed the intrinsic nature of the test of significance beyond that seen by Fisher and his colleagues. They demonstrated that the methods associated with the test of significance constitute not an assertion, or an induction, or a conclusion calculus, but a decision- or risk-evaluation calculus. Fisher has reacted to the decision-theory approach in polemic style, suggesting that its advocates were like "Russians [who] are made familiar with the ideal that research in pure science can and should be geared to technological performance, in the comprehensive organized effort of a five-year plan for the nation." He also suggested an American "ideological" orientation: "In the U. S. also the great importance of organized technology has I think made it easy to confuse the process appropriate for drawing correct conclusions, with those aimed rather at, let us say, speeding production, or saving money" (1955: 70).[5] But perhaps a more reasonable way of looking at this is to regard the decision-theory school to have explicated what was already implicit in the work of the Fisher school.

What then is our alternative, if the test of significance is really of such limited appropriateness? At the very least it would appear that we would be much better off if we were to attempt to estimate the magnitude of the parameters in the populations; and recognize that we then need to make other inferences concerning the psychological phenomena which may be manifesting themselves in these magnitudes. In terms of a statistical approach which is an alternative, the various methods associated with the theorem of Bayes', referred to earlier, may be appropriate; and the paper by Edwards, Lindman, and Savage (1963) and the book by Schlaifer (1959) are good starting points. However, what is expressed in the theorem of Bayes alludes to the more general process of inducing propositions concerning the nonmanifest (which is what the population is a special instance of) and ascertaining the way in which what is manifest (of which the sample is a special instance) bears on it. This is what the scientific method has been about for centuries. However, if the reader who might be sympathetic to the considerations set forth in this paper quickly goes out and reads some of the material on the Bayesian approach with the hope that thereby he will find a new basis for automatic inference, this paper will have misfired, and he will be disappointed.

What we have indicated in this paper in connection with the test of significance in psychological research may be taken as an instance of a kind of

5. For a reply to Fisher, see Pearson (1955).

essential mindlessness in the conduct of research which may be related to the presumption of the nonexistence of mind in the subjects of psychological research. Karl Pearson once indicated that higher statistics was only common sense reduced to numerical appreciation. However, that base in common sense must be maintained with vigilance. When we reach a point where our statistical procedures are substitutes instead of aids to thought, and we are led to absurdities, then we must return to common sense. Tukey (1962) has very properly pointed out that statistical procedures may take our attention away from the data, which constitute the ultimate base for any inferences which we might make. Schlaifer (1959) has dubbed the error of the misapplication of statistical procedures the "error of the third kind," the most serious error which can be made. Berkson has suggested the use of "the interocular traumatic test, you know what the data mean when the conclusion hits you between the eyes" (Edwards *et al.*, 1963). We must overcome the myth that if our treatment of our subject matter is mathematical it is therefore precise and valid. We need to overcome the handicap associated with limited competence in mathematics, a competence that makes it possible for us to run tests of significance while it intimidates us with a vision of greater mathematical competence if only one could reach up to it. Mathematics can serve to obscure as well as reveal.

Most important, we need to get on with the business of generating psychological hypotheses and proceed to do investigations and make inferences which bear on them, instead of, as so much of our literature would attest, testing the statistical null hypothesis in any number of contexts in which we have every reason to suppose that it is false in the first place.

26. Theory Testing in Psychology and Physics: A Methodological Paradox[*]

PAUL E. MEEHL

THE PURPOSE OF THE PRESENT PAPER is not so much to propound a doctrine or defend a thesis (especially as I should be surprised if either psychologists or statisticians were to disagree with whatever in the nature of a "thesis" it advances), but to call the attention of logicians and philosophers of science to a puzzling state of affairs in the currently accepted methodology of the behavior sciences which I, a psychologist, have been unable to resolve to my satisfaction. The puzzle, sufficiently striking (when clearly discerned) to be entitled to the designation "paradox," is the following: *In the physical sciences, the usual result of an improvement in experimental design, instrumentation, or numerical mass of data is to increase the difficulty of the "observational hurdle" which the physical theory of interest must successfully surmount; whereas, in psychology and some of the allied behavioral sciences, the usual effect of such improvement in experimental precision is to provide an easier hurdle for the theory to surmount.* Hence what we would normally think of as improvements in our experimental method tend (when predictions materialize) to yield *stronger* corroboration of the theory in physics, since to remain unrefuted the

Reprinted from *Philosophy of Science*, 34 (June, 1967), 103–115, by permission of the author and the publisher. Copyright © 1967 by the Philosophy of Science Association.

* The views and examples of Dr. David T. Lykken have influenced the form of the argument in this paper. For an application of these and allied considerations to a specific example of poor research in psychology, see Lykken (1966, 1968 [Chapter 27]).

theory must have survived a more difficult test; by contrast, such experimental improvement in psychology typically results in a *weaker* corroboration of the theory, since it has now been required to survive a more lenient test (Bunge, 1964; Popper, 1959, 1962).

Although the point I wish to make is one in logic and methodology of science and, as I think, does not presuppose adoption of any of the current controversial viewpoints in technical statistics, a brief exposition of the process of statistical inference as we usually find it in the social sciences is necessary. (The philosopher who is unfamiliar with this subject matter may be referred to any good standard text on statistics, such as the widely used book by Hays [1963] which includes a clear and succinct treatment of the main points I shall briefly summarize here.)

On the basis of a substantive psychological theory T in which he is interested, a psychologist derives (often in a rather loose sense of "derive") the consequence that an observable variable x will differ as between two groups of subjects. Sometimes, as in most problems of clinical or social psychology, the two groups are defined by a property the individuals under study already possess, e.g. social class, sex, diagnosis, or measured I.Q. Sometimes, as is more likely to be the case in such fields as psychopharmacology or psychology of learning, the contrasted groups are defined by the fact that the experimenter has subjected them to different experimental influences, such as a drug, a reward, or a specific kind of social pressure. Whether the contrasted groups are specified by an "experiment of nature" where the investigator takes the organisms as he finds them, or by a true "experiment" in the more usual sense of the word, is not crucial for the present argument; although, as will be seen, the implications of my puzzle for theory testing are probably more perilous in the former kind of research than in the latter.

According to the substantive theory T, the two groups are expected to differ on variable x, but it is recognized that errors of (a) measurement and (b) random sampling will, in general, produce *some* observed difference between the averages of the groups studied, even if their total population did not differ in the true value of $\bar{x}$ ($=$ mean of x).

Example: We are interested in the question whether girls are brighter than boys (i.e. that $\mu_g - \mu_b = \delta_{gb} > 0$). We do not have perfectly reliable measures of intelligence, and we are furthermore not in a position to measure the intelligence of all boys and girls in the hypothetical population about which we desire to make a general statement. Instead we must be content with fallible I.Q. scores, and with a sample of school children drawn from the hypothetical population. Each of these sources of error, measurement error and random sampling error, contributes to an untrustworthiness in the computed value we obtain for the average intelligence $\bar{x}_b$ of the boys and also for $\bar{x}_g$, that of the girls. If we observe a difference of, say $\bar{d} = 5$ I.Q. points in a sample of 100 boys and 100 girls, we must have some method to infer whether this obtained observational difference between the two groups reflects a real difference or one which is merely apparent, i.e. due to the combined effect of errors of measurement and sampling. We do this by means of a "statistical significance test," the mathematics of which is not relevant here, except to say that by combining the principles of probability with a characterization of the procedure by which the samples were constituted and quantifying the variation in observed intelligence score *within* each of the two groups being contrasted, it is possible to employ a formula which utilizes the observed averages together with the observed variations and sample sizes so as to answer certain relevant kinds of questions. Among such questions is the following: "If there were, in fact, no real difference in average I.Q. between the population of boys and girls, with what relative frequency would an investigator find a difference—in relation to the observed intra-group variation—of the magnitude our observations have actually found?"

The statistical hypothesis, that there is no population difference between boys and girls in I.Q., which is called the "null hypothesis" [$H_0 : \delta = 0$] is used to generate a random sampling distribution of the statistic ("t-test") employed in testing the presence of a significant difference. If the observed data would be very improbable on the hypothesis that H_0 obtained, we abandon H_0 in favor of its alternative. We conclude that since H_0 is false, its alternative, i.e. that there exists a real average difference between the sexes, obtains. In the past, it was customary to deal with what may be called the "point-null" hypothesis, which says that there is zero difference between the two averages in the populations. In recent years it has been more explicitly recognized that what is of theoretical interest is not the mere presence of *difference* (i.e. that H_0 is false, i.e. that $\mu_b \neq \mu_g$) but rather the presence of a difference *in a certain direction* (in this case, that $\mu_g > \mu_b$). It is therefore increasingly frequent that the behavioral scientist employs the so-called "directional null hypothesis," say H_2, instead of the point-null hypothesis H_0. If our substantive theory T involves the prediction that the average I.Q. of girls in the entire population exceeds that of boys, we test the alternative to this statistical hypothesis about the population, i.e. that *either the average I.Q. of boys exceeds that of girls* (H_2) *or that there is no difference* (H_0). That is, we adopt for statistical test (with the anticipation of refuting it) a disjunction of the old-fashioned point-null hypothesis H_0 with the hypothesis H_2 that H_0 is false and it is false in a direction *opposite* to that implied by our substantive theory. However, this directional null hypothesis ($H_{02} : \mu_g \leqslant \mu_b$), unlike the old-fashioned point-null hypothesis ($H_0 : \mu_g = \mu_b$), does not generate a theoretically expected distribution, because it is not precise, i.e. it does not specify a point-value for the unknown parameter $\bar{\delta} = \mu_{girls} - \mu_{boys}$). However, we can employ it as we do the point-null hypothesis, by reasoning that *if* the point-null hypothesis H_0

obtained in the state of nature, *then* an observed difference (in the direction that our substantive theory predicts) of such-and-such magnitude has a calculable probability; and that calculable probability is an upper bound upon the desired (but unknown) probability based on $H_{02} : \mu_g \leqslant \mu_b$. That is to say, if the probability of observed girl-over-boy difference ($d_{gb} = \bar{x}_g - \bar{x}_b$) arising through random error is p, given the point-null hypothesis $H_0 : \mu_g = \mu_b$, then the probability of the observed difference arising randomly given any of the point-hypotheses constituting $H_2 : \mu_g < \mu_b$ will of course be less than p. Hence p is an upper bound on this probability for the inexact directional null hypothesis ($H_{02} : \mu_g \leqslant \mu_b$). Proceeding in this way directs our interest to only one tail of the theoretical random sampling distribution instead of both tails, which has given rise to a certain amount of controversy among statisticians, but that controversy is not relevant here. (For an excellent clarifying discussion, see Kaiser [1960].) Suffice it to say that having formulated a directional null hypothesis H_{02} which is the alternative to the statistical hypothesis of interest, H_1, and which includes the point-null hypothesis H_0 as one (very unlikely) possibility for the state of nature, we then carry out the experiment with the anticipation of *refuting* this directional null hypothesis, thereby confirming the alternative statistical hypothesis of interest (H_1) and, since H_1 in turn was implied by the substantive theory T, of corroborating T.

In such a situation we know in advance that we are in danger of making either of two sorts of "errors," not in the sense of committing scientific mistakes but in the sense of (rationally) inferring what is objectively a false conclusion. If the null hypothesis (point or directional) is in fact true, but due to the combination of measurement and sampling errors we obtain a value which is so improbable upon H_2 or H_0 that we decide in favor of their alternative, H_1, we will then have committed what is known as an *error of the first kind* or *Type I error*. An error of the first kind is a statistical inference that the null hypothesis is false, when in the state of nature it is actually true. This means we will have concluded in favor of a statistical statement H_1 which flowed as a consequence of our substantive theory T, and therefore we will believe ourselves to have obtained empirical support for T, whereas in reality this statistical conclusion is false and, consequently, such support for the substantive theory is objectively lacking. Measurement and sampling error may, of course, also result in a sampling deviation in the opposite direction; or, the true difference $\bar{\delta}$ may be so small that even if our sample values were to coincide exactly with the true ones, the sheer algebra of the significance test would not enable us to reach the prespecified level of statistical significance. If we conclude until further notice that the directional null hypothesis H_{02} is tenable, on the grounds that we have failed to refute it by our investigation, then we have failed to support its statistical alternative H_1, and therefore failed to confirm one of the predictions of the substantive theory T. Retention of the null hypothesis H_{02} when it is in fact false is known as an *error of the second kind* or *Type II error*.

In the biological and social sciences there has been widespread adoption of the probabilities .01 or .05 as the allowable theoretical frequency of Type I errors. These values are called the 1 percent and 5 percent "levels of significance." It is obvious that there is an inverse relationship between the probabilities of the two kinds of errors, so that if we adopt a significance level which increases the frequency of Type I errors, such a policy will lead to a greater number of claims of statistically significant departure from the null hypothesis; and, therefore, in whatever unknown fraction of all experiments performed the null hypothesis is in reality false, we will more often (correctly) conclude its falsity, i.e. we will thereby be reducing the proportion of Type II errors.

Suppose we hold fixed the theoretically calculable incidence of Type I errors. Thus, we determine that *if* the null hypothesis is in fact true in the state of nature, we do not wish to risk erroneously concluding that it is false more than, say, five times in 100. Holding this 5 percent significance level fixed (which, as a form of scientific strategy, means leaning over backward not to conclude that a relationship exists when there isn't one, or when there is a relationship in the wrong direction), we can decrease the probability of Type II errors by improving our experiment in certain respects. There are three general ways in which the frequency of Type II errors can be decreased (for fixed Type I error-rate), namely, (a) by improving the logical structure of the experiment, (b) by improving experimental techniques such as the control of extraneous variables which contribute to intragroup variation (and hence appear in the denominator of the significance test), and (c) by increasing the size of the sample. Given a specified true difference in the range of H_1, the complement $(1 - p)$ of the probability of a Type II error is known as the *power*, and an improvement in the experiment by any or all of these three methods yields an increase in power (or, to use words employed by R. A. Fisher, the experiment's "sensitiveness" or "precision"). For many years relatively little emphasis was put upon the problem of power, but recently this concept has come in for a good deal of attention. Accordingly, up-to-date psychological investigators are normally expected to include some preliminary calculations regarding power in designing their experiments. We select a logical design and choose a sample size such that it can be said in advance that if one is interested in a true difference provided it is at least of a specified magnitude (i.e. if it is smaller than this we are content to miss the opportunity of finding it), the probability is high (say, 80 percent) that we will successfully refute the null hypothesis. (See, for example, Jacob Cohen's literature sampling [1962] on the problem of power.) For an incisive critique of the whole approach, a critique which has been given far less respectful attention than it deserves (conspiracy of silence?), I recommend Rozeboom's excellent contribution (1960 [Chapter 24]). But I should emphasize that my argument in this paper does not hinge upon the reader's agreement with Rozeboom's very strong attack (although I, myself, incline to go along with him).

It is important to keep clear the distinction between the *substantive theory* of interest and the *statistical hypothesis* which is derived from it (Bolles, 1962). In the I.Q. example there was almost no substantive theory or a very impoverished one; i.e. the question being investigated was itself stated as a purely statistical question about the average I.Q. of the two sexes. In the great majority of investigations in psychology the situation is otherwise. Normally, the investigator holds some substantive theory about unconscious mental processes, or physiological or genetic entities, or perceptual structure, or about learning influences in the persons' past, or about current social pressures, which contains a great deal more content than the mere statement that the population parameter of an observational variable is greater for one group of individuals than for another. While no competent psychologist is unaware of this obvious distinction between a substantive psychological theory T and a statistical hypothesis H implied by it, in practice there is a tendency to conflate the substantive theory with the statistical hypothesis, thereby illicitly conferring upon T somewhat the same degree of support given H by a successful refutation of the null hypothesis. Hence the investigator, upon finding an observed difference which has an extremely small probability of occurring on the null hypothesis, gleefully records the tiny probability number "$p < .001$," and there is a tendency to feel that the extreme smallness of this probability of a Type I error is somehow transferable to a small probability of "making a theoretical mistake." It is as if, when the observed statistical result would be expected to arise only once in a thousand times through a Type I statistical error given H_{02}, therefore one's substantive theory T, which entails the alternative H_1, has received some sort of direct quantitative support of magnitude around .999 $[= 1 - .001]$.

To believe this literally would, of course, be an undergraduate mistake of which no competent psychologist would be guilty; I only want to point to the fact that there is subtle tendency to "carry over" a very small probability of a Type I error into a sizable resulting confidence in the truth of the substantive theory, even among investigators who would never make an explicit identification of the one probability number with the complement of the other.

One reason why the directional null hypothesis ($H_{02} : \mu_g \leqslant \mu_b$) is the appropriate candidate for experimental refutation is the universal agreement that the old point-null hypothesis ($H_0 : \mu_g = \mu_b$) is [quasi-] always false in biological and social science. Any dependent variable of interest, such as I.Q., or academic achievement, or perceptual speed, or emotional reactivity as measured by skin resistance, or whatever, depends mainly upon a finite number of "strong" variables characteristic of the organisms studied (embodying the accumulated results of their genetic makeup and their learning histories) plus the influences manipulated by the experimenter. Upon some complicated, unknown mathematical function of this finite list of "important" determiners is then superimposed an indefinitely large number of essentially "random" factors which contribute to the intragroup variation and therefore boost the error term of the statistical significance test. In order for two groups which differ in some identified properties (such as social class, intelligence, diagnosis, racial or religious background) to differ not at all in the "output" variable of interest, it would be necessary that all determiners of the output variable have precisely the same average values in both groups, or else that their values should differ by a *pattern of amounts of difference* which precisely counterbalance one another to yield a net difference of zero. Now our general background knowledge in the social sciences, or, for that matter, even "common sense" considerations, make such an exact equality of all determining variables, or a precise "accidental" counterbalancing of them, so extremely unlikely that no psychologist or statistician would assign more than a negligibly small probability to such a state of affairs.

Example: Suppose we are studying a simple perceptual-verbal task like rate of color-naming in school children, and the independent variable is father's religious preference. Superficial consideration might suggest that these two variables would not be related, but a little thought leads one to conclude that they will almost certainly be related by *some* amount, however small. Consider, for instance, that a child's reaction to any sort of school-context task will be to some extent dependent upon his social class, since the desire to please academic personnel and the desire to achieve at a performance (just because it is a *task*, regardless of its intrinsic interest) are both related to the kinds of sub-cultural and personality traits in the parents that lead to upward mobility, economic success, the gaining of further education, and the like. Again, since there is known to be a sex difference in color-naming, it is likely that fathers who have entered occupations more attractive to "feminine" males will (on the average) provide a somewhat more feminine father-figure for identification on the part of their male offspring, and that a more refined color vocabulary, making closer discriminations between similar hues, will be characteristic of the ordinary language of such a household. Further, it is known that there is a correlation between a child's general intelligence and his father's occupation, and of course there will be *some* relation, even though it may be small, between a child's general intelligence and his color vocabulary, arising from the fact that *vocabulary in general* is heavily saturated with the general intelligence factor. Since religious preference is a correlate of social class, all of these social class factors, as well as the intelligence variable, would tend to influence color-naming performance. Or consider a more extreme and faint kind of relationship. It is quite conceivable that a child who belongs to a more liturgical religious denomination would be somewhat more color-oriented than a child for whom bright colors were not associated with the religious life. Everyone familiar with psychological research knows that numerous "puzzling, unexpected" correlations pop up all the time, and that it requires only a moderate amount of motivation-plus-ingenuity to construct very plausible alternative theoretical explanations for them.

These armchair considerations are borne out by the finding that in psychological and sociological investigations involving very large numbers of subjects, it is regularly found that almost all correlations or differences between means are statistically significant. See, for example, the papers by Bakan (1966 [Chapter 25]) and Nunnally (1960). Data currently being analyzed by Dr. David Lykken and myself, derived from a huge sample of over 55,000 Minnesota high school seniors, reveal statistically significant relationships in 91 percent of pairwise associations among a congeries of 45 miscellaneous variables such as sex, birth order, religious preference, number of siblings, vocational choice, club membership, college choice, mother's education, dancing, interest in woodworking, liking for school, and the like. The 9 percent of nonsignificant associations are heavily concentrated among a small minority of variables having dubious reliability, or involving arbitrary groupings of nonhomogeneous or nonmonotonic sub-categories. The majority of variables exhibited significant relationships *with all but three of the others*, often at a very high confidence level ($p < 10^{-6}$).

This line of reasoning is perhaps not quite as convincing in the case of true *experiments*, where the subjects are randomly assigned by the investigator to different experimental manipulations. If the reader is disinclined to follow me here, my overall argument will, for him, be applicable to those kinds of research in social science which study the correlational relationships or group differences between subjects "as they come," but not to the type of investigation which constitutes an experiment in the usual scientific sense. However, I myself believe that even in the strict sense of "experiment," the argument is still strong, although the quantitative departures from the point-null H_0 would be expected to run considerably lower on the average. Considering the fact that "everything in the brain is connected with everything else," and that there exist several "general state-variables" (such as arousal, attention, anxiety, and the like) which are known to be at least *slightly* influenceable by practically any kind of stimulus input, it is highly unlikely that *any* psychologically discriminable stimulation which we apply to an experimental subject would exert literally *zero* effect upon any aspect of his performance. The psychological literature abounds with examples of small but detectable influences of this kind. Thus it is known that if a subject memorizes a list of nonsense syllables in the presence of a faint odor of peppermint, his recall will be facilitated by the presence of that odor. Or, again, we know that individuals solving intellectual problems in a "messy" room do not perform quite as well as individuals working in a neat, well-ordered surrounding. Again, cognitive processes undergo a detectable facilitation when the thinking subject is concurrently performing the irrelevant, noncognitive task of squeezing a hand dynamometer. It would require considerable ingenuity to concoct experimental manipulations, except the most minimal and trivial (such as a very slight modification in the word order of instructions given a subject) where one could have confidence that the manipulation would be utterly without effect upon the subject's motivational level, attention, arousal, fear of failure, achievement drive, desire to please the experimenter, distraction, social fear, etc. So that, for example, while there is no very "interesting" psychological theory that links hunger drive with color-naming ability, I myself would confidently predict a significant difference in color-naming ability between persons tested after a full meal and persons who had not eaten for 10 hours, provided the sample size were sufficiently large and the color-naming measurements sufficiently reliable, since one of the effects of the increased hunger drive is heightened "arousal," and anything which heightens arousal would be expected to affect a perceptual-cognitive performance like color-naming. Suffice it to say that there are very good reasons for expecting at least *some* slight influence of almost any experimental manipulation which would differ sufficiently in its form and content from the manipulation imposed upon a control group to be included in an experiment in the first place. In what follows I shall therefore assume that the point-null hypothesis H_0 is, in psychology, [quasi-] always false.

Let us now conceive of a large "theoretical urn" containing counters designating the indefinitely large class of actual and possible substantive theories concerning a certain domain of psychology (e.g. mammalian instrumental learning). Let us conceive of a second urn, the "experimental-design" urn, containing counters designating the indefinitely large set of possible experimental situations which the ingenuity of man could devise. (If anyone should object to my conceptualizing, for purposes of methodological analysis, such a heterogeneous class of theories or experiments, I need only remind him that such a class is universally presupposed in the logic of statistical significance testing.) Since the point-null hypothesis H_0 is [quasi-] always false, almost every one of these experimental situations involves a non-zero difference on its output variable (parameter). Whichever group we (arbitrarily) designate as the "experimental" group and the "control" group, in half of these experimental settings the true value of the dependent variable difference (experimental minus control) will be positive, and in the other half negative.

It may be objected that this is a use of the Principle of Insufficient Reason and presupposes one particular answer to some disputed questions in statistical theory (as between the Bayesians and the Fisherians). But I must emphasize that I have said nothing about the *form* or *range* or other parametric characteristics of the distribution of true differences. I have merely said that the point-null hypothesis H_0 is always false, and I have then *assigned*, in a strictly random fashion, the names "experimental" and "control" to the two groups which a given experimental setup treats in two different ways. That is, it makes no difference here whether a group of subjects learning nonsense syllables while squeezing a hand dynamometer is called the experimental group, or whether we call "experimental" the group that learns the nonsense syllables without such squeezing. Hence my use of the Principle of Insufficient Reason is one of those legitimate, noncontroversial uses following directly when the basic principles of probability are applied to a specification of procedure for random assignment.

We now perform a random pairing of the counters from the "theory" urn with the counters from the "experimental" urn, and arbitrarily stipulate—quite irrationally—that a "successful" outcome of the experiment means that the difference favors the experimental group $[\mu_E - \mu_C > 0]$. This preposterous model, which is of course much worse than anything that can exist even in the most primitive of the social sciences, provides us with a lower bound for the expected frequency of a theory's successfully predicting the direction in which the null hypothesis fails, *in the state of nature* (i.e. we are here not considering sampling problems, and therefore we neglect errors of either the first or the second kind). It is obvious that if the point-null hypothesis H_0 is [quasi-] always false and there is no logical connection between our theories and the direction of the experimental outcomes, then if we arbitrarily assign one of the two directional hypotheses H_1 or H_2 to each theory, that hypothesis will be correct half of the time, i.e. in half of the arbitrary urn-counter-pairings. Since even my late, uneducated grandmother's common-sense psychological theories had nonzero verisimilitude, we can safely say that the value $p = \frac{1}{2}$ is a lower bound on the success-frequency of experimental "tests," assuming our experimental design had perfect power.

Countervailing the unknown increment over $p = \frac{1}{2}$, which arises from the fact that the experimental and theoretical counters are not thus drawn randomly (since our theories do possess, on the average, at least some tiny amount of verisimilitude), there is the statistical factor that among the counter-pairings which are accidentally "successful" (in the sense that the state of nature falsifies the null hypothesis in the expected direction), we will sometimes fail to refute it because of measurement and sampling errors, since our experiments will always, in practice, have less than perfect power. Even though the point-null hypothesis H_0 is always false, so that the directional null hypothesis H_{02} is false in the (theoretically pseudo-predicted) direction half the time, we will sometimes fail to discover this because of Type II errors. Without making illegitimate prior-probability assumptions concerning the actual distribution of true differences in the whole vast world of psychological experimental contexts, one cannot say anything definite about the extent to which this countervailing influence of Type II errors will wash out (or even overcome) the fact that our theories tend to have non-negligible verisimilitude. But by setting aside this latter fact, i.e. by assuming counterfactually that there is *no connection whatever between our theories and our experimental designs* (the two-urn idealization), thereby fixing the expected frequency of successful refutations of the directional null hypothesis H_{02} at $p = \frac{1}{2}$ for experiments of *perfect power*, it follows that, as the power of our experimental designs and significance tests is increased by any of the three methods described above, we approach $p = \frac{1}{2}$ as the limit of our expected frequency of "successful outcomes," i.e. of attaining statistically significant experimental results in the theoretically predicted direction.

I conclude that the effect of increased precision, whether achieved by improved instrumentation and control, greater sensitivity in the logical structure of the experiment, or increasing the number of observations, is to yield a probability approaching $\frac{1}{2}$ of corroborating our substantive theory by a significance test, *even if the theory is totally without merit.* That is to say, the ordinary result of improving our experimental methods and increasing our sample size, proceeding in accordance with the traditionally accepted method of theory-testing by refuting a directional null hypothesis, yields a prior probability $p \simeq \frac{1}{2}$ and very likely somewhat above that value by an unknown amount. It goes without saying that successfully negotiating an experimental hurdle of this sort can constitute only an extremely weak corroboration of any substantive theory, *quite apart from currently disputed issues of the Bayesian type regarding the assignment of prior probabilities to the theory itself.*

So far as I am able to discern, this methodological truth is either unknown or systematically ignored by most behavioral scientists. I do not know to what extent this is attributable to confusion between the substantive theory T and the statistical hypothesis H_1, with the resulting mis-assignment of the probability $(1 - p_I)$ complementary to the significance level p_I attained, to the "probability" of the substantive theory; or to what extent it arises from insufficient attention to the truism that the point-null hypothesis H_0 is [quasi-] always false. It seems unlikely that most social science investigators would think in their usual way about a theory in meteorology which "successfully predicted" that it would rain on the 17th of April, given the antecedent information that it rains (on the average) during half the days in the month of April!

But this is not the worst of the story. Inadequate appreciation of the extreme weakness of the test to which a substantive theory T is subjected by merely predicting a directional statistical difference $\bar{d} > 0$ is then compounded by a truly remarkable failure to recognize the logical asymmetry between, on the one hand, (formally invalid) "confirmation" of a theory via affirming the consequent in an argument of the form: $[T \supset H_1, H_1, \text{infer } T]$, and on the other hand the deductively tight *refutation* of the theory *modus tollens* by a falsified prediction, the logical form being: $[T \supset H_1, \sim H_1, \text{infer } \sim T]$.

While my own philosophical predilections are somewhat Popperian, I daresay any reader will agree that no full-fledged Popperian philosophy of science is presupposed in what I have just said. The destruction of a theory *modus tollens* is, after all, a matter of deductive logic; whereas that the "confirmation" of a theory by its making successful predictions involves a much weaker kind of inference. This much would be conceded by even the most anti-Popperian "inductivist." The writing of behavioral scientists often reads as though they assumed—what it is hard to believe anyone would explicitly assert if challenged—that successful and unsuccessful predictions are practically on all fours in arguing for and against a substantive theory. Many experimental articles in the behavioral sciences, and, even more strangely, review articles which purport to survey the current status of a particular theory in the light of all available evidence, treat the confirming instances and the disconfirming instances with equal methodological respect, as if one could, so to speak, "count noses," so that if a theory has somewhat more confirming than disconfirming instances, it is in pretty good shape evidentially. Since we know that this is already grossly incorrect on purely formal grounds, it is a mistake *a fortiori* when the so-called "confirming instances" have themselves a prior probability, as argued above, somewhere in the neighborhood of $\frac{1}{2}$, quite apart from any theoretical considerations.

Contrast this bizarre state of affairs with the state of affairs in physics. While there are of course a few exceptions, the usual situation in the experimental testing of a physical theory at least involves the prediction of a *form* of function (with parameters to be fitted), or, more commonly, the prediction of a quantitative magnitude (point-value). Improvements in the accuracy of determining this experimental function-form or point-value, whether by better instrumentation for control and making observations or by the gathering of a larger number of measurements, has the effect of narrowing the band of tolerance about the theoretically predicted value. What does this mean in terms of the significance-testing model? It means: *In physics, that which corresponds, in the logical structure of statistical inference, to the old-fashioned point-null hypothesis* H_0 *is the value which flows as a consequence of the substantive theory* T; so that an increase in what the statistician would call "power" or "precision" has the methodological effect of stiffening the experimental test, of setting up a more difficult observational hurdle for the theory T to surmount. Hence, in physics the effect of improving precision or power is that of *decreasing* the prior probability of a successful experimental out-come if the theory lacks verisimilitude, that is, precisely the reverse of the situation obtaining in the social sciences.

As techniques of control and measurement improve or the number of observations increases, the methodological effect in physics is that a successful passing of the hurdle will mean a greater increment in corroboration of the substantive theory; whereas in psychology, comparable improvements at the experimental level result in an empirical test which can provide only a pro-gressively weaker corroboration of the substantive theory.

In physics, the substantive theory predicts a point-value, and when physi-cists employ "significance tests," their mode of employment is to compare the theoretically predicted value x_0 with the observed mean $\bar{x}_0$, asking whether they differ (in either direction!) by more than the "probable error" of determination of the latter. Hence $H : H_0 = \mu_x$ functions as a point-null hypothesis, and the prior (logical, antecedent) probability of its being correct in the absence of theory approximates zero. As the experimental error asso-ciated with our determination of $\bar{x}_0$ shrinks, values of $\bar{x}_0$ consistent with x_0 (and hence, compatible with its implicans, T) must lie within a narrow range. In the limit (zero probable error, corresponding to "perfect power" in the significance test) any nonzero difference $(\bar{x}_0 - x_0)$ provides a *modus tollens* refutation of T. If the theory has negligible verisimilitude, the logical proba-bility of its surviving such a test is negligible. Whereas in psychology, the result of perfect power (i.e. *certain* detection of any nonzero difference in the predicted direction) is to yield a prior probability $p = \frac{1}{2}$ of getting experimental results compatible with T, because perfect power would mean guaranteed detection of whatever difference exists; and a difference [quasi-] always exists, being in the "theoretically expected direction" half the time if our substantive theories were all of negligible verisimilitude (two-urn model).

This methodological paradox would exist for the psychologist even if he played his own statistical game fairly. The reason for its existence is obvious, namely, that most psychological theories, especially in the so-called "soft" fields such as social and personality psychology, are not quantitatively devel-oped to the extent of being able to generate point-predictions. In this respect, then, although this state of affairs is surely unsatisfactory from the methodo-logical point of view and stands in great need of clarification (and, hopefully, of constructive suggestions for improving it) from logicians and philosophers of science, one might say that it is "nobody's fault," it being difficult to see just how the behavioral scientist could extricate himself from this dilemma without making unrealistic attempts at the premature construction of theories which are sufficiently quantified to generate point-predictions for refutation.

However, there are five social forces and intellectual traditions at work in the behavioral sciences which make the research consequences of this situation even worse than they may have to be, considering the state of our knowledge. In addition to (a) failure to recognize the marked evidential asymmetry between confirmation and *modus tollens* refutation of theories, and (b) inadequate appreciation of the extreme weakness of the hurdle provided by the mere directional significance test, there exists among psychologists (c) a fairly widespread tendency to report experimental findings with a liberal use of *ad hoc* explanations for those that didn't "pan out." This last methodological sin is especially tempting in the "soft" fields of (personality and social) psychology, where the profession highly rewards a kind of "cuteness" or "cleverness" in experimental design, such as a hitherto untried method for inducing a desired emotional state or a particularly "subtle" gimmick for detecting its influence upon behavioral output. The methodological price paid for this highly-valued "cuteness" is, of course, (d) an unusual ease of escape from *modus tollens* refutation. For, the logical structure of the "cute" component typically involves use of complex and rather dubious auxiliary assumptions, which are required to mediate the original prediction and are therefore readily available as (genuinely) plausible "outs" when the prediction fails. It is not unusual that (e) this *ad hoc* challenging of auxiliary hypotheses is repeated in the course of a series of related experiments, in which the auxiliary hypothesis involved in Experiment 1 (and challenged *ad hoc* in order to avoid the latter's *modus tollens* impact on the theory) becomes the focus of interest in Experiment 2, which in turn utilizes further plausible but easily challenged auxiliary hypotheses, and so forth. In this fashion a zealous and clever investigator can slowly wend his way through a tenuous nomological network, performing a long series of related experiments which appear to the uncritical reader as a fine example of "an integrated research program," *without ever once refuting or corroborating so much as a single strand of the network.* Some of the more horrible examples of this process would require the combined analytic and reconstructive efforts of Carnap, Hempel, and Popper to unscramble the logical relationships of theories and hypotheses to evidence. Meanwhile our eager-beaver researcher, undismayed by logic-of-science considerations and relying blissfully on the "exactitude" of modern statistical hypothesis-testing, has produced a long publication list and been promoted to a full professorship. In terms of his contribution to the enduring body of psychological knowledge, he has done hardly anything. His true position is that of a potent-but-sterile intellectual rake, who leaves in his merry path a long train of ravished maidens but no viable scientific offspring.

Detailed elaboration of the intellectual vices (a)–(e) and their scientific consequences must be left for another place, as must constructive suggestions for how the behavioral scientist can improve his situation. My main aim here has been to call the attention of logicians and philosophers of science to what, as I think, is an important and difficult problem for psychology, or for any science which is largely in a primitive stage of development such that its theories do not give rise to point-predictions.[1]

1. Since the readers of this article cannot, by and large, be expected to possess familiarity with the field of psychology or the contributions of various psychologists, and since quantitative empirical documentation of these admittedly impressionistic comments is still in preparation for subsequent presentation elsewhere, it is perhaps neither irrelevant nor in bad taste to present a few biographical data. Lest the philosophical reader wonder (quite appropriately) whether these impressions of the psychological literature ought perhaps to be dismissed as mere "sour grapes" from an embittered, low-publication psychologist *manqué*, it may be stated that the author (a past president of the American Psychological Association) has published over 70 technical books or articles in both "hard" and "soft" fields of psychology, is a recipient of the Association's Distinguished Scientific Contributor Award, also of the Distinguished Contributor Award of the Division of Clinical Psychology, has been elected to Fellowship in the American Academy of Arts and Sciences, and is actively engaged in both theoretical and empirical research at the present time. He's not mad at anybody—but he is a bit distressed at the state of psychology.

27. Statistical Significance in Psychological Research

DAVID T. LYKKEN

IN A RECENT JOURNAL ARTICLE Sapolsky (1964) developed the following substantive theory: Some psychiatric patients entertain an unconscious belief in the "cloacal theory of birth" which involves the notions of oral impregnation and anal parturition. Such patients should be inclined to manifest eating disorders: compulsive eating in the case of those who wish to get pregnant and anorexia in those who do not. Such patients should also be inclined to see cloacal animals, such as frogs, on the Rorschach. This reasoning led Sapolsky to predict that Rorschach frog responders show a higher incidence of eating disorders than patients not giving frog responses. A test of this hypothesis in a psychiatric hospital showed that 19 of 31 frog responders had eating disorders indicated in their charts, compared to only 5 of the 31 control patients. A highly significant Chi-square was obtained.

It will be an expository convenience to analyze Sapolsky's article in considerable detail for purposes of illustrating the methodological issues which are the real subject of this paper. My intent is not to criticize a particular author but rather to examine a kind of epistemic confusion which seems to be endemic in psychology, especially, but by no means exclusively, in its "softer" precincts. One would like to demonstrate this generality with

Reprinted from *Psychological Bulletin*, 70 (September, 1968), 151–159, by permission of the author and the publisher. Copyright © 1968 by the American Psychological Association, Inc.

multiple examples. Having just combed the latest issues of four well-known journals in the clinical and personality areas, I could undertake to identify several papers in each issue wherein, because they were able to reject a directional null hypothesis at some high level of significance, the authors claimed to have usefully corroborated some rather general theory or to have demonstrated some important empirical relationship. To substantiate that these claims are overstated and that much of this research has not yet earned the right to the reader's overburdened attentions would require a lengthy analysis of each paper. Such profligacy of space would ill become an essay one aim of which is to restrain the swelling volume of the psychological literature. Therefore, with apologies to Sapolsky for subjecting this one paper to such heavy-handed scrutiny, let us proceed with the analysis.

Since I regarded the prior probability of Sapolsky's theory (that frog responders unconsciously believe in impregnation *per os*) to be nugatory and its likelihood unenhanced by the experimental findings, I undertook to check my own reaction against that of 20 colleagues, most of them clinicians, by means of a formal questionnaire. The 20 estimates of the prior probability of Sapolsky's theory, which these psychologists made before being informed of his experimental results, ranged from 10^{-6} to 0.13 with a median value of 0.01, which can be interpreted to mean, roughly, "I don't believe it." Since the prior probability of many important scientific theories is considered to be vanishingly small when they are first propounded, this result provides no basis for alarm. However, after being given a fair summary of Sapolsky's experimental findings, which "corroborate" the theory by confirming the operational hypothesis derived from it with high statistical significance, these same psychologists attached posterior probabilities to the theory which ranged from 10^{-5} to 0.14, with the median unchanged at 0.01. I interpret this consensus to mean, roughly, "I still don't believe it." This finding, I submit, *is* alarming because it signifies a sharp difference of opinion between, for example, the consulting editors of the journal and a substantial segment of its readership, a difference on the very fundamental question of what constitutes good (i.e. publishable) clinical research.

The thesis of the present paper is that Sapolsky and the editors were in fact following, with reasonable consistency, our traditional rules for evaluating psychological research, but that, as the Sapolsky paper exemplifies, at least two of these rules should be reconsidered. One of the rules examined here asserts roughly the following: "When a prediction or hypothesis derived from a theory is confirmed by experiment, a nontrivial increment in one's confidence in that theory should result, especially when one's prior confidence is low." Clearly, my 20 colleagues were violating this rule here since their confidence in the frog responder-cloacal birth theory was not, on the average, increased by the contemplation of Sapolsky's highly significant Chi-square.

From their comments it seems that they found it too hard to accept that a belief in oral impregnation could lead to frog responding merely because the frog has a cloacus. (One must, after all, admit that few patients know what a cloacus is or that a frog has one and that those few who do know probably will also know that the frog's eggs are both fertilized and hatched externally so neither oral impregnation nor anal birth are in any way involved. Hence, *neither* the average patient *nor* the biologically sophisticated patient should logically be expected to employ the frog as a symbol for an unconscious belief in oral conception.) My colleagues, on the contrary, found it relatively easy to believe that the observed association between frog responding and eating problems might be due to some other cause entirely (e.g. both symptoms are immature or regressive in character; the frog, with its disproportionately large mouth and voice may well constitute a common orality totem and hence be associated with problems in the oral sphere; "squeamish" people might tend both to see frogs and to have eating problems; and so on).

Assuming that this first rule *is* wrong in this instance, perhaps it could be amended to allow one to make exceptions in cases resembling this illustration. For example, one could add the codicil: "This rule may be ignored whenever one considers the theory in question to be overly improbable or whenever one can think of alternative explanations for the experimental results." But surely such an amendment would not do. ESP, for example, could never become scientifically respectable if the first exception were allowed, and one consequence of the second would be that the importance attached to one's findings would always be inversely related to the ingenuity of one's readers. The burden of the present argument is that this rule is wrong not only in a few exceptional instances *but as it is routinely applied to the majority of experimental reports in the psychological literature.*

Corroborating Theories by Experimental Confirmation of Theoretical Predictions[1]

Most psychological experiments are of three kinds: (*a*) studies of the effect of some treatment on some output variables, which can be regarded as a special case of (*b*) studies of the difference between two or more groups of individuals with respect to some variable, which in turn are a special case of (*c*) the study of the relationship or correlation between two or more variables within some specified population. Using the bivariate correlation design as paradigmatic, then, one notes first that the strict null hypothesis must always be assumed to be false (this idea is not new and has recently been illuminated by Bakan,

1. Much of the argument in this section is based upon ideas developed in certain unpublished memoranda by P. E. Meehl (personal communication, 1963) and in a recent article (Meehl, 1967 [Chapter 26]).

1967 [Chapter 25]). Unless one of the variables is wholly unreliable so that the values obtained are strictly random, it would be foolish to suppose that the correlation between any two variables is identically equal to 0.0000 . . . (or that the effect of some treatment or the difference between two groups is exactly *zero*). The molar dependent variables employed in psychological research are extremely complicated in the sense that the measured value of such a variable tends to be affected by the interaction of a vast number of factors, both in the present situation and in the history of the subject organism. It is exceedingly unlikely that any two such variables will not share at least some of these factors and equally unlikely that their effects will exactly cancel one another out.

It might be argued that the more complex the variables the smaller their average correlation ought to be since a larger pool of common factors allows more chance for mutual cancellation of effects in obedience to the Law of Large Numbers. However, one knows of a number of unusually potent and pervasive factors which operate to unbalance such convenient symmetries and to produce correlations large enough to rival the effects of whatever causal factors the experimenter may have had in mind. Thus, we know that (*a*) "good" psychological and physical variables tend to be positively correlated; (*b*) experimenters, without deliberate intention, can somehow subtly bias their findings in the expected direction (Rosenthal, 1963); (*c*) the effects of common method are often as strong as or stronger than those produced by the actual variables of interest (e.g. in a large and careful study of the factorial structure of adjustment to stress among officer candidates, Holtzman and Bitterman [1956] found that their 101 original variables contained five main common factors representing, respectively, their rating scales, their perceptual-motor tests, the McKinney Reporting Test, their GSR variables, and the MMPI); (*d*) transitory state variables, such as the subject's anxiety level, fatigue, or his desire to please, may broadly affect all measures obtained in a single experimental session.

This average shared variance of "unrelated" variables can be thought of as a kind of ambient noise level characteristic of the domain. It would be interesting to obtain empirical estimates of this quantity in our field to serve as a kind of Plimsoll mark against which to compare obtained relationships predicted by some theory under test. If, as I think, it is not unreasonable to suppose that "unrelated" molar psychological variables share on the average about 4 percent to 5 percent of common variance, then the expected correlation between any such variables would be about .20 in absolute value and the expected difference between any two groups on some such variable would be nearly 0.5 standard deviation units. (Note that these estimates assume zero measurement error. One can better explain the near-zero correlations often observed in psychological research in terms of unreliability of measures than in terms of the assumption that the true scores are in fact unrelated.)

Suppose now that an investigator predicts that two variables are positively correlated. Since we expect the null hypothesis to be false, we expect his prediction to be confirmed by experiment with a probability of very nearly 0.5; by using a large enough sample, moreover, he can achieve any desired level of statistical significance for this result. If the ambient noise level for his domain is represented by correlations averaging, say, .20 in absolute value, then his chances of finding a statistically significant confirmation of his prediction with a reasonable sample size will be quite high (e.g. about one in four for $N = 100$) even if there is no truth whatever to the theory on which the prediction was based. Since most theoretical predictions in psychology, especially in the areas of clinical and personality research, specify no more than the direction of a correlation, difference, or treatment effect, we must accept the harsh conclusion that a single experimental finding of this usual kind (confirming a directional prediction), no matter how great its statistical significance, will seldom represent a large enough increment of corroboration for the theory from which it was derived to merit very serious scientific attention. (In the natural sciences, this problem is far less severe for two reasons: (a) theories are powerful enough to generate point predictions or at least predictions of some narrow range within which the dependent variable is expected to lie; and (b) in these sciences, the degree of experimental control and the relative simplicity of the variables studied are such that the ambient noise level represented by unexplained and unexpected correlations, differences, and treatment effects is often vanishingly small.)

The Significance of Large Correlations

It might be argued that, even where only a weak directional prediction is made, the obtaining of a result which is not only statistically significant but large in absolute value should constitute a stronger corroboration of the theory. For example, although Sapolsky predicted only that frog responding and eating disorders would be positively related, the fourfold point correlation (phi coefficient) between these variables in his sample was about .46, surely much larger than the average relationship expected between random pairs of molar variables on the premise that "everything is related to everything else." Does not such a large effect therefore provide stronger corroboration for the theory in question?

One difficulty with this reasonable sounding doctrine is that, in the complex sort of research considered here, *really large* effects, differences, or relationships are not usually to be expected and, when found, may even argue *against* the theory being tested. To illustrate this, let us take Sapolsky's theory seriously and, by making reasonable guesses concerning the unknown base rates involved, attempt to estimate the actual size of the relationship between frog responding and eating disorders which the theory should lead us to expect. Sapolsky found that 16 percent of his control sample showed eating disorders; let us take this value as the base rate for this symptom among patients who do not hold the cloacal theory of birth. Perhaps we can assume that all patients who do hold this theory will give frog responses but surely not all of these will show eating disorders (any more than will all patients who believe in vaginal conception be inclined to show coital or urinary disturbances); it seems a reasonable assumption that no more than 50 percent of the believers in oral conception will therefore manifest eating problems. Similarly, we can hardly suppose that the frog response *always* implies an unconscious belief in the cloacal theory; surely this response can come to be emitted now and then for other reasons. Even with the greatest sympathy for Sapolsky's point of view, we could hardly expect more than, say, 50 percent of frog responders to believe in oral impregnation. Therefore, we might reasonably predict that 16 of 100 nonresponders would show eating disorders in a test of this theory, 50 of 100 frog responders would hold the cloacal theory and half of these show eating disorders, while 16 percent or 8 of the remaining 50 frog responders will show eating problems too, giving a total of 33 eating disorders among the 100 frog responders. Such a finding would produce a significant Chi-square but the actual degree of relationship as indexed by the phi coefficient would be only about .20. In other words, if one considers the supplementary assumptions which would be required to make a theory compatible with the actual results obtained, it becomes apparent that the finding of a really strong association may actually embarrass the theory rather than support it (e.g. Sapolsky's finding of 61 percent eating disorders among his frog responders is *significantly larger* [$p < .01$] than the 33 percent generously estimated by the reasoning above).

Multiple Corroboration

In the social, clinical, and personality areas especially, we must expect that the size of the correlations, differences, or effects which might reasonably be predicted from our theories will typically not be very large relative to the ambient noise level of correlations and effects due solely to the "all-of-a-pieceness of things." The conclusion seems inescapable that the only really satisfactory solution to the problem of corroborating such theories is that of *multiple corroboration*, the derivation and testing of a number of separate, quasi-independent predictions. Since the prior probability of such a multiple corroboration may be on the order of $(0.5)^n$, where n is the number of independent[2]

---

2. Tests of predictions from the same theory are seldom strictly independent since they often share some of the same supplementary assumptions, are made at the same time on the same sample, and so on.

predictions experimentally confirmed, a theory of any useful degree of predictive richness should in principle allow for sufficient empirical confirmation through multiple corroboration to compel the respect of the most critical reader or editor.

The Relation of Experimental Findings to Empirical Facts

We turn now to the examination of a second popular rule for the evaluation of psychological research, which states roughly that "When no obvious errors of sampling or experimental method are apparent, one's confidence in the general proposition being tested (e.g. Variables A and B are positively correlated in Population C) should be proportional to the degree of statistical significance obtained." We are following this rule when we say, "Theory aside, Sapolsky has at least demonstrated an empirical fact, namely, that frog responders have more eating disturbances than patients in general." This conclusion means, of course, that in the light of Sapolsky's highly significant findings we should be willing to give very generous odds that any other competent investigator (at another hospital, administering the Rorschach in his own way, and determining the presence of eating problems in whatever manner seems reasonable and convenient for him) will also find a substantial positive relationship between these two variables.

Let us be more specific. Given Sapolsky's fourfold table showing 19 of 31 frog responders to have eating disorders (61 percent), it can be shown by Chi-square that we should have 99 percent confidence that the true population value lies between 13/31 and 25/31 (between 42 percent and 81 percent). With 99 percent confidence that the population value is at least 13 in 31, we should have $.99(99) = 98$ percent confidence that a new sample from that population should produce at least 6 eating disorders among each 31 frog responders, assuming that 5 of each 31 nonresponders show eating problems also, as Sapolsky reported. That is, we should be willing to bet $98 against only $2 that a replication of this experiment will show *at least as many* eating disorders among frog responders as among nonresponders. The reader may decide for himself whether his faith in the "empirical fact" demonstrated by this experiment can meet the test of this gambler's challenge.

Three Kinds of Replication

If, as suggested above, "demonstrating an empirical fact" must involve a claim of confidence in the replicability of one's findings, then to clearly understand the relation of statistical significance to the probability of a "successful" replication it will be helpful to distinguish between three rather different methods of replicating or cross-validating an experiment. *Literal replication,* of course, would involve exact duplication of the first investigator's sampling procedure, experimental conditions, measuring techniques, and methods of analysis; asking the original investigator to simply run more subjects would perhaps be about as close as we could come to attaining literal replication and even this, in psychological research, might often not be close enough. In the case of *operational replication*, on the other hand, one strives to duplicate exactly just the sampling and experimental procedures given in the first author's report of his research. The purpose of operational replication is to test whether the investigator's "experimental recipe"—the conditions and procedures he considered salient enough to be listed in the "Methods" section of his report—will in other hands produce the results that he obtained. For example, replication of the "Clever Hans" experiment revealed that the apparent ability of that remarkable horse to add numbers had been due to an uncontrolled and unsuspected factor (the presence of the horse's trainer within his field of view). This factor, not being specified in the "methods recipe" for the result, was omitted in the replication which for that reason failed. Operational replication would be facilitated if investigators would accept more responsibility for specifying what they believe to be the minimum essential conditions and controls for producing their results. Psychologists tend to be inconsistently prolix in describing their experimental methods; thus, Sapolsky tabulates the age, sex, and diagnosis for each of his 62 subjects. Does he mean to imply that the experiment will not work if these details are changed? Surely not, but then why describe them?

In the quite different process of *constructive replication*, one deliberately avoids imitation of the first author's methods. To obtain an ideal constructive replication, one would provide a competent investigator with *nothing more than* a clear statement of the empirical "fact" which the first author would claim to have established—for example, "psychiatric patients who give frog responses on the Rorschach have a greater tendency toward eating disorders than do patients in general"—and then let the replicator formulate his own methods of sampling, measurement, and data analysis. One must keep in mind that the data, the specific results of a particular experiment, are only seldom of any real interest in themselves. The "empirical facts" which we value so highly consist usually of confirmed conceptual or constructive (not operational) hypotheses of the form "Construct A is positively related to Construct B in Population C." We are interested in the *construct* "tendency toward eating disorders," not in the *datum* "has reference made to overeating in the nurse's notes for May 15th." An operational replication tests whether we can duplicate our findings using the same methods of measurement and sampling; a constructive replication goes further in the sense of testing the validity of these methods.

Thus, if I cannot confirm Sapolsky's results for patients from my hospital, assessing eating disorders by means of informant interviews, say, or actual measurements of food intake, then clearly Sapolsky has *not* demonstrated any "fact" about eating disorders among psychiatric patients in general. I could then revert to an operational replication, assessing eating problems from the psychiatric notes as Sapolsky did and selecting my sample to conform with the age, sex, and diagnostic properties of his, although I might not regard this endeavor to be worth the effort since, under these circumstances, even a successful operational replication could not establish an empirical conclusion of any great generality or interest. Just as a reliable but invalid test can be said to measure something, but not what it claimed to measure, so an experiment which replicates operationally but not constructively could be said to have demonstrated something, but not the relation between meaningful constructs, generalizable to some broad reference population, which the author originally claimed to have established.[3]

Relation of the Significance Test to the Probability of a "Successful" Replication

The probability values resulting from significance testing can be directly used to measure one's confidence in expecting a "successful" literal replication only. Thus, we can be 98 percent confident of finding at least 6 of 31 frog responders to have eating problems only if we reproduce all of the conditions of Sapolsky's experiment with absolute fidelity, something that he himself could not undertake to do at this point. Whether we are entitled to anything approaching such high confidence that we could obtain such a result from an operational replication depends entirely upon whether Sapolsky has accurately specified all of the conditions which were in fact determinative of his results. That he did not in this instance is suggested by the fact that, investigating the feasibility of replicating his experiment at the University of Minnesota Hospitals, I found that I should have to review several thousand case records in order to turn up a sample of 31 frog responders like his. Although he does not indicate how many records he examined, one strongly suspects that the base rate of Rorschach frog responding must have been higher at Sapolsky's hospital, either because of some difference in the patient population or, more probably, because an investigator's being interested in some class of responses

---

3. This distinction between operational and constructive replication seems to have much in common with that made by Sidman (1960) between what he calls "direct" and "systematic" replication. However, in the operant research context to which Sidman directs his attention, "replication" means to run another animal or the same animal again; thus, direct replication involves maintaining the same experimental conditions in detail whereas in systematic replication one allows all supposedly irrelevant factors to vary from one subject to the next in the hope of demonstrating that one has correctly identified the variables which are really in control of the behavior being studied.

will tend to subtly elicit such responses at a higher rate unless the testing procedure is very rigorously controlled. If the base rates for frog responding are so different at the two hospitals, it seems doubtful that the response can have the same correlates or meaning in the two populations and therefore one would be reckless indeed to offer high odds on the outcome of even the most careful operational replication. The likelihood of a successful constructive replication is, of course, still smaller since it depends on the additional assumptions that Sapolsky's samples were truly representative of psychiatric patients in general and that his method of assessing eating problems was truly valid, that is, would correlate highly with a different, equally reasonable appearing method.

Another Example

It is not my purpose, of course, to criticize statistical theory or method but rather to suggest ways in which these tools are sometimes misused or misinterpreted by writers or readers of the psychological literature. Nor do I mean to abuse a particular investigator whose research report happened to serve as a convenient illustration of the components of the argument. An abundance of articles can be found in the journals which exemplify these points quite as well as Sapolsky's but space limitations forbid multiple examples. As a compromise, therefore, I offer just one further illustration, showing how the application of these same critical principles might have increased a reader's—and perhaps even an editor's—skepticism concerning some research of my own.

The purpose of the experiment in question (Lykken, 1957) was to test the hypothesis that the "primary" psychopath has reduced ability to condition anxiety or fear. To segregate a subgroup in which such primary psychopaths might be concentrated, I asked prison psychologists to separate inmates already diagnosed as psychopathic personalities into one group that met 14 rather specific clinical criteria specified by Cleckley (1950: 355–392) and to identify another group which clearly did not fit some of these criteria. The normal control subjects were comparable to the psychopathic groups in age, IQ, and sex. Fear conditioning was assessed using the GSR as the dependent variable and a rather painful electric shock as the unconditioned stimulus (UCS). On the index used to measure rate of conditioning, the primary psychopathic group scored significantly lower than did the controls. By the usual reasoning, therefore, one might conclude that this result demonstrates that primary psychopaths are abnormally slow to condition the GSR, at least with an aversive UCS, and this empirical fact in turn provides significant support for the theory that primary psychopaths have defective fear-learning ability (i.e. a low "anxiety IQ").

But to anyone who has actually participated in research of this kind, this seemingly straightforward reasoning must appear appallingly oversimplified. It is quite impossible to obtain anything resembling a truly random sample of psychopaths (or of nonpsychopathic normals either, for that matter) and it is a matter of unquantifiable conjecture how a sample obtained by a different investigator using equally defensible methods might perform on the tests which I employed. Even with the identical sample, no two investigators are likely to measure the GSR in the same way, use the same conditioned stimulus (CS) and UCS or the same pattern of reinforced and CS-only trials. Given even the same set of protocols, there is no standard formula for obtaining an index of degree or rate of conditioning; the index I used was essentially arbitrary and whether it was a good one is a matter of opinion. My own evaluation of the methods used, together with a complex set of supplementary assumptions difficult to explicate, leads me to believe that these results increase the likelihood that primary psychopaths have slower GSR conditioning with an aversive UCS; I might now give odds of two to one that this empirical generalization is true and odds of three to two that another investigator would be able to confirm it by means of a constructive replication. But this already biased claim is far more modest than the one which is implicit in the significance testing operation, namely: "Such a mean difference would only be expected 5 times in 100 if the [generalization] is not true."

This empirical generalization, about GSR conditioning, is derivable from the hypothesis of interest, that psychopaths have a low anxiety IQ, by a chain of reasoning so complex and elliptical and so burdened with accessory assumptions as to be quite impossible to spell out in the detail required for rigorous logical analysis. Psychologists knowledgeable in the area can evaluate whether it is a reasonable derivation but their opinions will not necessarily agree. Moreover, even if the derivation could pass the scrutiny of some "Certified Public Logician," confirmation of the prediction about GSR conditioning should add only very slightly to our confidence in the hypothesis about fear conditioning. Even if this confirmation were made relatively more firm by, for example, constructive replication of the generalization, "aversive GSR conditioning is retarded in primary psychopaths," the hypothesis that these individuals have a low anxiety IQ could still be said to have passed only the weakest kind of test. This is so because such simple directional predictions about group differences have nearly a 50–50 chance of being true *a priori* even if our particular hypothesis is false. There are doubtless many possible explanations for low GSR conditioning scores in psychopaths other than the possibility of defective fear conditioning. Indeed, some of my subjects whose conditioning scores were nearly as low as those of the most extreme primary psychopaths seemed to me to be clearly neurotic with considerable anxiety and I attempted to account for their GSR performance with an *ad hoc* conjecture involving a kind of repression phenomenon, that is, a denial that a low GSR index implied poor fear conditioning in their cases.

A redeeming feature of this study was that two other related but distinguishable predictions from the same hypothesis were tested at the same time, namely, that primary psychopaths should do as well as normals on a learning task involving positive reward but less well on an avoidance learning problem, and that they should be more willing than normals to choose embarrassing or frightening situations in preference to alternatives involving tedium, frustration, physical discomfort, and the like. Tests of these predictions gave affirmative results also, thus providing some of the multiple corroboration necessary for the hypothesis to claim the attention of other experimenters.

Obviously, I do not mean to criticize the editor's decision to publish my (1957) paper. The tendency to evaluate research in terms of mechanical rules based on the results of the significance tests should not be replaced by equally rigid requirements concerning replication or corroboration. This study, like Sapolsky's or most others in this field, can be properly evaluated only by a qualified reader who can substitute his own informed judgment and scientific intuition for the rigorous reasoning and experimental control that is usually not achievable in clinical and personality research. As it happens, subsequent work has provided some encouraging support for my 1957 findings. The two additional predictions mentioned above have received operational replication (i.e. the same test methods used in a different context) by Schachter and Latené (1964). The prediction that psychopaths show slower GSR conditioning with an aversive UCS has been constructively replicated (i.e. independently tested with no attempt to copy my procedures) by Hare (1965a). Finally, two additional predictions from the theory that the primary psychopath has a low anxiety IQ have been tested with affirmative results (Hare, 1965b; 1966). All told, then, this hypothesis can now boast of having led to at least five quasi-independent predictions which have been experimentally confirmed and three of which have been replicated. The hypothesis is therefore entitled to serious consideration, although one would be rash still to regard it as proven. At least one alternative hypothesis, that the psychopath has an unusually efficient mechanism for inhibiting emotional arousal, can account equally well for the existing findings so that, as is usually the case, further research is called for.

Conclusions

The moral of this story is that the finding of statistical significance is perhaps the least important attribute of a good experiment: It is *never* a sufficient condition for concluding that a theory has been corroborated, that a useful empirical fact has been established with reasonable confidence—or that an experimental report ought to be published. The value of any research can be determined, not from the statistical results, but only by skilled, subjective evaluation of the coherence and reasonableness of the theory, the degree of experimental control employed, the sophistication of the measuring techniques, the scientific or practical importance of the phenomena studied, and so on. Ideally, all experiments would be replicated before publication but this goal is impractical. "Good" experiments will tend to replicate better than poor ones (and, when they do not, the failures will tend to be informative in themselves, which is not true for poor experiments) and should be published so that they may stimulate replication and extension by others. Editors must be bold enough to take responsibility for deciding which studies are good and which are not, without resorting to letting the p value of the significance tests determine this decision. There is little real danger that anything of value will be lost through this approach since the unpublished investigator can always resort to constructive replication to induce editorial acceptance of his empirical conclusions or to multiple corroboration to compel editorial respect for his theory. Since operational replication must really be done by an independent second investigator and since constructive replication has greater generality, its success strongly implying that an operational replication would have succeeded also, one should usually replicate one's own work constructively, using different sampling and measurement procedures within the purview of the same constructive hypothesis. If only unusually well done, provocative, and important research were published without such prior authentication, operational replication of such research by others would become correspondingly more valuable and entitled to the respect now accorded capable replication in the other experimental sciences.

PART FOUR
Criticism from Other Quarters

INTRODUCTION

T HE PAPERS in this section fall outside the parallel mainstreams of discussions of significance tests in sociology and psychology but are clearly in the critical spirit of many of the papers in the previous sections. Berkson's "Tests of Significance Considered as Evidence" [Chapter 28] is noteworthy because it shows that great dissatisfaction with the tests existed in other quarters many years before Selvin's [Chapter 9] major critique in sociology and Rozeboom's [Chapter 24] critique in psychology. Moreover, many of the points in Berkson's critique are the same as those independently made and re-made subsequently.

Berkson, a medical researcher, addresses the issues of meaning, power, and level by means of several concrete examples and relates these issues to the issues of form, process, and purpose in scientific inference. Like Hogben, he shows how the meaning that significance testing assigns to a rare event is not consonant with the inferential aims of the scientist. Nor, he claims, does the practice of rejecting no-difference null hypotheses help in achieving the positive knowledge about alternatives which science demands. Berkson demonstrates that the automatic features of inference with significance tests often lead to conclusions that are patently incorrect by any criterion of good judgment and common sense, and he clearly illustrates the sense in which failure to consider the power of a test can lead to questionable conclusions.

In his "Publication Decisions and Their Possible Effects on Inferences

Drawn from Tests of Significance—or Vice Versa" [Chapter 29], Sterling, a statistician, analyzes several psychology journals and shows that most of the articles employ significance tests, show significant ($p < .05$) results, and do not replicate previous studies. Sterling notes that many more studies are done than appear in professional journals, and he argues that many of these unpublished studies are likely to be unsuccessful ($p > .05$) duplications of published studies. He thus concludes that the findings reported in published studies actually have a much larger probability of unsuccessful replication than is implied in the significance level reported. Moreover, Sterling speculates that the publication of a significant finding tends greatly to decrease the probability of the study's replication in the future.

Tullock [Chapter 30], an economist, generally endorses Sterling's notions in a brief note of response, but he argues that the problem is probably not so much that an accumulation of unsuccessful replications exists or that a successful result tends to prevent replication as it is that the practice of replication is simply not a part of the research traditions of the social sciences. Sterling and Tullock do not question significance tests as a mode of inference for science, but clearly they raise a difficult issue about the inferential validity of current practices. Bakan [Chapter 25] and Lykken [Chapter 27] later discussed the same problems in the context of their more general critiques of the tests.

28. Tests of Significance Considered as Evidence

JOSEPH BERKSON, M.D.

*"After all, the higher statistics
are only common sense
reduced to numerical
appreciation."*—KARL PEARSON.

THERE WAS A TIME when we did not talk about tests of significance; we simply did them. We tested whether certain quantities were significant in the light of their standard errors, without inquiring as to just what was involved in the procedure or attempting to generalize it. In recent years tests of significance have been more broadly conceived as tests of hypotheses, and they have been generalized as t tests, F tests, and certain amplifications of these, such as analysis of variance or of covariance. It is hardly an exaggeration to say that statistics, as it is taught at present in the dominant school, consists almost entirely of tests of significance, though not always presented as such, some comparatively simple and forthright, others elaborate and abstruse. Behind this is a doctrine of analysis that consists of setting up what is called a "null hypothesis" and testing it. Indeed, in this conception not only does this procedure characterize the method of statistics, but it is considered to be the very essence of all experimental science. In his well known book, *The Design of Experiments*, R. A. Fisher wrote, "Every experiment may be said to exist only in order to give the facts a chance of disproving the null hypothesis (1935a, 2nd ed.: 19).

What is this null hypothesis procedure? I quote from a recent text.

Reprinted from the *Journal of the American Statistical Association*, 37 (September, 1942), 325–335, by permission of the author and the publisher.

We have just set up the hypothesis that our sample of 900, which has a mean of 15,071 miles, is a random sample drawn from the population having a known mean of 15,200 miles. . . . Such a hypothesis is called a *null* hypothesis since our computations undertake to nullify it. The procedure may be summarized into three steps: (1) Set up the hypothesis that the true difference is zero. (2) Upon the basis of this hypothesis determine the probability that such a difference as the one observed might occur because of sampling variations. (3) Draw a conclusion concerning the hypothesis. If such observed difference could hardly have occurred by chance, we have cast much doubt upon the hypothesis. We therefore abandon the hypothesis and conclude that the observed difference is significant (Croxton and Cowden, 1940: 310).

This I believe is a fair if abbreviated statement of the essential procedure as it is generally understood. If the experience at hand would occur only very infrequently in a given hypothesis, the hypothesis is considered disproved.

The argument has an apparent plausibility and for many years I adhered to it. However, set against experience with actual problems, reflection has led me to the conclusion that it is erroneous, and that a re-evaluation will lead to clearer comprehension in the application of tests of significance and also serve as a corrective of some of its misuses.

In the first place, the argument seems to be basically illogical. Consider it in symbolic form. It says "If A is true, B will happen sometimes; therefore if B has been found to happen, A can be considered disproved." There is no logical warrant for considering an event known to occur in a given hypothesis, even if infrequently, as disproving the hypothesis.

More to the present point, the argument does not seem to accord with what would be the mode of reasoning in ordinary rational discourse, nor with the rationale of usual procedures as they are observed in the scientific laboratory. Suppose I said, "Albinos are very rare in human populations, only one in fifty thousand. Therefore, if you have taken a random sample of 100 from a population and found in it an albino, the population is not human." This is a similar argument but if it were given, I believe the rational retort would be, "If the population is not human, what *is* it?" A question would be asked that demands an *affirmative* answer. In the null hypothesis schema we are trying only to nullify something: "The null hypothesis is never proved or established but is possibly disproved in the course of experimentation." But ordinarily evidence does not take this form. With the *corpus delicti* in front of you, you do not say, "Here is evidence against the hypothesis that no one is dead." You say, "Evidently someone has been murdered."

Nor do you find experimentalists typically engaged in disproving things. They are looking for appropriate evidence for affirmative conclusions. Even if the mediate purpose is the disestablishment of some current idea, the immediate objective of a working scientist is likely to be to gain affirmative evidence in favor of something that will refute the allegation which is under attack.

Does this mean that the application of tests of significance is in basic disaccord with rational scientific procedure? I am not sure. I think that there is a possibility of using them soundly, but the rule of inference on which they are supposed to rest has been misconceived, and this has led to certain fallacious uses.

Consider the objective of testing whether a distribution is normal. One could validly say, "If the distribution is normal and, the skewness of the sample, g_1, having been calculated, if a die of 100 faces, five of which are black, is thrown at random, a black face will occur only five times in 100." No one would suggest that the finding of a black face on a die following such a calculation is any reason for rejecting the null hypothesis that the distribution is normal. But when one says, "If the distribution is normal, a value of $g_1/S_{g_1} \geqslant 1.96$ will occur only five times in 100," the finding of such a value for g_1/S_{g_1} is taken as reason for rejecting the null hypothesis. What is the essential difference between the two situations? Following the procedures which were outlined for dealing with a null hypothesis, one should reject the hypothesis that the distribution is normal on the finding of a black face, for it is surely an event rare in the circumstance of the distribution being normal. The difference appears to be that we recognise that if the distribution actually were *abnormal* (skew), the occurrence of a black face still would not be expected, but a large value of g_1/S_{g_1} *would be expected*. The latter constitutes evidence *in favor* of skewness. We may discern, as operating in the realm of tests of significance, a principle that I suggest is generally operative in scientific inquiry. It is this: The finding of an event which is *frequent* under a hypothesis H_1 can be taken as evidence *in favor* of H_1. If H_0 is a contradictory alternative to H_1 for which the event would not be frequent then per corollary the finding of the event is, in so far, evidence in disfavor of H_0.

At this point I can imagine the question rising, "What difference does it make whether you say that you *reject* H_0 because for it the event is not frequent, or because you are *accepting* the alternative H_1 for which it is frequent?" To this the first answer must be that it would seem to be a sound idea to get one's head clear as to what are the principles on which one is really acting. If an event has occurred, the definitive question is not, "Is this an event which would be rare if H_0 is true?" but "Is there an alternative hypothesis under which the event would be relatively frequent?" If there is no plausible alternative at all, the rarity is quite irrelevant to a decision, and if there is such an alternative, the decisive question is, "Would the event be relatively frequent?" Secondly, the pursuit of a false principle for testing the null hypothesis will lead to false conclusions that will be avoided if one is consciously guided by the principle suggested here as being the correct one. I shall cite an example.

As an illustration of a test of linearity under the caption, "Test of straightness of regression line," R. A. Fisher utilizes data relating the temperature to the number of eye facets of *Drosophila melanogaster*, the facet number being measured in factorial units. An analysis of variance procedure is utilized for the test and, the calculations having been made, Fisher says,

> The deviations from linear regression are evidently larger than would be expected, if the regression were really linear, from the variations within the arrays. For the value of z we have 1.2434 while the 1 percent point is about .488. There can therefore be no question of the statistical significance of the deviations from the straight line . . . . the departure from linearity was markedly significant (1925a, 7th ed.: 259–265).

I have plotted the data of mean facet number in relation to temperature together with the least square line and they are shown in Figure 28.1. It was found by the significance test as applied that this regression was not straight, but on inspection it appears as straight a line as one can expect to find in biological material. What has betrayed the author is a faithful adherence to an unsound principle: to wit, reject the null hypothesis tested, in this case that the regression is linear, if the P of the test is small.

Figure 28.1. Mean number of eye facets of Drosophila melanogaster *raised at different temperatures and best fitting straight line by method of least squares*

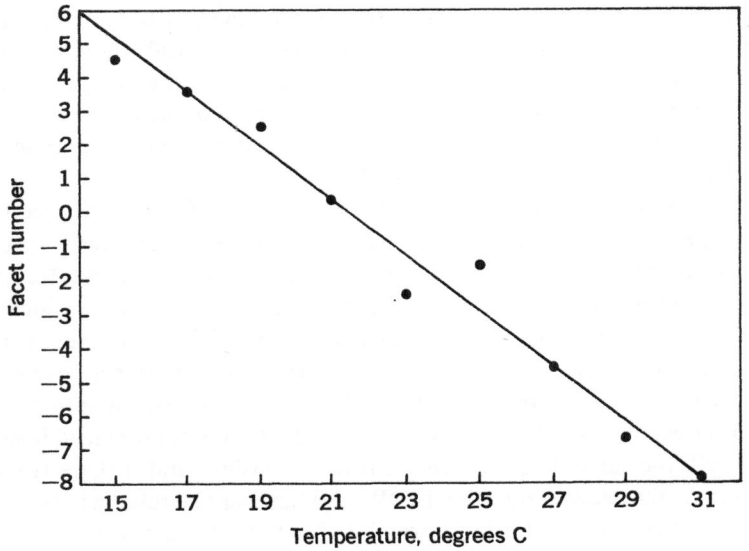

Source: Data from R. A. Fisher (1925a, 7th ed.: 260).

Let us consider the problem according to the principle advanced here. The event which has been found to have happened, in this case the small P, is to be considered as evidence in favor of any hypothesis under which it would be a frequent occurrence. Under what hypothesis would the P, considering its mode of calculation, be a frequent occurrence? If the regression were curvilinear, a small P is to be expected relatively frequently. In so far as this is so, a small P is evidence *in favor of curvilinearity* and because of this, and *primarily because of this*, a small P can be considered evidence in disfavor of its alternative, linearity. But also a small P is to be expected relatively frequently if the regression is linear and the variability heteroscedastic; hence a small P is also evidence in favor of linearity plus heteroscedasticity. Or again a small P is to be expected frequently if the regression is linear and a value of the abscissal variate, in this case the temperature, is not constant but subject to fluctuation. And there may be other conditions which, with linearity, would produce a small P relatively frequently. The small P is favorable evidence for any or several of these. Which of these shall be taken to have been demonstrated by the evidence of the small P will have to be determined by other evidence, possibly other statistical tests. In this case my own judgment would be, not that the regression is nonlinear, but that the temperature has varied during each or some of the experiments. At least that would explain the small P.

According to what is advocated here, we cannot lay down any pat axiomatic rules such as "A very small P disproves the hypothesis tested," or "Equally, a very high P disproves the hypothesis," for it is not primarily the infrequency of the P which gives the finding its meaning. Each test will have to be examined and the circumstances in which it is applied will have to be examined, to find out, as best we can, whether any particular regions of P will occur relatively frequently in the case of an alternative to the tested hypothesis. There are situations in which a very large P will be frequent in an alternative, and in these circumstances, but *only in these circumstances*, a very high P can be said to disfavor the null hypothesis. I cite an example.

If with $(n+1)$ observations from a frequency distribution of a variate x the quantity $ns^2/\bar{x}$ is calculated, where $\bar{x} = \Sigma x/(n+1)$ and $s^2 = \Sigma(x-\bar{x})^2/n$, it is known that the quantity is distributed in random samples as Chi-square for n degrees of freedom, if the distribution is Poisson. Small values of P, say $P \leqslant 0.05$, will occur with the small frequency of five times in 100. If, however, the distribution is what has been called supernormal, a distribution that is known to characterize certain physical situations, the variance σ^2 is greater than the mean μ, and in random samples large values of the quantity Chi-square, and correspondingly low values of $P \leqslant 0.05$ will be more frequent than five in 100. The finding of a $P \leqslant 0.05$ therefore can be taken as preponderant favorable evidence for the super-Poisson, and hence as unfavorable to the null hypothesis tested that the distribution is Poisson. Similarly, if the distribution is Poisson, large values of P, say $P \geqslant 0.95$, will occur with the small frequency, five times in 100. If, however, the distribution is Bernoullian-binomial or sub-Poisson, the variance σ^2 will be less than the mean μ, and small values of the quantity Chi-square and correspondingly large values of $P \geqslant 0.95$ will be more frequent than five times in 100. The finding of a $P \geqslant 0.95$ therefore can be taken as preponderant favorable evidence for the Bernoullian or sub-Poisson, and hence as unfavorable to the null hypothesis tested that the distribution is Poisson. Here then is a case in which either a very low value of P or a very high value can be considered as warrant for rejecting the null hypothesis. There are other such cases, but the rule is not general.

So much for the meaning of P's which are relatively frequent in the case of an alternative, and in so far, are evidence in disfavor of the null hypothesis tested. In the cases in which a very low P or very high P is evidence in favor of an alternative, what can we say of the finding of a middle value of P, say a P in the region 0.3 to 0.7? Statistical authors are not very clear about this. For the most part they merely confine themselves to statements that a low P disproves and one which is not low does not disprove. In some cases they say explicitly that a low P *disproves* but one which is not low does not *prove* the null hypothesis. What such a P should mean according to the principle advanced here is unequivocally clear. Since by definition such P's will occur frequently in the case in which the null hypothesis is true, the finding of one is to be taken as prima facie evidence *in favor* of the *null hypothesis*. That is in fact the way the statistician uses them, in contradistinction to the way he says they should be used when he describes the testing of the null hypothesis.

This was somewhat amusingly illustrated at one of our meetings. One of our most eminent members gave a paper presenting the application of the lambda test and used for illustration data designed to test a certain Mendelian hypothesis. The data having been examined and the test applied, a P of about 0.6 was found. "We can say therefore," he remarked, "that the results substantiate the hypothesis." He applied the test illustratively to several other sets of data successively and getting a P of considerable size, each time he said, "The results therefore substantiate the hypothesis." When he was finished, an equally eminent mathematical colleague rose to object and said, "You cannot say that the results of the test support the hypothesis; all you are able to say is that they have not in these data disproved it." The most interesting part of the colloquy is that the first mathematician accepted the correction!

This I find is rather typical. In the abstract the mathematical statistician insists that a middle value of P only fails to refute the hypothesis, but if he is dealing with real data and gets interested in the physical problem in hand, he forgets his statistical principles and relapses to the rules of inference applied generally in such problems.

That statisticians with real problems in hand do interpret a middle P as positive support for the null hypothesis can be readily illustrated by innumerable examples to be found in the literature. I shall cite one that is in a field in which I once did some work. "Student," in his classic paper on the error of count with a haemocytometer (1906–1907), used a series of data to examine whether the actual distribution in the haemocytometer followed the Poisson distribution, as it should on certain physical assumptions. He applied the Pearson Chi-square test to a number of series and finding the P's taken together fairly large, he concluded that the distribution was sensibly Poisson, and that therefore the variability could be taken as the square root of the average count. If this positive conclusion in favor of the null hypothesis tested was not obtained from the relatively high P's, then his statistical work was entirely irrelevant. Other examples of the use by statisticians of relatively high P's for demonstration of the null hypothesis are easily found if one keeps a weather eye open for them.

When I say that a middle value of P is to be considered valid evidence in favor of the null hypothesis, I have by no means resolved all the pertinent questions that may be asked regarding it. I do not say anything has been "proved" or "disproved." I leave to others the use of these words, which I think are quite inadmissible as applying to anything that can be accomplished by statistics. All I say is that what we have is in the nature of positive supporting evidence. Whether the evidence is of sufficient weight to be convincing is another matter.

The development of what should be taken to affect the weight of the evidence is beyond anything I wish to undertake but a few pertinent remarks I do wish to make. Whereas it can be said that the evidence provided by a small P correctly evaluated is broadly independent of the number in the sample from which it has been calculated, this is not true for such evidence as is provided by a P in the middle region, say 0.3 to 0.7. Consider Table 28.1 depicting the hypothetical results of a physician's judgments based on a

Table 28.1. Hypothetical results: determination of sex

Category	Experience 1			Experience 2		
	Total	Judgment of sex		Total	Judgment of sex	
		Correct	Incorrect		Correct	Incorrect
Expected by chance	10	5	5	1000	500	500
Physician's judgment	10	6	4	1000	505	495
P		0.38			0.38	

serological test, designed to ascertain the sex of a fetus *in utero*. Examine experience 1, divest yourself of formal rules, and consider what would be your reaction. I think I can fairly guess that it would be something like this: "We cannot say anything from this experience; it certainly does not present any convincing evidence that the physician can discriminate between the sexes. But I should not want to say either that he cannot discriminate. The experience is too small for any conclusion." With experience 2 I think you would say, or at any rate I should: "There is no question in my mind; quite evidently the physician does not possess any ability to discriminate by this serological test between the sexes. The experiment is quite large enough, and if he could discriminate to any significant degree we should see it in the results, which we do not."

Now for both experiences, the *P*, which is the probability of obtaining by chance as good a result as the one obtained, on the null hypothesis that the probability of either sex is a half, is the same, namely, 0.38. But the experience 2, being based on large numbers, is convincing positive evidence of the truth of the null hypothesis within practical limits. I do not intend to attempt to analyze what is the justification for the added conviction provided when the numbers are large, beyond suggesting that it has the same basis as what has been argued here is the general principle of inference which is operative throughout. When the numbers are small, a middle *P* will occur with considerable frequency if the null hypothesis is true or if an alternative is; with large numbers such a *P* will occur frequently in the case of the null hypothesis but not in the case of a practical alternative. Hence with large numbers, a middle *P* provides probative evidence in favor of the null hypothesis.

Here we have disclosed one fundamental weakness in the position of those who contend that small samples can be effectively utilized in statistical investigations if the calculations of the *P*'s are correctly made. If it were a fact that conclusions are drawn only when the *P* is very small and the null hypothesis disproved, then so far as concerns the main considerations here developed, there would be a certain validity to this view, for small *P*'s are more or less independent, in the weight of the evidence they afford, of the numbers in the sample. But if actually it is the fact that conclusions will be drawn from *P*'s which are not small, then only very considerable numbers in the sample are reliable.

If a test for the difference between means has yielded a large or middle *P*, it does not merely fail to disprove the null hypothesis that the true means are equal; it furnishes *affirmative evidence* that the means *are* substantially equal. If the numbers on which the test is based are large, the evidence will have convincing weight; otherwise not. Contrariwise a low *P* points affirmatively toward the alternative that the means are unequal. It is the merit of some kinds of tests that they indicate unequivocally the specific alternative toward which they point.[1] Such are tests for the difference between means or the difference between variances or tests for skewness. Other tests such as the frequency Chi-square or some applications of the analysis of variance do not have this characteristic. In Table 28.2 is presented an experience of mortalities following certain operations with and without the use of a vaccine for the prevention of peritonitis. Four tests are given for the "null hypothesis" that the true mortality rates are identical for patients with and without vaccine: (1) the probability of getting as many differences in the favorable direction as found; (2) the appropriate P for the Chi-square test of the four-fold table constituted by the totals; (3) the Fisher test of combining the value of Chi-square$= -2ln P_{x^2}$; (4) the summation of the Chi-square and degrees of freedom for the separate operations. The resulting P's are considerably different. In terms of the usual rationalization, each of these tests is equally valid for testing the null hypothesis. If the null hypothesis were true, that is, if the vaccine were ineffective and the mortality for any operation were the same whether the vaccine were used or not, the appropriate limiting value of each test function would occur only infrequently—one just as infrequently as the other. But the tests are differently sensitive to the presence of different *alternatives*. In terms of the Neyman–Pearson formulation they have different

Table 28.2. *Mortality rates for operations with and without use of vaccine: tests of significance of differences*

Type of operation	Vaccine			No vaccine			Mortality difference, percent
	Operations	Hospital deaths		Operations	Hospital deaths		
		Number	Percent		Number	Percent	
A	107	2	1.9	142	4	2.8	−0.9
B	28	3	10.7	60	9	15.0	−4.3
C	21	3	14.3	34	5	14.7	−0.4
D	21	4	19.0	34	8	23.5	−4.5
E	47	3	6.4	45	4	8.9	−2.5
F	21	1	4.8	26	2	7.7	−2.9
Total	245	16	6.53	341	32	9.38	−2.85

Test	P
1. Signs	0.016
2. Total difference mortality	0.11
3. Combination of P's—Fisher	0.91
4. Summation of Chi-square and D.F.	0.98

1. Elsewhere I have suggested that those tests are ones which in principle can be stated alternatively and equivalently in terms of an estimate and its confidence limits (Berkson, 1941).

powers for any particular alternative, and hence are likely to give different results in any particular case. How blind is the procedure of doing some test of significance, when there is no knowledge at hand as to whether it is likely to show a significant result or not show one, no matter how importantly different the facts may be from the hypothesis tested. The importance of this consideration is underscored when we realize that in practical applications the failure to show a significant result will be taken to corroborate the null hypothesis. It is an important but neglected task of mathematical statistics to investigate what alternatives are particularly pointed to by specified findings with different tests.

I should like to see the development of investigation of the finding of middle P's. I am not ready to say what this should be or just what it would lead to. But this is an example of what I mean. With the development that we now have, which emphasizes the low P's, we find such statements as the following in the literature, and it is typical of the essential procedure in many fields in which statistical tests are applied. A standard curve for estimating dosage from mortality has been established with its confidence zones, from a first set of data. A set of data for another drug is to be used for estimating the potency of a second drug. But realizing the possibility that the standard curve may not be applicable any more, the author counsels the use of some controls to see whether the standard curve still applies for the first drug. He says, "When the controls have been shown to agree with the standard of the regression line by the appropriate Chi-square or t test, the first curve can be used." Now what is meant by this is that if the test does not show a low P, the curve can be used, which is to say that if the test shows a middle P, the curve *will* be used. It should be clear on consideration that if there is a real discrepancy of a given size between the present conditions and the curve, a P which is not low will result with small numbers, while with the same discrepancy a low P will result if the numbers are large. The use of the suggested rule could easily be disastrous if drugs were standardized on the basis of it and small numbers were used. Investigation should be made which could result in a rule not such as just given, but rather of the following kind: "If the control is tested with data including so and so degrees of freedom and if the test results in a P of this amount or higher, the curve may be accepted as stable."

29. Publication Decisions and Their Possible Effects on Inferences Drawn from Tests of Significance—or Vice Versa

THEODORE D. STERLING

It has become commonplace to speak of a "level of significance" in reporting outcomes of experiments. This significance level refers to risks of rejecting the null hypothesis, H_0, erroneously, and, seemingly, has no other direct relationship to experimental work. The experimenter who uses so called tests of significance to evaluate observed differences usually reports that he has tested H_0 by finding the probability of the experimental results on the assumption that H_0 is true, and he does (or does not) ascribe some effect to experimental treatments. What with the shortage of publication space and the desire for objectivity it often seems that the responsibility for rejecting a hypothesis rests squarely on a crucial value in a table of probabilities.

The risk of choosing the incorrect inference from experimental observation depends on a stated risk of rejecting H_0 if true and on the risk of failing to do so if H_0 is not true. Here is a dilemma which is dealt with in practice by two conventions. As Savage (1954: 256) notes, publications tend to report the results of the test as well as that level of significance for which the corresponding test of the relevant family would be on the borderline between acceptance and rejection (in the view of the author). The individual reader now makes his own test at a level of significance appropriate to him. How much uncertainty such a reader is willing to tolerate in rejecting a hypothesis that might

Reprinted from the *Journal of the American Statistical Association*, 54 (March, 1959), 30–34, by permission of the author and the publisher.

be true will depend on his confidence in the methods of data collection, his views concerning the relevance of alternative hypotheses, or the weight he gives to evidence from other sources. In addition, scientific readers differ in fundamental strategies for games against nature and their tolerance for errors can hardly be expected to remain unchanged from one experimental problem to another. The type of reporting mentioned by Savage may well be most satisfactory for author and reader alike.

Some publications, notably of social science content, have adopted a some-what more extreme convention. Here a borderline between acceptance and rejection of H_0 is taken as a relatively fixed point, usually at $Pr\,(E|H_0) \leqslant .05$ or at that approximate region for which the probability, (Pr) of the outcome (E) of the experiment, calculated on the assumption that H_0 is true, is no larger than five in a hundred (Edwards, 1950; McNemar, 1955; Walker, 1947).[1] General adherence to such a rigid strategy is interesting by itself but might have no further consequences on the decisions reached. However, when a fixed level of significance is used as a critical criterion for selecting reports for dissemination in professional journals it may result in embarrassing and unanticipated results.

Table 29.1 shows that for psychological journals a policy exists under

Table 29.1. Outcomes of tests of significance for four psychology research journals

Journals: all issues from January to December	Total number of research reports (1)	Number of research reports using tests of significance (2)	Number of research reports that reject H_0 with $Pr(E|H_0) \leqslant .05$ (3)	Number of research reports that fail to reject H_0 (4)	Number of research reports that are replications of previously published experiments (5)
Experimental Psychology (1955)	124	106	105	1	0
Comparative and Physiological Psychology (1956)	118	94	91	3	0
Clinical Psychology (1955)	81	62	59	3	0
Social Psychology (1955)	39	32	31	1	0
Total	362	294	286	8	0

1. The fact that some tables present only the .05 and .01 levels of significance encourages the use of these two levels of significance (Walker, 1947: 292).

which the vast majority of published articles satisfy a minimum criterion of significance. The table summarizes the number of research articles in four publications. The journals were selected at random from four major areas of psychology. The table gives the distribution for the number of reports that used tests of significance to test H_0 and either rejected H_0 or failed to do so at $Pr(E|H_0) \leqslant .05$. In addition the table gives the number of experiments that were replications of previously published investigations. Column 1 gives the number of experimental research reports and column 2 gives the number of those reports that used tests of significance to choose among possible alternative hypotheses. Column 3 shows how many of the reports of column 2 managed to reject H_0 and column 4 counts the number of reports that failed to reject H_0 (either for the major hypothesis tested or for the majority of hypotheses under investigation).[2] Finally, column 5 gives the number of experiments representing a replication of work previously reported in the literature.

Table 29.2 shows the same distributions as proportions of columns 1 and 2. A glance at the tables is sufficient to show that most articles published

---

2. Some explanatory remarks concerning Table 29.1 are in order. Almost all of the 294 studies that used tests of significance were of a multivariable design. All evaluated observed differences against the assertion of H_0; however, H_0 was sometimes not rejected for all variables tested. The following rules were adopted in compiling Table 29.1:

    a. The attempt was made to determine the major variable or prediction tested by the research design. Such was usually clear from the author's preliminary remarks; the multivariable design was most frequently used to control for conditions not covered by the experimental procedure. The level of significance for which H_0 was rejected for the major prediction was noted (if H_0 was rejected at all).

    b. If the design tested two or more variables for which no unambiguous decision as to major importance could be made, the lowest level of significance for which at least half the variables rejected H_0 was noted. If H_0 was not rejected for at least half the variables, the article was placed in the class of studies for which H_0 was *not* rejected.

    c. Where results from more than one research design were reported, an attempt was made to determine the one study deemed most crucial by the author and the level of significance for that study was recorded if it rejected H_0.

    d. If all studies seemed of equal importance, the lowest level of significance for which at least half the reported studies rejected H_0 was recorded. If H_0 was not rejected for at least half the studies reported, the article was placed in the class of studies for which H_0 was *not* rejected. (This special provision in 2 and 4 was not really necessary since for no single article were less than half of the quoted results in the significant category.)

    e. Two studies that obtained $Pr(E|H_0) \leqslant .1$ were included because the authors had expressly pointed out that they rejected H_0 since the obtained significance level was close enough to the conventional .05 to suit their purposes.

Since the *Psychological Abstracts* essentially attempt to present an outline of the major points made in almost all research articles of interest to psychologists the procedure used here could be checked for reliability with that publication. Of 100 research articles selected at random from volumes covering 1952 to 1957, 94 reported positive results, five reported negative results, and one was a replication of a previous study. These proportions agreed by and large with the total proportions in Table 29.1. No comparison for use of tests of significance was made since that journal seldom reports results of statistical tests. However, the words "significantly different" were applied to most of the reported results.

during the year by the journals in question used tests of significance as aides in choosing among alternative experimental hypotheses and, at the same time, that nearly all managed to reject H_0 at the recommended level of certainty. It need not be assumed that the observed distributions are due to

Table 29.2. Percent of articles using tests of significance and percent of articles rejecting H_0

Journals: all issues from January to December	Percent of articles using tests of all articles published (2/1)	Percent of articles rejecting H_0 of all articles using tests (3/2)	Percent of articles not rejecting H_0 of all articles using tests (4/2)
Experimental Psychology (1955)	85.48	99.06	0.94
Comparative and Physiological Psychology (1956)	79.66	96.81	3.19
Clinical Psychology (1955)	76.54	95.16	4.84
Social Psychology (1955)	82.05	96.88	3.12
Total	81.22	97.28	2.72

explicit editorial rules. The single factor contributing most to the selection of articles in which H_0 is rejected may be implicit agreement among authors. The term "publication policy" will be used here largely as a matter of convenience. In fact, the distribution of articles in psychological journals in general appears to be similar to the ones shown in the table, and it seems likely that the authors' selection rather than editorial policy accounts for the observed profession-wide selection. Whatever the reasons, the tables indicate what gets printed with a high probability; namely, research reports that use tests of significance and at the same time reject H_0 for the effects of treatments in the design.[3] To state the above more concisely:

A_1 Experimental results will be printed with a greater probability if the relevant test of significance rejects H_0 for the major hypothesis with $Pr(E|H_0) \leqslant .05$ than if they fail to reject H_0 at that level.

A_2 The probability that an experimental design will be replicated becomes very small once such an experiment appears in print.

3. It is interesting that the *Journal of Experimental Psychology* appears to set the pace for the use of statistical tests as well as for the selection of articles that reject H_0. Some years ago the same journal was used (Lewis and Burke, 1949) to show that Chi-square was consistently misused by psychologists. The authors noted at the time that analyses in this journal would be typical for psychological publications in general and that the expectation of finding sound statistical treatments would be better in that journal than in others.

With respect to A_1, it is not known how many research results either reject H_0 or do not do so, or, are submitted or not submitted for publication. However, it does seem clear (American Psychological Association, 1957) that pressure exists which leads to the selection of a very small number of publications from a large number of submitted manuscripts. From a commonly admitted tendency to acknowledge only the most significant findings, and from perusal of statements concerning publication pressures (American Psychological Association, 1957), one could infer another reasonable assumption:

A_3 A great many more experiments are performed than appear in the pages of professional journals.

With respect to A_2, the lack of replication of experimentation in psychology has been noted elsewhere (Lubin, 1957). Replications are sometimes reported at professional meetings. Since such papers are rarely used as references unless they have been published they may be ignored as sources for widespread professional or scientific information.

The three assumptions are admittedly substantive in nature and strong supporting evidence for them, beyond that given here, is hard to come by. They may be taken as a fair statement of the prevailing conditions in which the scientific community is not equally aware of all experimental results. As a consequence, experiments for which $Pr(E|H_0)$ is large may well have a high frequency of replication by individuals who do not know that this particular comparison had been made previously, and that previous tests of significance had failed to reject H_0 at acceptable levels of significance. Once a study does result in a level of significance that meets this criterion, not only will it be published, but the likelihood of its ever being repeated appears to become very small. A picture emerges for which the number of possible replications of a test between experimental variates is related inversely to the actual magnitude of the differences between their effects. The smaller this difference the larger may be the likelihood of repetition. This chain is terminated apparently by an observation for which the relevant statistical test can reject H_0 with reasonable certainty. For any set of observed differences that are randomly variable (and which experimental observations are not?) a difference of some substance should then appear in print—*irrespective of the actual state of nature.* What credence can then be given to inferences drawn from statistical tests of H_0 if the reader is not aware of all experimental outcomes of a kind? Perhaps even more pertinent is the question: Can the reader justify adopting the same level of significance as does the author of a published study?

Two points are worth noting with respect to the last two questions. Both refer to the expectations a reader may form when he picks up an article in one of the journals of Table 29.1 (or in a journal following like practices).

First the reader's best expectation is that the author will reject H_0. The probability that he will commit a Type II error (accepting the null hypothesis when it is false) if he adopts the author's conclusion is, in consequence, extremely small. In fact, from Table 29.1 it appears that this risk is scarcely more than zero. One may therefore conclude that any and all tests used by authors are of equally high power for the reader. This obviously was not true for the individual investigator who attempted to choose the most powerful test in the first place.

There is also another side to this problem. The reader's expectations are that H_0 will be rejected. What risks does he take in making a Type I error by rejecting H_0 with the author? The author intended to indicate the probability of such a risk by stating a level of significance. On the other hand, the reader has to consider the selection that may have taken place among a set of similar experiments for which the one that obtained large differences by chance had the better opportunity to come under his scrutiny. The problem simply is that a Type I error (rejecting the null hypothesis when it is true) has a fair opportunity to end up in print when the correct decision is the acceptance of H_0 for a particular set of experimental variables. Before the reader can make an intelligent decision he must have some information concerning the distribution of outcomes of similar experiments or at least the assurance that a similar experiment has never been performed. Since the latter information is unobtainable he is in a dilemma. One thing is clear, however. The risk stated by the author cannot be accepted at its face value once the author's conclusions appear in print. It may be safe to conclude that pursuing statistical analyses under the conditions outlined here may have considerably less merit than psychologists like to ascribe to statistics in experimental design.

It would be unfair to close with the impression that the malpractices discussed here are the private domain of psychology. A few minutes of browsing through experimental journals in biology, chemistry, medicine, physiology, or sociology show that the same usages are widespread through other sciences. Some onus appears to be attached to reporting negative results. Certainly such results occur with lesser frequency in the literature than they may reasonably be expected to happen in the laboratory—even if it is assumed that all experimenters are outstandingly clever in selecting hypotheses. Perhaps the trend of our time is exemplified by the editors of a cancer journal who in a recent announcement took action to change the name of their yearly supplement from "Negative Data . . ." to ". . . Screening Data" (American Association for Cancer Research, 1958: 619).[4]

4. This was pointed out to me by Charles Stevens of the Kettering Laboratory, University of Cincinnati.

30. Publication Decisions and Tests of Significance: A Comment

GORDON TULLOCK

THE PURPOSE OF THIS COMMENT is not to criticize Sterling's important article (1959 [Chapter 29]), but to suggest that it has somewhat wider application than he implies. The specific type of error to which Dr. Sterling refers (1959: 33 [Chapter 29: 299]) is the repetition of some given experiment by different investigators until one of them obtains a significant result, which is then published. I do not think this is at all a frequent occurrence. If a hypothesis had been tested by a number of independent investigators, all but one of whom had obtained negative results, the publication of the positive results of the single investigator who had had the good fortune to obtain a "significant" result would normally bring the others out in the open. A man who has devoted a good deal of time to testing a hypothesis with a given experimental design and has been forced to conclude that the hypothesis is incorrect will normally simply file his results. If, however, he sees an article reporting a similar experiment with significant results, he will at least write the editor of the journal in which it appears. Since this type of comment on articles is rare, I judge that the duplication of experimental designs is rare.

But while this indicates that investigators have minds sufficiently independent that they seldom duplicate others' experiments by accident, it does not reduce the importance of Sterling's basic point. The necessary restrictions

Reprinted from the *Journal of the American Statistical Association*, 54 (September, 1959), 593, by permission of the author and the publisher.

imposed on articles by the scarcity of publication resources does result in greatly reducing the confidence we can put in statistical tests of significance as reported in the literature. There is little difference, statistically speaking, between the likelihood of a false correlation from 20 investigators all investigating independently the same false hypothesis with the same experimental design, or from the same 20 investigators investigating 20 false hypotheses. The same number of results of a given level of significance would be expected from either "design."

The moral of these considerations would appear to be clear. The tradition of independent repetition of experiments should be transferred from physics and chemistry to the areas where it is now a rarity.[1] It should be realized that repeating an experiment, while not necessarily showing great originality of mind, is nevertheless an important function. Journals should make space for brief reports of such repetitions, and foundations should undertake their support. Academics in the social sciences should learn to feel no more embarrassment in repeating someone else's experiment than their colleagues in the physics and chemistry departments do now.

1. A good illustrative example of the practice in the physical sciences is A. B. Mah's article (1957). Dr. Mah repeated a series of careful measurements made by several other scientists. Since his work simply confirmed theirs it was reported in the form of a brief note rather than in a full article.

PART FIVE
Epilogue

31. Significance Tests in Behavioral Research:
Skeptical Conclusions and Beyond

DENTON E. MORRISON and
RAMON E. HENKEL

O UR AIM HERE is not to convince readers of our positions on the issues but briefly to state what we are convinced of, provide a summary of our rationale,[1] and point out the readings we consider most relevant in arriving at our conclusions. As indicated in our "Introduction" *statistical issues* revolve around the assumptions of mathematical theory on which the tests are based and around conventional practices that affect applications or interpretations of the tests. *Philosophy of science* issues question the model of science implied in using the tests.

The Statistical Issues

1. *Sampling: the issue of how the method by which the data are generated is related to the applicability of significance tests to the data.* There is no basis in the statistical theory on which the tests are founded for any view other than that the data must be generated by a random procedure (either random sampling or randomization) for the tests to be legitimately used (Hogben [Chapters 2 and 3]; Camilleri [Chapter 16]). Often, particularly in non-experimental research, the assumption of a random procedure is not met at all when significance tests are employed, and typically the assumption is

---

1. Detailed justification of many of our views is contained in our essay on "Significance Tests Reconsidered" (1969 [Chapter 21]).

poorly approximated, as in cluster samples (Kish, 1957), attrition in surveys, and so on. The assumption of a random sampling or assignment procedure is required regardless of whether parametric or nonparametric tests are used.

Proponents of using the tests in circumstances where neither random sampling nor randomization is utilized justify their position on one or more of three grounds. The first is the argument that there is always measurement error even in a total enumeration or in a nonrandom selection of cases, and such error can be assumed to be random. The second is that any group of cases can be considered a random sample from a hypothetical infinite population of all possible samples that could have been generated under equivalent circumstances (Hagood [Chapter 4]). The third is that the theoretical sampling distribution upon which significance tests are based can be thought of as a model to compare with any given result from any given set of cases to "screen" the data to see if one needs to search for anything other than a random process as an explanation of the data (Gold [Chapter 20]; Winch and Campbell [Chapter 22]). All of these views are justified solely on the basis of the desire to use the statistical inference model; none of them can be defended by reference to the empirically demonstrable random generation procedure that the model formally and unequivocally requires for its application. The information value of such applications is, at best, completely unassessable, since the extent to which the assumption of randomness is met is unknown.

2. *Population: the issue of the population to which inferences based on significance tests apply*. The population to which significance tests apply is the population from which the random sample was drawn or the random assignment made (Camilleri [Chapter 16]; Bakan [Chapter 25]; McGinnis [Chapter 14]). The parameters of any population from which the sample cannot be demonstrated to be selected by a random procedure cannot be the subject matter of inference with the use of significance tests. This does not mean that only samples that are generated by a random procedure or that only populations from which such samples can be drawn are of relevance for scientific inference.

3. *Meaning: the issue of the meaning of the probability statements resulting from significance tests*. The probability indicated in a significance level is *auxiliary* probability, the relative frequency with which a given or larger difference or relationship would occur on repeated random generation from the population involved if the null hypothesis in question were true for that population. In contrast is *intrinsic* probability, the relationship between variables expressed in probabilistic terms (Camilleri [Chapter 16]). To argue that statistical significance provides a meaningful "screen" for substantive significance (Gold [Chapter 20]; Winch and Campbell [Chapter 22] is to commit two errors: first, to define substantive importance in terms of the

magnitude of intrinsic probability, and second, to confuse auxiliary and intrinsic probability by letting the former determine what will be minimally meaningful or acceptable for the latter. In addition, such a screening process means nothing in operational terms outside the consideration of power and level, and, as indicated above, is also advocated under inappropriate sampling conditions.

4. *Level: the issue of why and how a given significance level is chosen.* There is a basis for choosing a certain level of significance only when a decision about a hypothesis is required, as in the situation in which action will or will not be taken according to the outcome of the test. The mathematical theory underlying significance tests provides no clue whatsoever as to the appropriate choice of level. A level can be set rationally only when the cost of a wrong decision about a null hypothesis can be calculated. Such calculation is not typically possible in behavioral science, even in its more applied aspects (Rozeboom [Chapter 24]; Camilleri [Chapter 16]). Strong and arbitrary convention is reflected in the near universal choice of the .05 and .01 levels, rather than the knowledge needs of the researcher, or the existing knowledge on the subject. Findings that do not meet conventional levels are usually not published. Since by definition a certain proportion of null hypotheses will be falsely rejected, such practice doubtless provides an erroneous notion of the veracity of a given hypothesis (Sterling [Chapter 29]; Tullock [Chapter 30]).

5. *Power: the issue of the role of the power of significance tests in statistical inference.* The ability of a given significance test to reject a given null hypothesis at a given level of significance is directly proportional to the power of the test, and power is directly proportional to the size of the sample. While being rigid, arbitrary, and conservative on level, consideration of the power of a test is seldom made in behavioral science. The failure to consider power means that the choice of level is practically without meaning (Meehl [Chapter 26]; Chandler [Chapter 23]). As our samples become larger (the trend) the rejection of our null hypotheses is increasingly guaranteed (Meehl [Chapter 26]).

6. *Technique: the issue of the role of computation and data analysis techniques in significance testing.* Several sins that fall under this issue often operate singly or together to make reported level of significance of doubtful validity. Some common examples are: searching for and then reporting only the significant findings from a large set of findings (Selvin [Chapter 9]), collapsing categories or establishing cutting points to obtain significance (Selvin [Chapter 9]), failure to consider the joint significance of a set of findings, and the use of formulas for simple random samples on probability samples of other types (Kish, 1957).

7. *Causality: the issue of the role of significance tests in assessing causality.* Randomization (thus implicitly the theoretical basis of statistical inference)

is useful to increase confidence in causal inference in experiments, whether or not decisions based on significance tests are subsequently used on the results. Randomization is, however, only one of the features of the experimental design and execution relevant to assessing causality (Winch and Campbell [Chapter 22]). None of the other issues involved in the use of significance tests is avoided in experimental research. In nonexperimental research significance tests do not play the same role as in experiments because random sampling does not, like randomization, maximize assurance that correlated biases are controlled (Selvin [Chapter 9]). The absence of controls in non-experimental analysis does not, however, influence the applicability or relevance of significance tests (McGinnis [Chapter 14]).

The Philosophy of Science Issues

8. *Scope: the issue of whether significance tests contribute to obtaining the desired population scope of scientific inference.* The population scope of inference in basic science is the infinite conceptual universe, i.e. all past, present, and future instances of the phenomena in question wherever they occur under conditions specified in the theory under consideration (Camilleri [Chapter 16]). Of course, all tests of hypotheses must operate on some set of cases that is available. However, the use of significance tests makes a contribution to scientific inference only under the assumption that the population in question is itself a random sample from the infinite conceptual universe (Camilleri [Chapter 16]). This assumption can never be demonstrated empirically and seldom can any argument at all be offered for the assumption. Thus tests of significance give information of unassessable relevance for making inferences of the scope demanded of basic scientific research. While a case can possibly be made that in basic science there is no *harm* in using the tests to make precise inferences for the particular population studied (i.e. statistical inference), the benefits of this procedure for basic science are difficult to demonstrate, given the opportunity for misinterpreting the scope and meaning of such inferences, and in terms of the issues of form, process, and purpose discussed below. More important than random selection or the use of significance tests in basic scientific work is evidence that the cases are selected to meet or at least explore theoretical conditions for the validity of the hypothesis.

9. *Form: the issue of whether the logical characteristics of the hypotheses involved in significance testing measure up to the requirements of hypotheses that characterize basic science.* In a developed science the hypotheses are stated exactly (i.e. intrinsic probabilities are stated precisely), the hypotheses are systematically related to each other, and the conditions under which the hypotheses will hold are specified. These are some of the main formal

characteristics of hypotheses that constitute scientific theory. The hypotheses involved in significance testing as practiced in behavioral science are atheoretical in form in several ways (none of which are inherent formal characteristics of the statistical hypotheses involved in significance testing). (1) The simple directional and nondirectional alternative hypotheses most common in behavioral science specify no intrinsic probabilities and thus have a high *a priori* probability of being supported, just as the null hypotheses are generally known in advance to be false (Meehl [Chapter 26]). (2) The relative certainty of statistically significant associations leads to emphasis on magnitude of association as a more discriminating basis for judging the worth of individual hypotheses rather than to an emphasis on precise and deductively based prediction with a set of interrelated hypotheses. (3) Further, the conditions under which the hypotheses will hold are usually not specified in any way. Partly, the way significance testing is practiced in behavioral science simply reflects the impoverished state of both our theory and our approach to theory development. But the influence that significance testing has had in behavioral science has also contributed to the impoverished state of our theory. At any rate, significance testing in behavioral research is deeply implicated in our false search for empirical association, rather than a search for hypotheses that explain.

10. *Process: the issue of whether the process by which hypotheses are evaluated in significance testing is appropriate to the creation of scientific knowledge.* 11. *Purpose: the issue of the purpose of making inferences with significance tests as compared with the purpose of scientific inference.* These issues are inextricably linked and must thus be considered together. The use of significance tests involves the researcher in the process of making firm "reject" or "accept" decisions on each test of each null hypothesis on the basis of a formal, firm, and frequently arbitrary criterion, the significance level. This *decision making process* is antithetical to the *information accumulation process* of scientific inference. In the latter process the hypothesis of interest (the equivalent of the alternative hypothesis, in contrast with the null) is given credence in the scientific community to the extent that it satisfies the general and informal criterion of consistent support over replication (Rozeboom [Chapter 24]; Lykken [Chapter 27]).[2] Further, in science, the deductive strength and simplicity of the argument that led to a particular hypothesis as well as the fruitfulness of its concepts in other hypotheses inevitably enter into its assessment (Camilleri [Chapter 16]).

2. It is also worth noting that scientific inference is generally a more positive information producing process than statistical inference. In the latter, even a seemingly positive result, i.e. a rejected null hypothesis, doesn't actually produce positive knowledge, since the tone of the interpretation typically emphasizes the evidence against the null hypothesis and not the evidence for the hypothesis of interest, the alternative hypothesis.

Just as scientific inference involves a cognitive process, it involves a cognitive purpose: to allow the scientific community appropriately to adjust its degree of *belief* about hypotheses (Rozeboom [Chapter 24]). In contrast, significance testing involves a decision process and a decision purpose: to allow the researcher to make a decision about his result on a given hypothesis so that *action* may subsequently be guided by the decision. In the vast majority of instances in which significance tests are reported in behavioral research no firm decision is required because no specific actions are to be guided by the decisions. What is worse, obedience to the rituals of significance testing often means that results are reported in such a way as to prevent the reader from making a more appropriate interpretation in terms of the goals of scientific inference.

In summary, significance tests have severe restrictions and difficult requirements for use in any research endeavor, and both the opportunities for and the actualities of misuse are great. The tests are, however, clearly applicable and useful for research contexts in which: (1) random data generation from a specific population is possible, (2) knowledge about the particular population in question is of central import, (3) comparison of the relative frequency with which a statistic would occur if a null hypothesis were true with some set frequency (level) provides sufficient and substantively meaningful information for making the decision required about the population, (4) the costs of a wrong decision can be calculated so that level can be rationally and not simply arbitrarily set, (5) the power desired of the test can be rationally determined, (6) the methods of calculation and of data analysis are such that reported level of significance is accurate, (7) causal inference is either irrelevant for the research purpose or unequivocal from the research design.

These are not, however, the conditions for the use of the tests in basic science. Moreover, the conditions are not and cannot be met in most behavioral research, nor, in our view, should they be since most behavioral research is implicitly basic in its orientation. Basic science involves a different scope, form, process, and purpose of inference than statistical inference.

Beyond Significance

Full and systematic consideration of the question "What do we do without significance tests?" is not our purpose. Our purpose will be accomplished if researchers simply raise this question more often. We will have achieved our aim to the extent that we have caused researchers to obtain a critical perspective on the tests. We believe that such a perspective will free researchers from the restrictive and often irrelevant bonds of significance testing, and, moreover, that their research will be better rather than worse for this regardless of our not providing specific instructions on how to do research without the tests. Such concrete guidelines are, in fact, not available. No step by step procedures and no formulas for scientific inference exist. General clues and guidelines are, of course, given in several of the readings.

We realize that many researchers, including those who would claim that their orientation is to basic research, will regard the abandonment of the tests a threat to the very foundations of empirical behavioral research. In fact, our experience (among sociologists) has been that many researchers accept all or most of our arguments on rational grounds, but keep using significance tests as before simply because use is a strong norm in the discipline. They seem to be blind to the fact that use of the tests is a relatively recent norm in certain of the behavioral sciences and, at least in broad historical perspective, has not been a norm in science at all. Herein lies an important indicator of the contributions of the tests, one that should give pause to those who uncritically follow the normative procedures of their discipline in their research.

Major developments in the physical and natural sciences—developments that the behavioral sciences have yet to approach—took place long before significance tests were available, and major developments continue in these disciplines without the heavy reliance on the tests that generally characterizes behavioral science. Among the behavioral sciences there are great variations in the extent to which the tests are employed in the various disciplines, with practically no correlation between the extent of use and the degree to which the discipline is theoretically developed or "scientific." Economists, for example, use the tests far less than sociologists or psychologists, yet few would maintain that either of the latter disciplines rank above economics in theoretical development. (Quite appropriately, the tests are used much more in the applied fields of economics, e.g. business research.) Within each of the behavioral disciplines important contributions to basic science are often made without the tests, very seldom because of the tests, and sometimes in spite of the tests.

What we do without the tests, then, has always in some measure been done in behavioral science and needs only to be done more and better: the application of imagination, common sense, informed judgment, and the appropriate remaining research methods to achieve the scope, form, process, and purpose of scientific inference.

References

AMERICAN ASSOCIATION FOR CANCER RESEARCH (1958). *Cancer Research*, 18, Chicago: The University of Chicago Press.

AMERICAN PSYCHOLOGICAL ASSOCIATION (1957). *Publication Manual* (rev.), Washington, D.C.: American Psychological Association.

ANDERSON, R. L., and BANCROFT, T. A. (1952). *Statistical Theory in Research*, New York: McGraw-Hill.

ANSCOMBE, F. J. (1951). "Mr. Kneale on Probability and Induction," *Mind*, 60, 299–317.

——— (1948). "The Validity of Comparative Experiments," *Journal of the Royal Statistical Society*, Series A, 111, 181–200.

BABBAGE, CHARLES (1832). *On the Economy of Machinery and Manufacture*, London: C. Knight.

——— (1830). *Reflections on the Decline of Science in England and Some of Its Causes*, London: B. Fellowes.

BAKAN, DAVID (1967). "The Test of Significance in Psychological Research," *On Method*, San Francisco: Jossey-Bass, 1–29 (later version of Bakan, 1966).

——— (1966). "The Test of Significance in Psychological Research," *Psychological Bulletin*, 66 (December), 423–437 (earlier version of Bakan, 1967).

BARNARD, G. A. (1954). "Sampling Inspection and Statistical Decisions," *Journal of the Royal Statistical Society*, Series B, 16, 151–165.

——— (1948). "Discussion on Anscombe's 'The Validity of Comparative Experiments'," *Journal of the Royal Statistical Society*, Series A, 111, 201–202.

BATESON, WILLIAM (1909). *Mendel's Principles of Heredity*, Cambridge: Cambridge University Press.

313

BAYES, THOMAS (1763). "Essay Towards Solving a Problem in Doctrine of Chance," *Philosophical Transactions of the Royal Society of London*, 53, 376–398.

BERELSON, BERNARD, LAZARSFELD, PAUL F., and MCPHEE, WILLIAM N. (1954). *Voting: A Study of Opinion Formation in a Presidential Campaign*, Chicago: The University of Chicago Press.

BERGER, JOSEPH, COHEN, BERNARD P., SNELL, J. LAURIE, and ZELDITCH, MORRIS (1962). *Types of Formalization in Small Group Research*, Boston: Houghton-Mifflin.

BERKSON, JOSEPH (1942). "Tests of Significance Considered as Evidence," *Journal of the American Statistical Association*, 37, 325–335.

——— (1941). "Comments on Dr. Madow's 'Note on Tests of Departure from Normality' with Some Remarks Concerning Tests of Significance," *Journal of the American Statistical Association*, 46 (December), 539–541.

——— (1938). "Some Difficulties of Interpretation Encountered in the Application of the Chi-Square Test," *Journal of the American Statistical Association*, 33, 526–542.

BERNARD, CLAUDE (1957). *Introduction to the Study of Experimental Medicine* (tr. by Henry Copley Green), New York: Dover Publications.

BERTRAND, J. L. F. (1889). *Calcul Des Probabilités*, Paris: Gauthier-Villars.

BESHERS, JAMES (1958). "On 'A Critique of Tests of Significance in Survey Research'," *American Sociological Review*, 23 (April), 199.

BINDER, ARNOLD (1963). "Further Considerations on Testing the Null Hypothesis and the Strategy and Tactics of Investigating Theoretical Models," *Psychological Review*, 70, 107–115.

BLALOCK, HUBERT (1960). *Social Statistics*, New York: McGraw-Hill.

BOLLES, ROBERT C. (1962). "The Difference Between Statistical Hypotheses and Scientific Hypotheses," *Psychological Reports*, 11, 639–645.

BORING, E. G. (1950). *A History of Experimental Psychology* (2nd ed.), New York: Appleton-Century-Crofts.

BRADLEY, JAMES V. (1968). *Distribution Free Statistical Tests*, Englewood Cliffs, N. J.: Prentice-Hall.

BRAITHWAITE, R. B. (1953). *Scientific Explanation*, Cambridge: Cambridge University Press.

BUNGE, MARIO (ed.) (1964). *The Critical Approach to Science and Philosophy: Essays in Honor of Karl R. Popper*, New York: Free Press.

CAMILLERI, S. F. (1962). "Theory, Probability, and Induction in Social Research," *American Sociological Review*, 27 (April), 170–178.

CAMPBELL, DONALD T. (1969). "Reforms as Experiments," *American Psychologist*, 24 (April), 409–429.

——— (1968). "Quasi-experimental Design," in David L. Sills (ed.), *International Encyclopedia of the Social Sciences*, Vol. 5, New York: Macmillan & Free Press, 259–263.

——— (1957). "Factors Relevant to the Validity of Experiments in Social Settings," *Psychological Bulletin*, 54, (July), 297–312.

CAMPBELL, D. T., and STANLEY, J. C. (1963). "Experimental and Quasi-experimental Designs for Research," in N. L. Gage (ed.), *Handbook of Research on Teaching*, Chicago: Rand McNally, 171–246.

CHANDLER, ROBERT E. (1957a). "A Review of Guilford's *Fundamental Statistics in Psychology and Education* (3rd ed.)," *Personnel Psychology*, 10, 272–273.

—— (1957b). "The Statistical Concepts of Confidence and Significance," *Psychological Bulletin*, 54 (September), 429–430.

CHAPIN, F. STUART (1955). *Experimental Designs in Sociological Research* (2nd ed.), New York: Harpers.

CHUNG, J. H., and FRASER, D. A. S. (1958). "Randomization Tests for a Multivariate Two-sample Problem," *Journal of the American Statistical Association*, 53 (September), 729–735.

CICOUREL, AARON V. (1964). *Method and Measurement in Sociology*, New York: Free Press.

CLECKLEY, H. (1950). *The Mask of Sanity*, St. Louis: C. V. Mosby.

COCHRAN, WILLIAM G. (1953). *Sampling Techniques*, New York: Wiley.

COCHRAN, WILLIAM G., and COX, GERTRUDE M. (1950). *Experimental Designs*, New York: Wiley.

COCHRAN, WILLIAM G., MOSTELLER, FREDERICK, and TUKEY, JOHN W. (1953). "Statistical Problems of the Kinsey Report," *Journal of the American Statistical Association*, 48, 673–716.

COHEN, JACOB (1962). "The Statistical Power of Abnormal-Social Psychological Research: A Review," *Journal of Abnormal and Social Psychology*, 65 (September), 145–153.

COHEN, LILLIAN (1954). *Statistical Methods for Social Scientists*, Englewood Cliffs, N. J.: Prentice-Hall.

CORNFIELD, JEROME (1954). "Statistical Relationships and Proof in Medicine," *American Statistician*, 8 (December), 19–21.

CORNFIELD, JEROME, and TUKEY, JOHN W. (1956). "Average Values of Mean Squares in Factorials," *Annals of Mathematical Statistics*, 27 (December), 907–949.

COX, D. R. (1958). "Some Problems Connected with Statistical Inference," *Annals of Mathematical Statistics*, 29 (June), 357–372.

CRAMÉR, H. (1946). *Mathematical Methods of Statistics*, Princeton, N. J.: Princeton University Press.

CRONBACH, L. J. (1957). "The Two Disciplines of Scientific Psychology," *American Psychologist*, 12, 671–684.

CROXTON, F. E., and COWDEN, D. J. (1940). *Applied General Statistics*, New York: Prentice-Hall.

DAVIS, JAMES (1958). "Review of Robert K. Merton et al., *The Student Physician*," *American Journal of Sociology*, 63 (January), 445–446.

DEMING, W. EDWARDS (1950). *Some Theory of Sampling*, New York: Wiley.

DIXON, W. J., and MASSEY, F. J., Jr. (1957). *Introduction to Statistical Analysis* (2nd ed.), New York: McGraw-Hill.

DUBS, HOMER H. (1930). *Rational Induction*, Chicago: The University of Chicago Press.

DUGGAN, THOMAS J., and DEAN, CHARLES W. (1968). "Common Misinterpretations of Significance Levels in Sociology Journals," *The American Sociologist*, 3 (February), 45–46.

DUNCAN, DAVID B. (1955). "Multiple Range and Multiple F Tests," *Biometrics*, 11 (March), 1–42.

EDWARDS, A. L. (1954). *Statistical Methods for the Behavioral Sciences*, New York: Rinehart.

———— (1950). *Experimental Design in Psychological Research*, New York: Rinehart.

EDWARDS, W., LINDMAN, H., and SAVAGE, L. J. (1963). "Bayesian Statistical Inference for Psychological Research," *Pyschological Review*, 70, 193–242.

ETZIONI, AMITAI (1965). "Mathematics for Sociologists?" *American Sociological Review*, 30, 943–945.

EZEKIEL, M. (1941). *Methods of Correlation Analysis*, New York: Wiley.

FELLER, WILLIAM (1950). *An Introduction to Probability Theory and Its Applications*, Vol. 1., New York: Wiley.

FERGUSON, L. (1959). *Statistical Analysis in Psychology and Education*, New York: McGraw-Hill.

FICHTER, JOSEPH H. (1954). *Social Relations in the Urban Parish*, Chicago: The University of Chicago Press.

FISHER, R. A. (1955). "Statistical Methods and Scientific Induction," *Journal of the Royal Statistical Society*, Series B, 17, 69–78.

———— (1950). *Contributions to Mathematical Statistics*, London: Chapman and Hall.

———— (1945). "The Logical Inversion of the Notion of the Random Variable," *Sankhya*, 7, 129–132.

———— (1939). "The Comparison of Samples with Possibly Unequal Variances," *Annals of Eugenics*, 9, Part II, 174–180.

———— (1937). "On a Point Raised by M. S. Bartlett on Fiducial Probability," *Annals of Eugenics*, 7, 370–375.

———— (1936). "Uncertain Inference," *Proceedings of the American Academy of Arts and Sciences*, 71, 4, 245–258.

———— (1935a). *The Design of Experiments* (subsequent editions 1937, 1942, 1947, 1949, 1951), London: Oliver and Boyd.

———— (1935b). "The Logic of Inductive Inference," *Journal of the Royal Statistical Society*, 98, Part I, 39–54.

———— (1930). "Inverse Probability," *Proceedings of the Cambridge Philosophical Society*, 26, 528–535.

———— (1925a). *Statistical Methods for Research Workers* (subsequent editions 1928, 1930, 1932, 1934, 1936, 1938, 1941, 1944, 1946, 1950, 1954, 1958), London: Oliver and Boyd.

———— (1925b). "Theory of Statistical Estimation," *Proceedings of the Cambridge Philosophical Society*, 22, Part 5, 700–725.

———— (1921). "On the Mathematical Foundations of Theoretical Statistics," *Philosophical Transactions of the Royal Society of London*, Series A, 222, 309–368.

FORGE, ROLFE L. (1967). "Confidence Intervals or Tests of Significance in Scientific Research?" *Psychological Bulletin*, 68 (December), 446–447.

FURFEY, PAUL HANLY (1955). "A Review of Joseph H. Fichter's *Social Relations in the Urban Parish*," *American Sociological Review*, 20 (June), 354.

GALTUNG, JOHAN (1967). "On the Use of Statistical Tests," in *Theory and Methods of Social Research*, New York: Columbia University Press, 358–388.

GAVARRET, JULES (1840). *Principles Generaux de Statistique Medicale*, Paris: [No publisher given].

GOLD, DAVID (1969). "Statistical Tests and Substantive Significance," *The American Sociologist*, 4 (February), 42–46.

——— (1964). "Some Problems in Generalizing Aggregate Associations," *The American Behavioral Scientist*, 8 (December), 16–18.

——— (1958). "Comment on 'A Critique of Tests of Significance'," *American Sociological Review*, 23 (February), 85–86.

——— (1957). "A Note on Statistical Analysis in the *American Sociological Review*," *American Sociological Review*, 22 (June), 332–333.

GOODMAN, LEO, and KRUSKAL, WILLIAM H. (1954). "Measures of Association for Cross Classifications," *Journal of the American Statistical Association*, 49, 732–764.

GOSSART, ERNEST (1919). "Adolphe Quetelet et le Prince Albert de Saxe-Cobourge (1836–1861)," *Bulletin de la Classe des Lettres et de Sciences Morales et Politique*, Brussels: Academe Royale de Belgique, 211–254.

GREENBERG, B. G. (1929). *The Story of Evolution*, New York: Garden City.

GUILFORD, J. P. (1936). *Psychometric Methods*, New York: McGraw-Hill.

HAGOOD, MARGARET J. (1941). *Statistics for Sociologists*, New York: Reynal & Hitchcock.

HAGOOD, MARGARET J., and PRICE, DANIEL O. (1952). *Statistics for Sociologists* (rev. ed.), New York: Henry Holt.

HARE, R. D. (1966). "Psychopathy and Choice of Immediate Versus Delayed Punishment," *Journal of Abnormal Psychology*, 71, 25–29.

——— (1965a). "Acquisition and Generalization of a Conditioned Fear Response in Psychopathic and Non-Psychopathic Criminals," *Journal of Psychology*, 59, 367–370.

——— (1965b). "Temporal Gradient of Fear Arousal in Psychopaths," *Journal of Abnormal Psychology*, 70, 442–445.

HAYS, WILLIAM L. (1963). *Statistics for Psychologists*, New York: Holt, Rinehart, and Winston.

HEMPEL, CARL G. (1956). "A Logical Appraisal of Operationism," in Phillip G. Frank (ed.), *The Validation of Scientific Theories*, New York: Collier Books, 56–59.

HIRSCHI, TRAVIS, and SELVIN, HANAN C. (1967). "Statistical Inference," in *Delinquency Research: An Appraisal of Analytic Methods*, New York: Free Press, 216–234.

HODGES, J. L., and LEHMAN, E. L. (1954). "Testing the Approximate Validity of Statistical Hypotheses," *Journal of the Royal Statistical Society*, Series B, 16, 261–268.

HOEL, P. G. (1954). *Introduction to Mathematical Statistics* (2nd ed.), New York: Wiley.

HOGBEN, LANCELOT T. (1957). *Statistical Theory: The Relationship of Probability, Credibility, and Error. An Examination of the Contemporary Crisis in Statistical Theory from a Behaviourist Viewpoint*, New York: W. W. Norton (re-issued, 1968).

HOLTZMAN, W. H., and BITTERMAN, M. E. (1956). "A Factorial Study of Adjustment to Stress," *Journal of Abnormal and Social Psychology*, 52, 179–185.

HOVLAND, CARL I., LUMSDAINE, A. A., and SHEFFIELD, F. D. (1949). *Experiments on Mass Communication*, Princeton, N. J.: Princeton University Press.

HYMAN, HERBERT H. *et al.* (1955a). *Interviewing in Social Research*, Chicago: The University of Chicago Press.

—— (1955b). *Survey Design and Analysis*, Glencoe, Ill.: Free Press.

IRWIN, J. O. (1948). "Discussion on Anscombe's 'The Validity of Comparative Experiments'," *Journal of the Royal Statistical Society*, Series A, 111, 200–201.

JEFFREYS, HAROLD (1939). *Theory of Probability*, Oxford: At the Clarendon Press.

JONES, CARADOG D. (1921). *A First Course in Statistics*, London: Bell & Sons.

KAISER, H. F. (1960). "Directional Statistical Decision," *Psychological Review*, 67, 160–167.

KATZ, DANIEL (1955) "A Review of Bernard Berelson *et al.*, *Voting: A Study of Opinion Formation in a Presidential Campaign*," *Public Opinion Quarterly*, 19, 328.

KEMPTHORNE, OSCAR (1955). "The Randomization Theory of Experimental Inference," *Journal of the American Statistical Association*, 50 (September), 946–967.

KENDALL, M. G. (1952). "Regression, Structure and Functional Relationship, *Biometrika*, 39 (June), 96–108.

—— (1951). "Regression, Structure and Functional Relationship," *Biometrika*, 38 (June), 12–25.

—— (1949). "On the Reconciliation of Theories of Probability," *Biometrika*, 36, 101–116.

KENDALL, M. G. and BUCKLAND, W. R. (1957). *A Dictionary of Statistical Terms* (rev., 1960), London: Oliver and Boyd.

KENDALL, PATRICIA (1957). "Note on Significance Tests," Appendix C in Robert K. Merton, George C. Reader, and Patricia Kendall (eds.), *The Student Physician*, Cambridge, Mass: Harvard University Press, 301–305.

—— (1954). *Conflict and Mood: Factors Affecting the Stability of Response*, Glencoe, Ill.: Free Press.

KEYFITZ, NATHAN (1953). "A Factorial Arrangement of Comparisons of Family Size," *American Journal of Sociology*, 53 (March), 470.

KEYNES, J. M. (1921). *Treatise on Probability*, London: Macmillan.

KINSEY, ALFRED C., POMEROY, WARDELL, and MARTIN, CLYDE E. (1948). *Sexual Behavior in the Human Male*, Philadelphia: Saunders.

KISH, LESLIE (1965). *Survey Sampling*, New York: Wiley.

—— (1959). "Some Statistical Problems in Research Design," *American Sociological Review*, 24 (June), 328–338.

—— (1957). "Confidence Intervals for Clustered Samples," *American Sociological Review*, 22 (April), 154–165.

LABOVITZ, SANFORD (1968). "Criteria for Selecting a Significance Level: A Note on the Sacredness of .05," *The American Sociologist*, 3 (August), 220–222.

LACEY, O. L. (1953). *Statistical Methods in Experimentation*, New York: Macmillan.

LAPLACE, P. S. (1812). *Theorie Analytique des Probabilités*, Paris: Courcier.

LAZARSFELD, PAUL F. (1955). "Interpretation of Statistical Relations as a Research Operation," in Paul F. Lazarsfeld and Morris Rosenberg (eds.), *The Language of Social Research*, Glencoe, Ill.: The Free Press, 115–125.

LEWIS, DON, and BURKE, C. J. (1949), "The Use and Misuse of the Chi-square Test," *Psychological Bulletin*, 46 (November), 433–489.

LI, C. C. (1956). "The Concept of Path Coefficient and Its Impact on Population Genetics," *Biometrics*, 12 (June), 190–209.

LINDQUIST, E. F. (1940). *Statistical Analysis in Educational Research*, Boston: Houghton-Mifflin.

LIPSET, SEYMOUR M., TROW, MARTIN, and COLEMAN, JAMES S. (1956). "Statistical Problems," Appendix I-B in *Union Democracy*, Glencoe, Ill.: Free Press, 427–432.

LUBIN, A. (1957). "Replicability as a Publication Criterion," *American Psychologist*, 8, 519–520.

LUNDBERG, GEORGE (1947). *Can Science Save Us?* New York: Longmans, Green.

LYKKEN, DAVID (1968). "Statistical Significance in Psychological Research," *Psychological Bulletin*, 70 (September), 151–159.

——— (1966). "Statistical Significance in Psychiatric Research," *Reports from the Research Laboratories of the Department of Psychiatry, University of Minnesota*, Report No. PR–66–9 (December).

——— (1957). "A Study of Anxiety in the Sociopathic Personality," *Journal of Abnormal and Social Psychology*, 55, 6–10.

LYND, ROBERT S. (1939). *Knowledge For What?* Princeton, N. J.: Princeton University Press.

McCORMICK, THOMAS C. (1937). "Sampling Theory in Sociological Research," *Social Forces*, 16 (October), 67–74.

McGINNIS, ROBERT (1958). "Randomization and Inference in Sociological Research," *American Sociological Review*, 23 (August), 408–414.

McNEMAR, QUINN (1960). "At Random: Sense and Nonsense," *American Psychologist*, 15, 295–300.

——— (1955). *Psychological Statistics*, New York: Wiley.

MAH, A. B. (1957). "Heats of Formation of Alumina, Molybdenum Dioxide, and Molybdenum Trioxide," *Journal of Physical Chemistry*, 61, 1572.

MATHER, KENNETH (1942). *Statistical Analysis in Biology* (rev., 1946), London: Methuen.

MEEHL, PAUL E. (1967). "Theory Testing in Psychology and Physics: A Methodological Paradox," *Philosophy of Science*, 34 (June), 103–115.

MELTON, A. W. (1962). "Editorial," *Journal of Experimental Psychology*, 64, 553–557.

MERTON, ROBERT K. (1957). *Social Theory and Social Structure*, Glencoe, Ill.: Free Press.

MILTON, T. E. (1956). "A Review of Edwards' *Statistical Methods for the Behavioral Sciences*," *Journal of the American Statistical Association*, 51, 382.

MOOD, A. M. (1950). *Introduction to the Theory of Statistics*, New York: McGraw-Hill.

MORRISON, DENTON E., and HENKEL, RAMON E. (1969). "Significance Tests Reconsidered," *The American Sociologist*, 4 (May), 131–140.

MUELLER, JOHN, and SCHUESSLER, KARL (1961). *Statistical Reasoning in Sociology*, Boston: Houghton-Mifflin.

NAGEL, ERNEST (1955). *Principles of the Theory of Probability, International Encyclopedia of Unified Science*, Vol. I, Part II, Chicago: The University of Chicago Press.

NEYMAN, JERZY (1957). "'Inductive Behavior' as a Basic Concept of Philosophy of Science," *Review of the Mathematical Statistics Institute*, 25, 7–22.

——— (1952). *Lectures and Conferences on Mathematical Statistics and Probability*, Washington, D.C.: Graduate School of The United States Department of Agriculture, 143–154.

——— (1937). "Outline of a Theory of Statistical Estimation Based on the Classical Theory of Probability," *Philosophical Transactions of the Royal Society*, Series A, 236, 333–380.

NEYMAN, JERZY, and PEARSON, E. S. (1933). "On the Problem of the Most Efficient Tests of Statistical Hypotheses," *Philosophical Transactions of the Royal Society*, Series A, 231, 289–337.

——— (1932). "The Testing of Statistical Hypotheses in Relation to Probabilities a priori," *Proceedings of the Cambridge Philosophical Society*, 29, 492–516.

NUNNALLY, JUM C. (1960). "The Place of Statistics in Psychology," *Educational and Psychological Measurement*, 20, 641–650.

PARSONS, TALCOTT (1964). "The Sibley Report on Training in Sociology," *American Sociological Review*, 29, 747–748.

PEARSON, E. S. (1955). "Statistical Concepts in Their Relation to Reality," *Journal of the Royal Statistical Society*, Series B, 17, 204–207.

——— (1948a). "Discussion of Dr. Wishart's Paper 'The Teaching of Statistics'," *Journal of the Royal Statistical Society*, Series A, 111, 212–229.

——— (1948b). "Discussion on Anscombe's 'The Validity of Comparative Experiments'," *Journal of the Royal Statistical Society*, Series A, 111, 203–204.

——— (1947). "The Choice of Statistical Tests Illustrated on the Interpretation of Data Classed in a 2×2 Table," *Biometrika*, 34, 139–167.

PEARSON, KARL (1911). "Probability that Two Independent Distributions of Frequency are Really Samples from the Same Population," *Biometrika*, 8, 250–254.

——— (1900). "On the Criterion that a Given System of Deviations from the Probable in the Case of Correlated Systems of Variables is Such that it Can Reasonably be Supposed to Have Arisen from Random Sampling," *Philosophical Magazine*, Series V, 1, 157–175.

——— (1895). "Skew Variation in Homogeneous Material," *Philosophical Transactions of the Royal Society*, Series A, 343–414.

——— (1892). *The Grammar of Science*, London: W. Scott.

POPPER, KARL R. (1962). *Conjectures and Refutations*, New York: Basic Books.

——— (1959). *The Logic of Scientific Discovery*, New York: Basic Books.

QUETELET, L. A. J. (1835). *Sur L'Homme et le Developpment de ses Facultés, ou l'essai de Physique Sociale*, Paris: [No publisher given].

REYNOLDS, ROBERT (1969). "Replication and Substantive Support: A Critique on the Use of Statistical Inference in Social Research," *Sociology and Social Research*, 53 (April), 299–310.

RILEY, MATILDA WHITE (1963). *Sociological Research*, New York: Harcourt, Brace and World.

ROBERTS, C. L., and WIST, E. (n.d.). "An Empirical Side-light on the Aggregate-General Distinction," Unpublished manuscript.

ROBINSON, WILLIAM S. (1951). "The Logical Structure of Analytic Induction," *American Sociological Review*, 16 (December), 812–818.

ROETHLISBERGER, F. J., and DICKSON, W. J. (1939). *Management and the Worker*, Cambridge: Harvard University Press.

ROSENTHAL, R. (1963). "On the Social Psychology of the Psychological Experiment: The Experimentor's Hypothesis as Unintended Determinant of Experimental Results," *American Scientist*, 51, 268–283.

ROSENTHAL, R., and GAITO, J. (1963). "The Interpretation of Levels of Significance by Psychological Researchers," *Journal of Psychology*, 55, 33–38.

ROZEBOOM, WILLIAM W. (1960). "The Fallacy of the Null Hypothesis Significance Test," *Psychological Bulletin*, 57 (September), 416–428.

RYAN, T. A. (1959). "Multiple Comparisons in Psychological Research," *Psychological Bulletin*, 56 (January), 26–47.

SAPOLSKY, A. (1964). "An Effort at Studying Rorschach Content Symbolism: The Frog Response," *Journal of Consulting Psychology*, 28, 469–472.

SAVAGE, I. RICHARD (1957). "Nonparametric Statistics," *Journal of the American Statistical Association*, 52 (September), 332–333.

SAVAGE, L. J. (1954). *The Foundations of Statistics*, New York: Wiley.

SCHACHTER, S., and LATENÉ, B. (1964). "Crime, Cognition, and the Autonomic Nervous System," *Nebraska Symposium on Motivation*, 12, 221–273.

SCHEFFÉ, HENRY (1953). "A Method for Judging all Contrasts in the Analysis of Variance," *Biometrika*, 40 (June), 87–104.

SCHLAIFER, R. (1959). *Probability and Statistics for Business Decisions*, New York: McGraw-Hill.

Schultz, Henry (1938). *The Theory and Measurement of Demand*, Chicago: The University of Chicago Press.

SELLTIZ, CLAIRE, JAHODA, MARIE, DEUTSCH, MORTON, and COOK, STUART W. (1961). *Research Methods in Social Relations*, New York: Holt, Rinehart and Winston.

SELVIN, HANAN C. (1968). "Survey Analysis: Methods of Survey Analysis," in David L. Sills (ed.), *International Encyclopedia of the Social Sciences*, Vol. 15, New York: Macmillan & Free Press, 411–419.

——— (1965). "Mathematics and Sociology," *American Sociological Review*, 30 (April), 264–265.

——— (1958a) "Reply to Beshers," *American Sociological Review*, 23 (April), 199.

——— (1958b). "Reply to Gold's Comment on 'A Critique of Tests of Significance'," *American Sociological Review*, 23 (February), 86.

——— (1957). "A Critique of Tests of Significance in Survey Research," *American Sociological Review*, 22 (October), 519–527.

———— (1956). "The Uses and Misuses of Significance Tests," Appendix C in "The Effects of Leadership Climate on the Nonduty Behavior of Army Trainees," unpublished Ph.D. dissertation, Columbia University.

SEWELL, WILLIAM H. (1952). "Infant Training and the Personality of the Child," *American Journal of Sociology*, 53 (September), 150–159.

SIBLEY, ELBRIDGE (1965). "Parsons on the Sibley Report," *American Sociological Review*, 30 (February), 110.

SIDMAN, M. (1960). *Tactics of Scientific Research*, New York: Basic Books.

SIEGEL, SIDNEY (1956). *Nonparametric Statistics for the Behavioral Sciences*, New York: McGraw-Hill.

SIMON, HERBERT A. (1956). *Models of Man*, New York: Wiley.

———— (1954). "Spurious Correlation: A Causal Interpretation," *Journal of the American Statistical Association*, 49 (September), 467–479.

SKIPPER, JAMES K., GUENTHER, ANTHONY L., and NASS, GILBERT (1967). "The Sacredness of .05: A Note Concerning the Uses of Statistical Levels of Significance in Social Science," *The American Sociologist*, 1 (February), 16–18.

SMITH, C. A. B. (1960). "Review of N. T. J. Bailey, *Statistical Methods in Biology*," *Applied Statistics*, 9, 64–66.

SNEDECOR, GEORGE W. (1937). *Statistical Methods Applied to Experiments in Agriculture and Biology* (subsequent editions 1938, 1940, 1946, 1956, 1957), Ames: Iowa State College Press.

STERLING, THEODORE D. (1959). "Publication Decisions and Their Possible Effects on Inferences Drawn From Tests of Significance—or Vice Versa," *Journal of the American Statistical Association*, 54 (March), 30–34.

STOUFFER, SAMUEL A. (1934). "Sociology and Sampling," in L. L. Bernard (ed.), *Fields and Methods of Sociology*, New York: Long and Smith, 476–487.

"STUDENT" [W. S. GOSSETT] (1906–7). "On the Error of Counting with a Haemacytometer," *Biometrika*, 5, 351–360.

TIPPETT, L. H. C. (1931). *The Methods of Statistics*, London: Williams & Norgate.

TUKEY, J. W. (1962). "The Future of Data Analysis," *Annals of Mathematical Statistics*, 33, 1–67.

———— (1954a). "Causation, Regression, and Path Analysis," in Oscar Kempthorne et al., *Statistics and Mathematics in Biology*, Ames: The Iowa State College Press, Chapter 3.

———— (1954b). "Unsolved Problems of Experimental Statistics," *Journal of the American Statistical Association*, 49 (December), 710.

———— (1949). "Comparing Individual Means in the Analysis of Variance," *Biometrics*, 5 (June), 99–114.

TULLOCK, GORDON (1959). "Publication Decisions and Tests of Significance—A Comment," *Journal of the American Statistical Association*, 54 (September), 593.

UNDERWOOD, B. J., DUNCAN, C. P., TAYLOR, J. A., and COTTON, J. W. (1954). *Elementary Statistics*, New York: Appleton-Century-Crofts.

VENN, JOHN (1888). "Cambridge Anthropometry," *Journal of the Anthropological Institute*, 18, 140–154.

———— (1866). *The Logic of Chance*, Cambridge: Cambridge University Press.

VON MISES, RICHARD (1939). *Probability, Statistics, and Truth* (tr. by J. Neyman, D. Scholl, and E. Rabinowitsch), New York: Macmillan.

WALD, A. (1955). *Selected Papers in Statistics and Probability*, New York: McGraw-Hill.

——— (1950). *Statistical Decision Functions*, New York: Wiley.

——— (1939). "Contributions to the Theory of Statistical Estimation and Testing Hypotheses," *Annals of Mathematical Statistics*, 10, 299–326.

WALKER, H. M. (1956). "A Review of Adams' *Basic Statistical Concepts*," *Educational and Psychological Measurement*, 16, 554–557.

——— (1947). *Elementary Statistical Methods*, New York: Henry Holt.

WIGHTMAN, W. P. D. (1953). *The Growth of Scientific Ideas*, New Haven, Conn.: Yale University Press.

WILK, MARTIN B., and KEMPTHORNE, OSCAR (1956). "Some Aspects of the Analysis of Factorial Experiment in a Completely Randomized Design," *Annals of Mathematical Statistics*, 27 (December), 950–985.

——— (1955). "Fixed, Mixed, and Random Models," *Journal of the American Statistical Association*, 50 (December), 1144–1167.

WILKS, SAMUEL S. (1943). *Mathematical Statistics* (rev., 1962, New York: Wiley), Princeton, N. J.: Princeton University Press.

WILLER, DAVID (1967). "Conditional Universals and Scope Sampling," in *Scientific Sociology*, Englewood Cliffs, N. J.: Prentice Hall, 97–115.

WILSON, K. V. (1961). "Subjectivist Statistics for the Current Crisis," *Contemporary Psychology*, 6, 229–231.

WILSON, WARNER, MILLER, HOWARD, and LOWER, JEROLD S. (1967). "Much Ado About the Null Hypothesis," *Psychological Bulletin*, 67 (March), 188–196.

WINCH, ROBERT, F., and CAMPBELL, DONALD T. (1969). "Proof? No. Evidence? Yes. The Significance of Tests of Significance," *The American Sociologist*, 4 (May), 140–143.

WINER, B. J. (1962). *Statistical Principles in Experimental Design*, New York: McGraw-Hill.

WOLD, HERMAN (1956). "Causal Inference from Observational Data," *Journal of the Royal Statistical Society*, Series A, 119, Part 1 (January), 28–50.

WRIGHT, SEWALL (1954). "The Interpretation of Multi-Variate Systems," in Oscar Kempthorne *et al.*, *Statistics and Mathematics in Biology*, Ames, Iowa: The Iowa State College Press, Chapter 2.

YATES, FRANK (1951). "The Influence of *Statistical Methods for Research Workers* on the Development of the Science of Statistics," *Journal of the American Statistical Association*, 46 (March), 32–33.

——— (1939). "An Apparent Inconsistency Arising from Tests of Significance Based on Fiducial Distributions of Unknown Parameters," *Proceedings of the Cambridge Philosophical Society*, 35, 579–591.

YOUNG, PAULINE V. (1939). *Scientific Social Surveys and Research: An Introduction to the Background, Content, Methods, and Analysis of Social Studies* (2nd ed., 1950), New York: Prentice-Hall.

YULE, G. UDNY (1911). *An Introduction to the Theory of Statistics* (nine subsequent editions through 1932), London: Griffin.

YULE, G. UDNY, and GREENWOOD, M. (1915). "The Statistics of Anti-typhoid and Anti-cholera Inoculations and the Interpretation of Such Statistics in General," *Proceedings of the Royal Society of Medicine*, 8, Part II, 113–194.

YULE, G. UDNY, and KENDALL, M. G. (1937). *An Introduction to the Theory of Statistics* (11th ed. and three subsequent editions through 1950), London: Griffin.

ZEISEL, HANS (1955). "The Significance of Insignificant Differences," *Public Opinion Quarterly*, 17 (Fall), 319–321.

ZETTERBERG, HANS (1954). *On Theory and Verification in Sociology*, Stockholm: Almqvist & Wiksell.

Name Index

326

Subject Index

[Note: Designed to be used in conjunction with "A Tabular Guide to Authors and Issues," p. xiv.]

Backward Look, 4, 29
Bayesian approach, 4, 15, 18, 55, 250
Bias, correlated, 61, 98-137 *passim; see also* Causal inference vs. statistical inference

Calculus, types of, 3-23 *passim*
Causal *vs.* statistical inference, 61-63, 67-70, 85-86, 94-137, 144-147, 195-196; *see also* Statistical issues, causality; Randomization
Chi-square, *see* Significance tests
Confidence intervals, see Significance tests, and confidence intervals

Decision tests, *see* Significance tests, as decision tests

Errors, 10-29 *passim,* 98-137 *passim,* 148-149, 174, 183-198
random *vs.* systematic, 98-103, 174; *see also* Randomization
Type I and II compared, 156-160, 168, 214-215
see also Level of significance; Power

Fisher approach
criticized, 41-56
and decision theory compared, 32-39, 249-250
described, 32-33, 147, 233-234
Forward Look, 4, 30
Frequency interpretation of tests, 42*ff.*

Generalization, *see* Significance tests, and generalization

Hypotheses, general types of, 121-122, 184, 191
null and alternative, *see* Null hypotheses
probable validity of, *see* Probability, inverse or inductive
scientific, *see* Inference, scientific *vs.* statistical
tests of, *see* Significance tests
see also Theory, deterministic *vs.* probabilistic

Induction, 142-154
to aggregate *vs.* general, 175, 211, 244-248
analytic, 142
see also Inference; Probability, inverse or inductive
Inference
automatic, 211, 242-244, 283
causal, *see* Causal *vs.* statistical inference
fiducial, *see* Probability, fiducial
scientific *vs.* statistical, vii, x-xiii, 3-7, 8-56 *passim,* 63, 142-154 *passim,* 186-198 *passim,* 210-212, 221-279 *passim,* 285-294 *passim; see also* Philosophy of science issues
statistical, *see* Causal *vs.* statistical inference; Significance tests
uncertain, 8-21 *passim*
see also Induction; Population; Probability, inverse or inductive; Significance tests, and generalization
Interocular traumatic test, 251
Isotropic tables, 105

330